DATE DUE FOR RETURN

This book may be recalled before the above date.

Magnetic Cell Separation

LABORATORY TECHNIQUES IN BIOCHEMISTRY AND MOLECULAR BIOLOGY

Series Editors

P.C. van der Vliet—*Department for Physiological Chemistry, University of Utrecht, Utrecht, The Netherlands*

and

S. Pillai—*MGH Cancer Center, Boston, Massachusetts, USA*

Volume 32

ELSEVIER

AMSTERDAM • BOSTON • HEIDELBERG • LONDON • NEW YORK • OXFORD
PARIS • SAN DIEGO • SAN FRANCISCO • SINGAPORE • SYDNEY • TOKYO

MAGNETIC CELL SEPARATION

Edited by

M. Zborowski
Dept. of Biomedical Engineering
Cleveland Clinic
Cleveland, Ohio, USA

J. J. Chalmers
Dept. of Chemical and Biomolecular Engineering
The Ohio State University
Columbus, Ohio, USA

ELSEVIER

AMSTERDAM • BOSTON • HEIDELBERG • LONDON • NEW YORK • OXFORD
PARIS • SAN DIEGO • SAN FRANCISCO • SINGAPORE • SYDNEY • TOKYO

Elsevier
Radarweg 29, PO Box 211, 1000 AE Amsterdam, The Netherlands
Linacre House, Jordan Hill, Oxford OX2 8DP, UK

First edition 2008

Library of Congress Cataloging-in-Publication Data
A catalog record for this book is available from the Library of Congress

British Library Cataloguing in Publication Data
A catalogue record for this book is available from the British Library

ISBN: 978-0-444-52754-7
ISSN: 0075-7535

For information on all Elsevier publications
visit our website at books.elsevier.com

Printed and bound in The Netherlands
07 08 09 10 11 10 9 8 7 6 5 4 3 2 1

Contents

See Color Plate Section in the back of this book

Preface

Cell separation is at the core of current methods in experimental biology and medicine. Its importance is illustrated by the large number of physical and biochemical principles that have been evaluated for application to cell separation. The development of cell separation methods is driven by the needs of biological and medical research, and the ever increasing demands for sensitivity, selectivity, yield, timeliness, and economy of the process. On the other hand, the current, rapid progress in cell separation methods would not have been possible without considerable technical innovation and contributions from diverse disciplines of material science, physics, and engineering. An important contribution comes from commercial entrepreneurs and innovators who bring the product to the market and make it available to a large number of users at an affordable price. As a result, today's investigator has access to a large selection of cell separation methods and devices, from benchtop microcentrifuges to large machines that may require separate rooms and facilities.

Magnetic cell separation, the topic of this volume, is a particularly instructive example of the power of interdisciplinary research and development because it successfully combines developments in immunocytochemistry, materials science, clinical research, physics, electrical engineering, process engineering, and commercial entrepreneurship. The antigen–antibody reaction provides the potential for addressing the cell at a molecular level, with the sensitivity and specificity unrivaled by other separation methods based on physical cell properties. The magnetic separation method builds on advances in immunochemistry, in particular, improved conjugation techniques initially developed for applications with fluorescent dyes. In parallel with the development of immuno fluorescent markers, instrumentation has also advanced including automated optical scanners and sorters which opened a new chapter in the development of cell analysis and separation methods, and established flow cytometry as a standard tool in cell biology and clinical research. It is against the performance of optically based technology that magnetic cell separation methods are being evaluated.

Magnetic cell separation combines simplicity of physical separation with the sensitivity and specificity afforded by immunocytochemistry. From the user's point of view, in its simplest implementation, it is nothing more than a simple procedure of mixing a cell suspension with a magnetic microparticle-antibody reagent, followed by exposure of the tube with the labeled cell suspension to a magnet that precipitates the cell–antibody–microparticle complex on the tube wall and thus removes the cells from the suspension. Unlike the optical methods of cell scanning and sorting, magnetic separation does not require highly engineered fluidic, optic, and electronic systems for operation, resulting in a significant advantage of lower capital and operating costs, and therefore greater availability to an individual laboratory. Typically, the cost of the immunomagnetic reagents exceeds the cost of the separation apparatus in the course of just a few separation runs, and the separation time is negligible as compared to the time that it takes to prepare the sample, process the separated fractions, and analyze the data. These attributes made magnetic separation an extremely attractive alternative to other immunoseparation methods, and contribute to its widespread use in research and clinical laboratories.

By the same token, magnetic separation lacks the analytical capabilities of optical methods, and its application to a poorly defined cell mixture is comparable to flying blindly. Because of this, magnetic separation relies heavily on the analytical capabilities of the optical methods used in flow cytometry such as determination of the composition of the starting cell mixture and the final, separated cell fractions. It performs best when applied to well-defined cell suspensions, in routine separations involving large cell volumes that need to be performed quickly. It takes a subordinate role to automated optical cell scanners and sorters in application to exploratory research. Therefore, it is typically used in laboratories that have the necessary infrastructure and support for a full evaluation of the cell mixtures. The increasing use of magnetic cell separation to routine, large volume separations sparked interest in automation of the separation process, including preparative stages of adding the immunomagnetic reagents. There is also interest in combining magnetic separation with optical analysis of the separated fraction in a single system for rare cell detection.

The editors of this volume were confronted with the task of attempting a review of a field that is evolving quickly in an environment that provides easy access to information about magnetic cell separation applications, products, and reagents online. There is a wealth of information about the types of applications, experimental protocols, immunomagnetic reagents, research publications with details of experimental procedures, and separation results that are published by companies selling magnetic separators and immunomagnetic reagents. A good source of information are the conferences that are devoted to the topic of magnetic separations, including cell separation, in biology and medicine, and conferences that include sessions on magnetic cell separation, organized yearly or on a biennial basis. The editors believe, however, that there is a genuine need for a basic description of the fundamental processes involved in magnetic cell separation that may help the user in navigating the wealth of information available online and in scientific publications. This is particularly true for a discipline that is at the crossroads of cell biology, clinical research, inorganic chemistry, biochemistry, chemical engineering, materials science, physics, and electrical engineering, to name a few.

Magnetic cell separation is driven by a difference in magnetic susceptibility between different cell subsets, typically brought out by selective binding of the magnetizable micro- or submicroparticles. Chapter 1 provides a description of how the magnetic susceptibility is measured and calculated. This seemingly routine topic is complicated by a large number of definitions of the magnetic susceptibility that are used on equal footing in the literature, and the use of different systems of units. This chapter provides examples of how to calculate the volume magnetic susceptibility, a fundamental quantity for calculating the magnetic force acting on a cell, from various types of magnetic susceptibilities available in literature.

There is a vast physics and engineering literature on magnetic fields, and a large number of excellent textbooks at an undergraduate and graduate level on the topic. Nevertheless, in Chapter 2 we introduce the elements of magnetostatics as they apply to cell magnetization and the magnetization of magnetic micro- and nanoparticles used for cell separation. These elements of magnetostatics are used in later chapters, and in particular in Chapter 6 on synthesis and characterization of magnetic particles.

The description of cell motion in a magnetic field is complicated by the fact that the static magnetic force acting on magnetizable matter is a second-order function of the applied field. This is different from gravitational and electrostatic forces, familiar from everyday life experience, which are first-order functions of the respective field intensities. The consequence of this difference is that magnetic field lines, generated from classical representation of magnetic field in physics and electrical engineering, have no bearing on the description of magnetized cell motion, which is a source of considerable confusion at meetings devoted to magnetic cell separation. Instead, in Chapter 3, we provide the description of the magnetic field with the use of Maxwell stress, which is a second-order function of the magnetic field and therefore ideally suited for visualization of the magnetically susceptible cell trajectory. The Maxwell stress description has been shown to be particularly useful for the prediction of the behavior of ferrofluids, and its application to cell motion analysis is a natural extension of the earlier work on continuum electromechanics.

The application of Maxwell stress and the stress gradients to determine magnetized cell trajectories and the accumulation of cell mass on magnetized solid supports is illustrated in Chapter 4. The geometries of the magnet surfaces represent typical magnet configurations found in contemporary laboratory and commercial magnetic cell separators and analyzers. Examples include ferromagnetic wires and submillimeter-sized steel spheres used for high-gradient magnetic separation (HGMS), and interpolar gaps and quadrupole arrangements used for open-gradient magnetic separation and motion analysis. The application of the Maxwell stress description of the magnetostatic field leads to a highly visual representation of the effect of magnetic force on the magnetized cells and should be helpful in developing intuitive understanding of the magnetic cell separation process.

The motion induced by the magnetostatic field acting on a magnetic dipole, such as a magnetic microparticle or a cell tagged with magnetic nanoparticles, has been dubbed "magnetophoresis." This is in obvious analogy with electrophoresis, important because of its role in separation of macromolecules in biology. Differential cell magnetophoresis is the necessary condition for a successful magnetic cell separation, yet its quantitative description has not been well developed. In Chapter 5, the parameters that determine cell magnetophoresis are described, in particular the dependence of the magnetophoretic mobility on the material properties of the particle and the fluid medium. The experimental determination of the magnetophoretic mobility and its applications to measuring magnetic particle binding to cells is described in later chapters, particularly in Chapter 8.

The synthesis and characterization of magnetic micro- and submicroparticles for application to cell separation is described in Chapter 6. The science and application of the magnetic particles to cell separation have been developing rapidly during the past 20 years, and the trend continues unabated. Particularly interesting types of microparticles are the multifunctional particles, such as those combining magnetic and optical properties, including fluorescence.

An often asked question about magnetic cell separation concerns the fate and the toxicity of the magnetic micro- and nanoparticles. This is particularly relevant for clinical applications of magnetic cell separation. The methods and issues related to magnetic particle toxicity are discussed in Chapter 7.

The laboratory and clinical research-related applications of magnetic cell separation are discussed in Chapters 8 and 9. These include application of cell tracking velocimetry using well-defined magnetic fields (in particular, an isodynamic field) to measure antibody-magnetic bead conjugate binding to cells and magnetic bead binding titration. The preparative applications include elimination of certain classes of lymphocytes (such as T cells) and the enrichment of blood progenitor cells.

The current, commercially available magnetic cell separation systems are reviewed in Chapter 10. The list is not exhaustive and the selection was made mainly to illustrate successful application of various magnetic field configurations discussed in Chapter 4. As already mentioned above, the worldwide web and the web search engines provide an excellent source for up-to-date information from commercial companies, and the interested reader is encouraged to retrieve information posted on their websites. The huge volume of available information concerning numerous combinations of cell types, applications, and reagents cannot be included in a single book. The rapidly developing applications and the market forces would make any such book obsolete in the course of a year or two. Rather, it is hoped that the review of selected magnetic cell separators included in this volume provides a useful background for subsequent searches for any particular application that the reader may have in mind.

In the same vein, a few prototypical magnetic cell separation protocols are described in Chapter 11. These have been adapted from the protocols posted on company websites, and serve the purpose of illustrating the steps typically involved in achieving magnetic cell separation. With the current choice of separators and reagents, there is almost an infinite number of possible combinations, and the interested reader is again referred to the information posted online for specifics concerning a particular cell mixture and the matching reagents. Based on the authors' own experience, the quality of information and the experimental protocols freely available on company websites are very good indeed.

The continuing research on improved methods of magnetic cell separation is described in Chapter 12. The choice of topics has been admittedly colored by the authors' own research interests. Particularly, exciting developments include application of microelectromechanical (MEMS) technology for miniaturization of magnetic cell separators for possible inclusion in lab-on-a-chip devices, the use of magnetic field-flow fractionation for characterization of magnetic particles, continuous magnetic cell sorting in a flowing stream for high-volume separations, and the application of intrinsic cell magnetization to the separation of unmanipulated cells. The list could certainly be made longer, as suggested by the list of references.

The Appendices provide basic references to nomenclature, definition of units, vector notation used throughout this volume, additional information on Maxwell stress on extended bodies such as a cell labeled with magnetic nanoparticles, and a table of magnetic susceptibilities of selected elements and compounds.

Finally, a word of caution regarding what this volume is not about. We did not attempt to provide a systematic introduction to the topic of electrodynamics and physics of the magnetic field, or the separation science. As already mentioned above, there is a large number of excellent textbooks dealing with those subjects at an undergraduate and a graduate levels, and they are referenced in the bibliography. Therefore, the order of chapters does not follow the typical sequence from definitions of basic electromagnetic quantities to applications. Rather, it jumps right to the expression for force acting on the magnetically susceptible material, and issues dealing with differences in definitions of magnetic susceptibilities and systems of units, encountered in the literatures. On the other hand, we strived to provide a self-contained text, so that the reader is provided with the necessary formulas and background information to follow examples provided in the volume. This may lead to a nonlinear reading, where getting acquainted with the later chapters first may help answer questions encountered in examples discussed in earlier chapters. In this sense, this volume is intended as more of a compendium of basic information about the magnetic cell separation techniques and related physical concepts rather than a textbook on magnetic separation. The editors will be grateful for comments regarding the volume.

The editors wish to acknowledge the help in preparation of this volume of their colleagues who critically read and commented on sections of the manuscript, and in particular the members of their laboratories: Francesca Carpino, Christina Lohr, Xiaoxia Jin, Dr. Leonardo Cinque, and Elizabeth Zborowska, who assisted in preparation of figures and references. Dr. P. Stephen Williams and Mr. Lee Moore reviewed the definitions and double-checked the calculations provided in the examples, and critically read the early versions of the manuscript. Dr. Aaron Fleischman has kindly helped in navigating the labyrinth of word processor formatting commands. The indirect contribution comes from the former students in our laboratories, who brought the curiosity, enthusiasm, and motivation to our investigations, and in particular Drs. Liping Sun, Kara McCloskey, Oscar Lara, Masayuki Nakamura, Huading Zhang, Ying Jing, Yang Zhao, Zhaodong Tong, and Mrs. Kristie Melnik, Guo-ha (Jennifer) Chen, and Mr. Bingbing Fan. One of us (M.Z.) is especially indebted to Prof. Jim H. P. Watson for his kind review of the early version of the introductory Chapters (1–5) and comments on the presentation of the physics principles. We also wish to acknowledge fruitful interactions with Dr. Paul S. Malchesky and Profs. Robert S. Brodkey, Paul Todd, and J. Fredrick Cornhill, who directly or indirectly shaped our research on magnetic cell separations. The editors, however, accept sole responsibility for the contents of this volume.

Maciej Zborowski
Jeffrey J. Chalmers

Laboratory Techniques in Biochemistry and Molecular Biology, Volume 32
Magnetic Cell Separation
M. Zborowski and J. J. Chalmers (Editors)

CHAPTER 1

Magnetic susceptibility

Maciej Zborowski

Department of Biomedical Engineering, Lerner Research Institute, Cleveland Clinic, Cleveland, OH 44195, USA

zMagnetic susceptibility is a material property that describes response to an applied magnetic field [1–7]. Experimentally it is measured by the amount of force exerted on a defined amount of the substance by a well-defined magnetic field. Every substance is responsive to the magnetic field, but in the majority of cases the response is too weak to be of practical importance. *Such substances are considered "nonmagnetic" for all intents and purposes. The magnetic susceptibility of the "nonmagnetic" materials becomes apparent in the presence of high magnetic fields and gradients that become increasingly accessible for biological and clinical applications [8]. Here we restrict the discussion to the static and quasistatic magnetic field effects, and therefore the static magnetic susceptibility.*

The simplest experimental configuration to measure magnetic susceptibility is that of the magnetic balance, Fig. 1.1, the design of which goes back to Faraday. The most sensitive methods of magnetic susceptibility measurement apply alternating current (AC) fields and sample vibration, and are reviewed elsewhere [9, 10]. The magnetic force exerted on a volume V of substance characterized by the volume magnetic susceptibility χ is expressed by Eq. (1.1):

$$F = \chi V H \frac{\mathrm{d}B_0}{\mathrm{d}x} \tag{1.1}$$

DOI: 10.1016/S0075-7535(06)32001-3

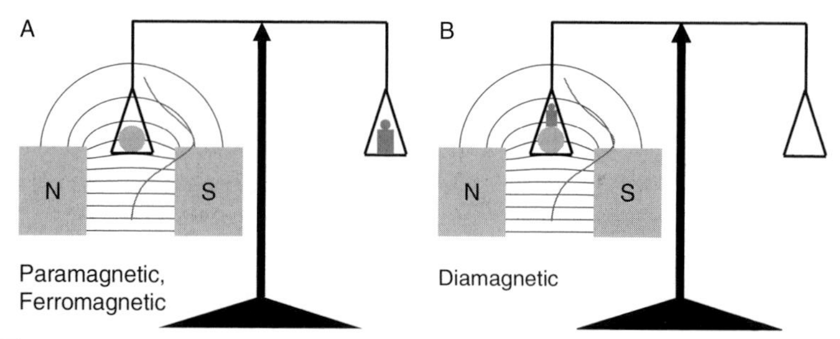

Fig. 1.1. Magnetic balance schematic. One arm of the balance holds a sample (circle) exposed to an inhomogeneous magnetic field at a point where the product of the magnetic field and the field gradient reaches maximum (as indicated by the vertical curve). The other arm of the balance (isolated from the magnetic field) holds a weight that keeps the balance at equilibrium. (A) For paramagnetic and ferromagnetic substances, the weight balances the magnetic force, acting on the sample. (B) For diamagnetic substances, the sample is pushed out from the magnet and therefore the weight (magnetically neutral) has to be added to the sample to keep the balance at equilibrium. The diamagnetic effects are much weaker than the para- and ferromagnetic ones.

where H and B_0 are magnitudes of the applied magnetic field vectors (for definitions see Chapter 2 and Appendices A, B, and C). The volume magnetic susceptibility is dimensionless. In particular, one notes that the magnetic force may add weight to the substance when placed on the Faraday balance ($\chi > 0$, paramagnetic substances) as well as decrease the weight of the substance ($\chi < 0$, diamagnetic substances, Fig. 1.1). The proper definition of the magnetic susceptibility is provided in Chapter 2. The same quantity, the volume magnetic susceptibility, is often represented by different symbols such as Greek "kappa" in lower case, κ, or uppercase, K. This is a common occurrence when reviewing literature across different disciplines, including electrical engineering, physical chemistry, and physics (for instance, compare notation between Purcell's physics textbook [1] and the chemistry and physics handbook [11]). In this volume, we adapt a convention of using the symbol χ to designate the volume magnetic susceptibility, and indicating the system of units in parentheses, thus (SI) for the SI System of Units and (CGS) for the CGS System of Units (where the system of units designation is dropped, it is to be understood as the SI). The two quantities differ by a constant 4π:

$$\chi(\mathrm{SI}) = 4\pi\chi(\mathrm{CGS}) \tag{1.2}$$

The susceptibility in the CGS units is often accompanied by the designation "EMU CGS" or "emu cgs," which stands for the "electromagnetic units centimeter-gram-second". This is to make a distinction from the alternative "ESU CGS" or "esu cgs" units (electrostatic units centimeter-gram-second), used in the electrostatics [3]. In the context of the magnetostatic literature, such a distinction is often considered superfluous and an abbreviated form "CGS" or "cgs" is used. It is worth remembering that the above designations only refer to the system of units as a reference for the susceptibility calculations but not to the actual units that often differ for different definitions of the magnetic susceptibility, reported in the literature. Reviews of the systems of units used in electrodynamics (part of which is the magnetostatics) are given in Becker [3]. In this volume, we adhere to the SI system of units, unless otherwise indicated. The review of the most commonly used expressions for the magnetic susceptibility is provided below.

1.1. Mass and volume magnetic susceptibilities

In dealing with solids, it is easier to determine the magnetic force acting on a known mass of the substance rather than its volume. This leads to a definition of the "mass susceptibility," also known as the "specific susceptibility," χ_g

$$F = \chi_g m_g H \frac{\mathrm{d}B_0}{\mathrm{d}x} \tag{1.3}$$

where m_g is the mass of the measured substance. By comparing the right-hand sides of Eqs. (1.1) and (1.3), one is able to calculate the volume magnetic susceptibility from the specific magnetic susceptibility providing that the density, ρ, of the given substance is known:

$$\chi = \frac{\chi_g m_g}{V} = \chi_g \rho \quad (1.4)$$

One observes that the specific susceptibility is not a dimensionless quantity like χ (as discussed in Chapter 2), but rather a compound quantity comprising two material constants, the density and the susceptibility proper. The following example illustrates the calculations:

Example Calculation of volume susceptibility from specific susceptibility.
The example concerns magnetic susceptibility of the protein part of the hemoglobin molecule, the globin. The magnetic properties of hemoglobin were studied by Pauling in 1930s [12] and are currently a basis for functional analysis by nuclear magnetic resonance imaging (MRI) [13, 14]. Magnetic separation of erythrocytes serves the useful purpose as a model of magnetic cell separation [15–17]. The contribution from the heme prosthetic group is discussed in detail in the later section.

$$\chi_{g,\text{globin}}(\text{CGS}) = -0.580 \times 10^{-6}\,\frac{\text{cm}^3}{\text{g}} \quad \text{the mass (or specific) susceptibility of the globin}$$

$$\rho_{\text{globin}} = 1.34\,\frac{\text{g}}{\text{cm}^3} \quad \text{the globin density}$$

$$\chi_{\text{globin}}(\text{CGS}) = \rho_{\text{globin}}\chi_{g,\text{globin}} = -0.774 \times 10^{-6}$$

$$\chi_{\text{globin}}(\text{SI}) = 4\pi\chi_{\text{globin}}(\text{CGS}) = -9.73 \times 10^{-6}$$

1.2. Molar magnetic susceptibility

In dealing with a compound substance whose chemical formula is known, another combination material constant is used, that of the "molar susceptibility," χ_m defined as follows:

$$F = \chi_m N H \frac{\mathrm{d}B_0}{\mathrm{d}x} \quad (1.5)$$

where N is the number of moles. One calculates the volume susceptibility from molar susceptibility by comparing Eqs. (1.1) and (1.5), yielding

$$\chi = \frac{\chi_m N}{V} \tag{1.6}$$

If exactly one mole of substance, $N = 1$, is used to measure the susceptibility, then the above formula is reworked to include the molar mass, M_w, and the density, ρ, of the substance by observing that $V_m = M_w/\rho$, where V_m is the molar volume of the substance, so that:

$$\chi = \frac{\chi_m \rho}{M_w} \tag{1.7}$$

In this context, the molar susceptibility χ_m is sometimes referred to as the "one gram formula weight susceptibility" [11].

1.3. Magnetic susceptibilities in CGS and SI systems of units

The same relationships hold for susceptibilities measured in the CGS system of units. In order to be consistent with the notation convention introduced with Eq. (1.2), for the calculations in the CGS units system, substitute χ (CGS) for χ in the Eqs. (1.3–1.7).

Examples of calculating volume susceptibility from one gram formula weight and molar susceptibilities are given below. The conversion factors between different types of susceptibilities are provided in Table 1.1.

Example Calculation of volume susceptibility from molar susceptibility.
Water

$$\chi_m(\text{CGS}) = -12.97 \times 10^{-6}\,\frac{\text{cm}^3}{\text{mol}}$$ one gram formula weight susceptibility (at 20 °C) [11]

TABLE 1.1
Magnetic susceptibilities (static magnetic fields)[a]

Susceptibility designation	Symbol	χ (EMU CGS units)	χ (SI units)	To obtain value in SI units, multiply value in EMU CGS unity by
Volume	χ	1	1	4π
Mass	χ_g	$\frac{\text{cm}^3}{\text{g}}$	$\frac{\text{m}^3}{\text{kg}}$	$\frac{4\pi}{1000}$
Molar	χ_m	$\frac{\text{cm}^3}{\text{mol}}$	$\frac{\text{m}^3}{\text{mol}}$ or $\frac{\text{m}^3}{\text{kmol}}$	$\frac{4\pi}{10^6}$ or $\frac{4\pi}{1000}$, respectively

[a]See Appendix A for other conversion factors.

$$\rho = 1 \frac{\text{g}}{\text{cm}^3} \quad \text{water density}$$

$$M_w = 18.02 \frac{\text{g}}{\text{mol}} \quad \text{water molar mass}$$

Susceptibility (CGS)

$$\chi(\text{CGS}) = \frac{\chi_m(\text{CGS})\rho}{M_w} = \frac{-12.97 \times 10^{-6}\left[\frac{\text{cm}^3}{\text{mol}}\right] \times 1\left[\frac{\text{g}}{\text{cm}^3}\right]}{18.02\left[\frac{\text{g}}{\text{mol}}\right]} = -0.720 \times 10^{-6} \tag{1.8}$$

Susceptibility (SI)

$$\chi(\text{SI}) = 4\pi\chi(\text{CGS}) = -9.04 \times 10^{-6} \tag{1.9}$$

Table 1.1 summarizes the conversion of various forms of susceptibility encountered in the literature. The numerical values of selected susceptibilities are listed in the Appendix D.

1.4. Bulk magnetic susceptibility of a mixture

If a given volume of substance consists of a mixture of n volumes of substances of different volume magnetic susceptibilities χ_i, $i = 1, 2, \ldots n$, then the bulk magnetic susceptibility χ is the weighed average of the individual susceptibilities:

$$\chi = \frac{V_1\chi_1 + V_2\chi_2 + \ldots + V_n\chi_n}{V} = \sum_{i=1}^{n} \frac{V_i}{V}\chi_i = \sum_{i=1}^{n} \phi_i\chi_i \qquad (1.10)$$

where ϕ_i is the volume fraction of ith component and $V = \sum_{i=1}^{n} V_i$. Note that the same applies to the mass and molar susceptibility providing that one substitutes volume fractions by the mass fractions or the molar fractions, respectively, as the weighing factors [by consulting Eqs. (1.4) and (1.7), correspondingly].

Equation (1.10) applies to well-mixed, single-phase physical solutions or colloidal suspensions whose components do not precipitate under the influence of the magnetic field [18].

Example Gadolinium solutions in water.
Gadolinium (Gd) is a member of the lanthanide series of elements [19]. It forms physical aqueous solutions as a trivalent cation or in the chelated form (as a DTPA Gd complex). It has a high atomic magnetic moment (7.95 Bohr magnetons, see below), which makes the solutions paramagnetic and therefore useful as an MRI contrast agent [20]. Gd and another high magnetic moment member of the lanthanide series, Erbium (Er), have been evaluated for applications to magnetic cell separation [21]. Paramagnetic Gd solutions are convenient standards for magnetic susceptibility measurements [16]. The magnetic susceptibility of 0.5 M Gd (as an ion, Gd^{3+}) in water is calculated below.

$$\chi_m = 0.027\frac{\text{cm}^3}{\text{mol}} = 27,000 \times 10^{-6}\frac{\text{cm}^3}{\text{mol}} \quad \text{molar susceptibility of Gd}$$

$$[\mathrm{Gd}] = 0.5\mathrm{M} = 0.5\frac{\mathrm{mol}}{\mathrm{liter}} \quad \text{molar concentration of Gd in water}$$

$$M_w = 157.25\frac{\mathrm{g}}{\mathrm{mol}} \quad \text{atomic weight}$$

$$\rho = 7.8\frac{\mathrm{g}}{\mathrm{cm}^3} \quad \text{density}$$

$$\chi(\mathrm{CGS}) = \frac{\chi_m\rho}{M_w} = \frac{0.027\left[\frac{\mathrm{cm}^3}{\mathrm{mol}}\right] \times 7.8\left[\frac{\mathrm{g}}{\mathrm{cm}^3}\right]}{157.25\left[\frac{\mathrm{g}}{\mathrm{mol}}\right]} = 0.00134 = 1340 \times 10^{-6}$$

$$\chi(\mathrm{SI}) = 4\pi\chi(\mathrm{CGS}) = 0.0168 = 16{,}800 \times 10^{-6}$$

$$V_m = \frac{M_w}{\rho} = \frac{157.25\left[\frac{\mathrm{g}}{\mathrm{mol}}\right]}{7.8\left[\frac{\mathrm{g}}{\mathrm{cm}^3}\right]} = 20.2\frac{\mathrm{cm}^3}{\mathrm{mol}}$$

$$= 0.0202\frac{\mathrm{liter}}{\mathrm{mol}} \quad \text{molar volume of Gd}$$

$$\varphi_{\mathrm{Gd}} = [\mathrm{Gd}]V_m = 0.5\left[\frac{\mathrm{mol}}{\mathrm{liter}}\right]0.0202\left[\frac{\mathrm{liter}}{\mathrm{mol}}\right] = 0.0101 \quad \text{Gd volume fraction in 0.5 M Gd solution}$$

$$\varphi_{\mathrm{H_2O}} = 1 - \varphi_{\mathrm{Gd}} = 0.99 \ \ \mathrm{H_2O} \text{ volume fraction in 0.5 M Gd solution}$$

$$\chi_{\mathrm{H_2O}}(\mathrm{SI}) = -9.04 \times 10^{-6} \quad \mathrm{H_2O} \text{ volume susceptibility}$$

$$\chi = \phi_{\mathrm{H_2O}}\chi_{\mathrm{H_2O}} + \phi_{\mathrm{Gd}}\chi_{\mathrm{Gd}} = 0.99 \times (-0.00000904) + 0.0101 \times 0.0168$$
$$= 0.000161 = 161 \times 10^{-6} \quad \text{magnetic susceptibility of 0.5 M Gd in water}$$

Commercial preparations used as MRI contrast agents are prepared as chelated compounds rather than pure ionic solutions. The empirical formula for one such preparation, Magnevist^{TM} (Berlex Laboratories, Bothell, WA), is $C_{28}H_{54}GdN_5O_{20}$[16].

1.5. Volume magnetic susceptibility and Bohr magneton

Bohr magneton, μ_B, is a natural unit of atomic and molecular magnetic dipole moments. The information about the molecular magnetic dipole moment allows one to calculate the magnetic susceptibility of the substance based on the theory developed by Paul Langevin for paramagnetic materials that are treated like an ideal gas [9]. The approximation works best for real gases (such as oxygen) and for dilute solutions. The elements of the Langevin model are presented in Chapter 2, in the context of sample magnetization. The examples of calculations of the susceptibilities from molecular magnetic moments are provided below.

Example Susceptibility of gadolinium chloride.
The atomic magnetic moment of Gd is 8 Bohr magnetons, $\mu_A = 8\,\mu_B$ (see below for numerical value), which remains unaffected by its chemical binding [9]. Therefore, as discussed in Chapter 2, the molar magnetic susceptibility of $GdCl_3$ at room temperature, $T = 293$ K, is:

$$\chi_m = \frac{N_A \mu_0 \mu_A^2}{3k(T - T_C)} = 0.346 \times 10^{-6} \frac{\text{m}^3}{\text{mol}} \tag{1.11}$$

where $T_C = 2.2$ K is the Curie temperature of $GdCl_3$ (N_A is the Avogadro number, k is the Boltzmann constant, and μ_0 is the magnetic permeability, see Appendix A). Interestingly, although the magnetic moment of Gd does not change with the chemical binding, the Curie temperature of $GdCl_3$ is significantly lower than that for the elemental Gd ($T_C = 290$ K). In the CGS system of units (Table 1.1)

$$\begin{aligned}\chi_m(\text{CGS}) &= \chi_m \frac{10^6}{4\pi} = 0.346 \times 10^{-6} \times \frac{10^6}{4\pi} = 0.0275 \\ &= 27,500 \times 10^{-6} \frac{\text{cm}^3}{\text{mol}}\end{aligned} \tag{1.12}$$

which is equal to one-gram formula weight susceptibility of $GdCl_3$ quoted in the CRC Handbook [11].

Example Susceptibility of ferritin.
Ferritin is an iron storage protein ubiquitous in the animal kingdom, found in bacteria and in mammals [22]. More about ferritin in Chapter 2. It can transport up to 4500 atoms of iron, with the corresponding ferritin molecular magnetic moment ranging from 150 to 400 Bohr magnetons [23]. It is a prototypical superparamagnetic molecule, with a magnetic dipole many times higher than that of iron (5.5 Bohr magnetons) or the lanthanides, that exhibits the highest atomic dipole moment (terbium, 12 Bohr magnetons) [9]. The magnetic moment of ferritin is referred to as the "effective" moment because it is calculated using Langevin approximation of an ideal gas in application to a crystalline structure of the ferritin iron core (without taking into account quantum mechanical effects). The ferritin loses its superparamagnetic properties below a blocking temperature, T_B, at which the ferritin behaves as an antiferromagnet, and above the Néel temperature, T_N, above which the ferritin behaves as a more conventional, paramagnetic substance. The native ferritin, such as isolated from horse spleen, has an average core consisting of 2000 iron atoms. Assuming the corresponding value of $\mu_{eff} = 100\ \mu_B$ and $T_B = 15$ K, one obtains for the molar susceptibility of ferritin in room temperature

$$\chi_m = \frac{N_A \mu_0 \mu_{\text{eff}}^2}{3k(T - T_B)} = 56.5 \times 10^{-6} \frac{\text{m}^3}{\text{mol}} \tag{1.13}$$

Converting to CGS system of units (Table 1.1), one obtains:

$$\chi_m(\text{CGS}) = \chi_m \frac{10^6}{4\pi} = 4.50 \frac{\text{cm}^3}{\text{mol}} = 4500 \times 10^{-3} \frac{\text{cm}^3}{\text{mol}} \tag{1.14}$$

which is somewhat lower than the value of 5900×10^{-3} cm^3/mol reported for the molar magnetic susceptibility of the ferritin iron core [24].

The corresponding volume susceptibility of ferritin is calculated from Eq. (1.7) by selecting the molecular weight of ferritin, $M_w = 560$ kg/mol, and density, $\rho = 2370$ kg/m^3, to obtain

$$\chi = \chi_m \frac{\rho}{M_w} = 0.000239 = 239 \times 10^{-6} \tag{1.15}$$

This value is almost three times higher than the ferritin susceptibility estimated for the native horse spleen ferritin [25], 87.9×10^{-6}. The difference may be related to the uncertainty as to the exact value of the ferritin molecular weight (the value as high as $M_w =$ 900,000 has been reported) and the ferritin density. The susceptibility is a square function of the effective magnetic moment, μ_A, Eq. (1.11), therefore the ferritin susceptibility strongly depends on the selection of the value of μ_A.

Example Susceptibility of the erythrocyte.
The red blood cell volume magnetic susceptibility is the weighted sum of the susceptibilities of its components, Eq. (1.10):

$$\chi_{\text{RBC}} = \phi_{\text{H}_2\text{O}}\chi_{\text{H}_2\text{O}} + (1 - S)\phi_{\text{Hb}}\chi_{\text{ferro}} + \phi_{\text{Hb}}\chi_{\text{globin}} \tag{1.16}$$

where $\phi_{\text{H}_2\text{O}}$ is the volume fraction of water in the erythrocyte, ϕ_{Hb} is the volume fraction of hemoglobin in the erythrocyte, $\chi_{\text{H}_2\text{O}}$ is the volume susceptibility of water, χ_{ferro} is the volume susceptibility of the four ferroheme groups in deoxyhemoglobin, ϕ_{globin} is the volume susceptibility of the protein (globin) part of the hemoglobin, and S is the oxygen saturation of oxyhemoglobin.

It is important to note that the heme group is only paramagnetic in the absence of bound oxygen (deoxygenated state); the heme's magnetic moment is zero in the presence of bound oxygen (oxygenated state) [12, 14, 26]. Equation (1.16) was adapted from Refs. [27] and [28] as modified by others groups [13, 14, 17, 29, 30]. The susceptibility values are given in the CGS system of units to facilitate comparison with the literature data; the conversion to values in the SI unit system requires multiplication by 4π.

The average magnetic moment of the ferroheme group in deoxyhemoglobin is $n_B = 5.46$ Bohr magnetons [12], therefore:

$$\mu_A = n_B \mu_B = 5.46\mu_B = 5.46 \times 9.274 \times 10^{-24}\,\frac{\mathrm{J}}{\mathrm{T}} \tag{1.17}$$

The molar susceptibility contributed by the four ferroheme groups of the deoxyhemoglobin at room temperature (293 K) is calculated as described in Chapter 2:

$$\chi_{m,\mathrm{ferro}} = 4\frac{N_A \mu_0 \mu_A^2}{3kT} = 0.0509 = 50{,}900 \times 10^{-6}\,\frac{\mathrm{cm}^3}{\mathrm{mol}} \tag{1.18}$$

The volume fraction taken up by water in the erythrocyte is determined from the volume fraction of the hemoglobin [compare with Eq. (1.10)]:

$$\phi_{\mathrm{H_2O}} = 1 - V_{m,\mathrm{Hb}} c_{\mathrm{Hb}} = 0.734 \tag{1.19}$$

where $V_{m,\mathrm{Hb}} = 48,277\ \mathrm{cm}^3/\mathrm{mol} = 48.277\ \mathrm{liter/mol}$ is the molar volume of hemoglobin, and $c_{\mathrm{Hb}} = 5.5\ \mathrm{mM} = 5.5 \times 10^{-3}\ \mathrm{mol/liter}$ is the intracellular hemoglobin concentration.

The conversion to volume susceptibilities is obtained by observing that

$$\begin{aligned} \phi_{\mathrm{Hb}} \chi_{\mathrm{ferro}} &= c_{\mathrm{Hb}} \chi_{m,\,\mathrm{ferro}} \\ \phi_{\mathrm{Hb}} \chi_{\mathrm{globin}} &= c_{\mathrm{Hb}} \chi_{m,\,\mathrm{globin}} \end{aligned} \tag{1.20}$$

The molar susceptibility of the globin is calculated from its mass (or specific) susceptibility by combining Eqs. (1.4) and (1.7) so that

$$\chi_{m,\,\mathrm{globin}} = M_{w,\,\mathrm{globin}} \chi_{g,\,\mathrm{globin}} = -0.0374 = -37,400 \times 10^{-6}\,\frac{\mathrm{cm}^3}{\mathrm{mol}} \tag{1.21}$$

where $M_{w,\,\mathrm{globin}} = 64,450\ \mathrm{g/mol}$ is the molecular weight of the globin, and $\chi_{g,\,\mathrm{globin}} = -0.580 \times 10^{-6}\mathrm{cm}^3/\mathrm{g}$ is the mass (or specific) susceptibility of the globin.

Note that the magnitude of the diamagnetic contribution of the globin portion, Eq. (1.21), is smaller than the magnitude of the paramagnetic contribution of the four ferroheme groups, Eq. (1.18).

By inserting terms from Eqs. (1.18) through (1.21) into Eq. (1.16), one arrives at the following function of the erythrocyte volume susceptibility as a function of the oxyhemoglobin saturation, S:

$$\chi_{RBC} = -0.4538 \times 10^{-6} - 0.2799 \times 10^{-6} \times S \tag{1.22}$$

where $0 \leq S \leq 1$.

The magnetic susceptibility of the erythrocyte depends on its oxygen saturation, in particular,

$\chi_{RBC} = -0.7337 \times 10^{-6}$ fully oxygenated erythrocyte, $S = 1$

$\chi_{RBC} = -0.4538 \times 10^{-6}$ fully deoxygenated erythrocyte, $S = 0$

Considering that the magnetic susceptibility (CGS) of water is -0.720×10^{-6} [Eq. (1.8)], one observes that

$\chi_{RBC} < \chi_{H_2O}$ fully oxygenated erythrocyte (diamagnetic contrast relative to water)

$\chi_{RBC} > \chi_{H_2O}$ fully deoxygenated erythrocyte (paramagnetic contrast relative to water)

The methemoglobin contribution to the erythrocyte magnetic susceptibility has a similar form as the one above:

$$\chi_{RBC} = \phi_{H_2O}\chi_{H_2O} + Z\phi_{Hb}\chi_{met\ Hb} + \phi_{globin}\chi_{globin} \tag{1.23}$$

where x_{metHb} and χ_{metHb} are the volume fraction and the magnetic susceptibility of methemoglobin, and Z is the fraction of hemoglobin converted to methemoglobin. Here it is assumed that hemoglobin exists in two forms only, as methemoglobin (ferrihemoglobin) or oxyhemoglobin. Also, $\phi_{Hb}\chi_{metHb} = c_{Hb}\chi_{m,metHb}$, where $\chi_{m,metHb} = 57{,}428\ \text{cm}^3/\text{mol}$ [30, 31].

Example Magnetic susceptibility of the white blood cell (approximate).

Assuming that the protein concentration of the white blood cell is similar to the globin concentration of the erythrocyte and that the magnetic susceptibility of the nuclear material and other organelles is not much different from that of the globin, one obtains that the susceptibility of the white blood cell is approximated by that of the oxygenated erythrocyte, χ (CGS) $= -0.7337 \times 10^{-6}$.

1.6. Multiphase magnetic suspensions

A necessary, but not sufficient, condition for achieving significant magnetic effects in microscale is that the magnetic field energy is greater than that caused by the random motion due to the thermal energy of the system:

$$\frac{\mu B_0}{kT} \geq 10 \tag{1.24}$$

where μ is the magnetic dipole moment (see Chapter 2 for additional explanation), B_0 is the applied magnetic field intensity, k is the Boltzmann constant, and T is the absolute temperature (in kelvin). The above condition determines only if the magnetic field is sufficient to maintain fixed orientation of the magnetic dipole against randomizing effects of thermal motion. As noted in the next section, the above condition is not sufficient for achieving magnetic separation as the magnetically induced particle displacement depends also on the magnetic field gradient. Therefore, Eq. (1.24) provides only a preliminary check before any further consideration is made of the use of the magnetic field for separation in microscale. The application of the above equation is illustrated for two cases: (1) colloidal suspensions and (2) microparticle suspensions.

Example Colloidal suspension of 50-nm gelatin particles doped with hematite and particulate suspension of 2-μm polystyrene microspheres covered with magnetite (see Chapter 6 for details about particle synthesis) and B_0 of 1 T.

The relationship between the magnetic moment, μ, the particle volume magnetization, M, and the particle mass magnetization, M_g is discussed in Chapter 2. For colloidal suspension of 50-nm particles one obtains:

$$R = 25 \text{ nm}$$

$$V = 6.54 \times 10^{-23}\,\text{m}^3$$

$$M_g = 40\,\frac{\text{emu}}{\text{g}}$$

$$\rho = 2\,\frac{\text{g}}{\text{cm}^3}$$

$$M = M_g\rho = 80\,\frac{\text{emu}}{\text{cm}^3} = 80{,}000\,\frac{\text{A}}{\text{m}}$$

$$\mu = MV = 5.24 \times 10^{-18}\,\frac{\text{J}}{\text{T}}$$

$$B_0 = 1 \text{ T}$$

$$\mu B_0 = 5.24 \times 10^{-18}\,\text{J}$$

As an aside, note that in the units of Bohr magnetons ($\mu_B = 9.274 \times 10^{-24}\,\text{J/T}$) the particle magnetic moment is equal to $\mu \approx 6 \times 10^5\,\mu_B$, orders of magnitude higher than that of the molecular ferritin ($\mu \approx 150$–$400\,\mu_B$) or the deoxyhemoglobin ($\mu = 5.46\,\mu_B$), discussed above.

The magnetic to the thermal energy ratio in room temperature (20 °C) and a high field of 1 T is therefore

$$k = 1.381 \times 10^{-23}\,\frac{\text{J}}{\text{K}}$$

$$T = 300 \text{ K}$$

$$kT = 4.14 \times 10^{-21}\,\text{J}$$

$$\frac{\mu B_0}{kT} = \frac{5.24 \times 10^{-18}}{4.14 \times 10^{-21}} \approx 10^3$$

Since the magnetic energy greatly predominates, the magnetic field effects on the orientation of the magnetic dipoles in colloidal suspension in 1-T field at room temperature are not negligible. The ratio of the magnetic to thermal energy is even higher for the 2-μm microspheres of magnetization 4 emu/g and density 1.5 g/cm^3, and equals to $\sim 6 \times 10^6$.

1.7. Magnetic phase separation

Magnetic separation requires not only the application of the field but also a suitably high magnetic field gradient. This is described in detail in the following chapters. Here general conditions of the feasibility of the magnetic separation are discussed, as they apply to the difference in the magnetic susceptibility between two phases. Such conditions must include not only the field and the field gradient, but also properties of the fluid (viscosity) and the residence time in the magnetic field, in order to be of practical value. The effects at the interface of immiscible continuous media differing in the magnetic susceptibility are described in detail elsewhere [32–34].

At first, the condition of magnetic separation is described for a single particle suspended in a stationary liquid. The description is based on a random walk model of diffusion and a motion of a small sphere in a viscous medium induced by the magnetic body force. Second, the condition is extended to a particle entrained in a laminar flow past the magnetic field source. For both the stationary and the laminar flow conditions, three different types of particles are considered: a submicron synthetic particle, an erythrocyte and a synthetic micron-sized particle.

1.7.1. Stationary fluid

The condition of the magnetic displacement greatly exceeding the randomizing thermal effects can be expressed by the following inequality:

$$x_m \gg \langle x \rangle \tag{1.25}$$

where x_m is the displacement caused by the magnetic field, and $\langle x \rangle$ is the root-mean-square distance due to the Brownian motion. The magnetic field displacement is calculated from the following expression:

$$x_m = mS_m t \tag{1.26}$$

where m is the magnetophoretic mobility, S_m is the magnetophoretic driving force (Maxwell pressure gradient, to be discussed in detail in later chapters and in Appendix C), and t is the residence time in the magnetic field.

The root-mean-square displacement due to thermal motion is given by the Einstein formula [35]:

$$\langle x \rangle = \sqrt{2Dt} \tag{1.27}$$

where D is the diffusion coefficient, determined from the Stokes–Einstein formula [36]

$$D = \frac{kT}{6\pi\eta R} \tag{1.28}$$

It appears that the condition in Eq. (1.25) could always be met, providing a sufficiently long residence time in the magnetic field has elapsed, however, there are obvious practical limitations to the observation time, t. In particular, if the particles are sufficiently large, the sedimentation effects may become more important than the randomizing thermal motion. In that case, one also requires that the magnetic displacement is significantly larger than that due to the sedimentation, or

$$x_m \gg x_g \tag{1.29}$$

where

$$x_g = s_g g t \tag{1.30}$$

Here g is the standard gravitational acceleration ($g = 9.81\mathrm{m/s^2}$) and s_g is the sedimentation coefficient [37]

$$s_g = \frac{\Delta\rho V}{6\pi\eta R} \tag{1.31}$$

In order to obtain more specific conditions for the separability of the magnetically susceptible species, one needs to consider the expression for m in Eq. (1.26). Here we have two cases, depending on whether the particle is linearly polarizable (paramagnetic and diamagnetic substances) or behaves as a saturated magnetic dipole (ferromagnetic substance).

1.7.1.1. Linearly polarizable particles (paramagnetic and diamagnetic substances)

The derivation of the magnetophoretic mobility, m, and the magnetophoretic driving force, S_m, are provided in Chapter 5. Here one substitutes the following expressions for m and S_m:

$$m = \frac{\Delta\chi V}{6\pi\eta R} \tag{1.32}$$

and

$$S_m = \frac{\mathrm{d}}{\mathrm{d}x}\left(\frac{1}{2}HB_0\right) = H\frac{\mathrm{d}B_0}{\mathrm{d}x} \quad (B_0 \text{ is the applied field}) \tag{1.33}$$

so that

$$x_m = mS_m t = \frac{\Delta\chi V}{6\pi\eta R} H \frac{\mathrm{d}B_0}{\mathrm{d}x} t \tag{1.34}$$

Here

$$\Delta\chi = \chi_2 - \chi_1 \tag{1.35}$$

is the magnetic susceptibility difference between the two phases. The parameter $\Delta\chi$ is related to the MRI contrast [20], and the parameter $\Delta\chi V$ is related to the "particle-field interaction parameter" [35] (also related to the magnetic polarizability [3]). The conditions of a magnetic separation described in Eqs. (1.25) and (1.29) will be illustrated using as an example the prototypical paramagnetic particle, the deoxygenated erythrocyte.

Example The deoxygenated erythrocyte.
In the context of cell separation, the temperature ranges typically from 4 °C to the room temperature, 23 °C, or 300 K. The solution viscosity depends strongly on temperature in that temperature range and for 155-mM NaCl solution, $\eta(4\,°\mathrm{C}) = 1.593$ mPa s, and $\eta(23\,°\mathrm{C}) = 0.948$ mPa s. Other parameters entering Eq. (1.34), characteristic of the field-induced displacement of the deoxygenated erythrocytes in the high gradient magnetic field, are as follows:

$R = 2\ \mu m$ the erythrocyte radius

$V = 3.35 \times 10^{-17}\,\mathrm{m}^3$ the erythrocyte volume

$\chi_{\mathrm{deoxy}}(\mathrm{CGS}) = -0.454 \times 10^{-6}$ deoxygenated erythrocyte susceptibility

$\chi_{\mathrm{H_2O}}(\mathrm{CGS}) = -0.720 \times 10^{-6}$ water susceptibility

$\Delta\chi(\mathrm{CGS}) = 0.266 \times 10^{-6}$ net susceptibility of deoxygenated erythrocyte in water

$\Delta\chi = 3.35 \times 10^{-6}(\mathrm{SI})$ as above but in the SI units system

$\rho = 1.15 \times 10^3\,\dfrac{\mathrm{kg}}{\mathrm{m}^3}$ the erythrocyte density

$\Delta\rho = 0.15 \times 10^3\,\dfrac{\mathrm{kg}}{\mathrm{m}^3}$ difference between the erythrocyte and water density

$D = 1.10 \times 10^{-13}\,\dfrac{\mathrm{m}^2}{\mathrm{s}}$ erythrocyte diffusion coefficient [from Eq. (1.28), at 293 K]

$\eta = 10^{-3}$ Pa s solution viscosity at room temperature, RT = 293 K

$B_0 = 1\,\mathrm{T}$ magnetic field

$\dfrac{\mathrm{d}B_0}{\mathrm{d}x} = 100\,\dfrac{\mathrm{T}}{\mathrm{m}}$ magnetic field gradient

$$H = \frac{B_0}{\mu_0} = 796{,}000 \frac{\text{A}}{\text{m}} \quad \text{magnetic field strength}$$

$$g = 9.81 \frac{\text{m}}{\text{s}^2} \quad \text{standard gravitational acceleration}$$

$$t = 10 \text{ min} = 600 \text{ s}$$

Substitution of the above numerical values to Eqs. (1.34), (1.27), and (1.30) yields the following erythrocyte displacements in solution due to the magnetic, thermal, and gravitational effects:

$$x_m = 140 \ \mu\text{m}$$

$$\langle x \rangle = 11 \ \mu\text{m}$$

$$x_g = 790 \ \mu\text{m}$$

One concludes that even in the presence of the strong magnetic field and gradient, the gravitational effects predominate, and that the magnetic field displacement is just about an order of magnitude higher than that caused by random thermal motion [15].

1.7.1.2. Saturated magnetic dipole (superparamagnetic iron oxide particle in saturating magnetic field)

Here the magnetic displacement is calculated for magnetic particles described in Section 1.6. As in the case of the deoxygenated erythrocyte, discussed above, it is assumed that the particles are suspended in aqueous solution, in the absence of paramagnetic ions, at room temperature. Typically, the magnetization of superparamagnetic particles saturates in low to moderate magnetic fields (as discussed in the subsequent chapters) and is orders of magnitude higher than that of the aqueous solution [38–40]. Therefore, two modifications are made in the expression for the particle magnetophoretic mobility in Eq. (1.32): (1) the contribution of water is dropped, and (2) the particle susceptibility, χ_p, is interpreted as the effective susceptibility, that is, the ratio of particle magnetization, M, to the applied magnetic field, H (Chapter 2):

$$m = \frac{\Delta\chi V}{6\pi\eta R} \approx \frac{\chi_p V}{6\pi\eta R} = \frac{MV}{6\pi\eta RH}$$
$$x_m = mS_m t = \frac{MV}{6\pi\eta R}\frac{\mathrm{d}B_0}{\mathrm{d}x}t \qquad (1.36)$$

In contrast to the paramagnetic particles, Eq. (1.34), the magnetic displacement of the saturated magnetic particles does not depend explicitly on the applied magnetic field H (the dependence on H is hidden in the assumption that the saturating magnetic field is applied to the particles).

In comparison, for the type of magnetic particles described in Section 1.6 and the magnetic field as in the example of the deoxygenated erythrocyte magnetophoresis, above, one obtains:

Example Colloidal, 50-nm hematite gelatin particles.

$$x_m = 670\ \mu\text{m}$$
$$\langle x \rangle = 100\ \mu\text{m}$$
$$x_g = 0.82\ \mu\text{m}$$

The displacement due to sedimentation is less than that due to the thermal motion, explaining the colloidal character of the particles. Although the magnetic displacement is significantly higher than that due to the sedimentation, it is only about seven times higher than the displacement caused by the thermal motion. This indicates only a weak magnetic effect on the particle displacement [compare inequality in Eq. (1.25)].

It is interesting to note that here the magnetic field effects predicted on the basis of the particle displacement, Eq. (1.25), are much weaker than the effects predicted on the basis of the ratio of the magnetic to thermal energy, Eq. (1.24), and discussed in Section 1.6. There, it was calculated that the ratio of magnetic to thermal energy of the nanoparticle is on the order of 10^3. The difference is explained by the fact that the field alone does not cause the magnetic particle displacement (but only the alignment of the magnetic dipole with the magnetic field lines, as discussed in the later chapters), and that the field gradient is necessary for the magnetic particle displacement. Moreover, the comparison based on the particle displacement, Eq. (1.25), includes energy dissipation due to friction in viscous media, which does not enter the Eq. (1.24).

Example 2-μm polystyrene microspheres covered with magnetite.

$$x_m = 80{,}000\ \mu\text{m}$$

$$\langle x \rangle = 160\ \mu\text{m}$$

$$x_g = 650\ \mu\text{m}$$

Here the displacement due to sedimentation is higher than that due to the thermal motion, resulting in the particle precipitation from the suspension (albeit at a slow rate). Also, the magnetic displacement greatly exceeds the thermal and gravitational effects, indicating highly magnetic properties of such particles.

1.7.2. Laminar flow

The superposition of laminar flow of the fluid medium on the magnetic, thermal, and gravitational effects adds yet another condition to be considered when determining the likelihood of separation of magnetic from nonmagnetic phase. In its simplest form, it can be stated as a requirement that the magnetic body force greatly exceeds the viscous drag of the suspending fluid medium, or

$$F_m \gg F_d \tag{1.37}$$

where the magnetic body force is obtained from Eq. (1.1) and the viscous drag is directly proportional to the fluid velocity, v, relative to the magnetic particle

$$F_d = 6\pi\eta R v \tag{1.38}$$

The combination of Eqs. (1.1), (1.37), and (1.38) leads to the condition

$$v \ll \frac{2}{9}\frac{\Delta\chi R^2}{\eta} H \frac{\mathrm{d}B_0}{\mathrm{d}x} \tag{1.39}$$

where the use was made of the relationship between the particle volume, V, and the particle radius, R, and the particle susceptibility, χ, was replaced by the difference between particle and fluid susceptibilities, $\Delta\chi$.

The effect of the flow on the separation of magnetic and non-magnetic phases is illustrated by calculating the maximum flow velocity, v_{max}, that meets the condition stated in Eq. (1.39). Here, as in previous examples, the sign "$\ll$" implies "at least one order of magnitude" smaller. The examples shown below concern the type of suspensions discussed above for stationary fluids.

Example An erythrocyte in the laminar flow.
For the numerical parameter values shown in Section 1.7.1.1, one obtains

$$v_{\mathrm{max}} \ll 0.28\ \mu\mathrm{m/s}$$

This obviously severely limits application of any type of flow that could be applied to the magnetically separated erythrocytes, without disturbing the phase separation.

Example Colloidal, 50-nm hematite gelatin particles.
Here, as in the section on stationary fluid, one replaces the expression for the difference between magnetic susceptibilities of the particle and fluid by the ratio of particle magnetization, M, to that of the applied field, H, in Eq. (1.39), leading to the condition

$$v \ll \frac{2}{9}\frac{MR^2}{\eta}\frac{\mathrm{d}B_0}{\mathrm{d}x} \tag{1.40}$$

where the magnetic field strength, H, is implicit in the condition of the particle saturation magnetization, M. Using the numerical values of parameters provided in Section 1.6, one obtains:

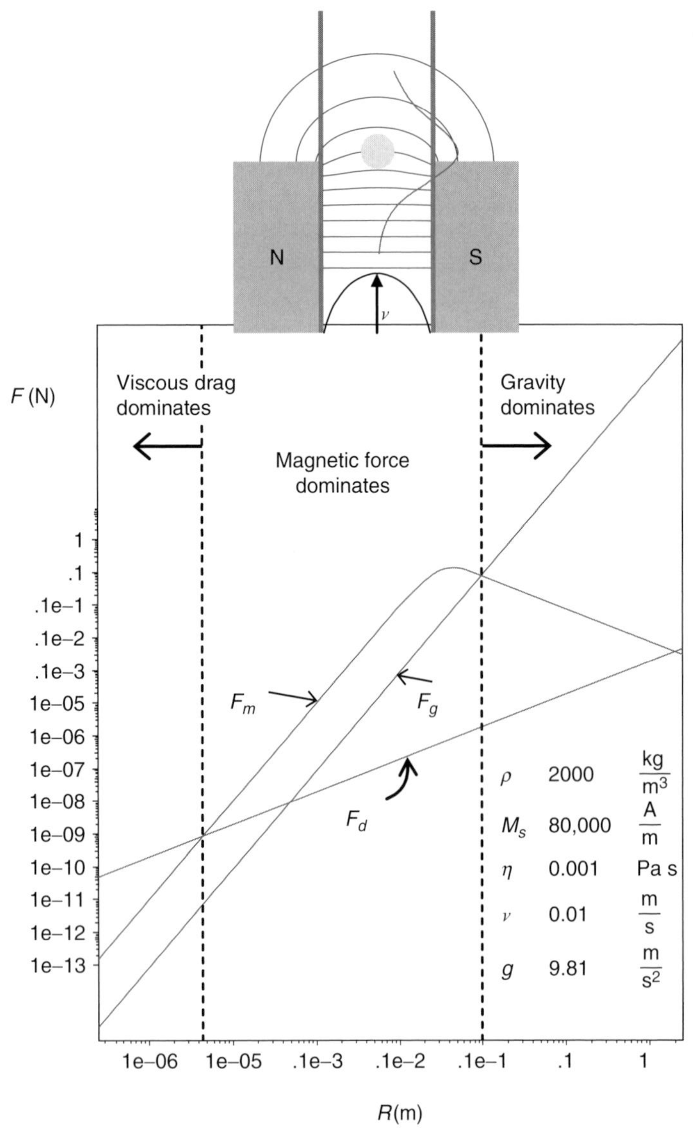

Fig. 1.2. Comparison of magnetic, F_m, to viscous drag, F_d, to gravitational, F_g, forces as a function of magnetic particle diameter, R, for a particle held in a

$$v_{max} \ll 1.1\,\mu m/s$$

This restricts the allowable flow almost as severely as in the case of the deoxygenated erythrocytes, discussed above.

Example 2-μm polystyrene microspheres covered with magnetite. Using the same particle, fluid, and field parameters as those discussed for the stationary fluids, one obtains:

$$v \ll 0.130\,mm/s$$

The right-hand side of the inequality corresponds to the average velocity of the volumetric flow rate of 1.5 ml/h when pumped through a capillary of 1 mm in diameter.

The examples provided above demonstrate that the viscous effects predominate when considering magnetic separation of micrometer-sized and submicron particles. This point is further emphasized in a summary provided in Fig. 1.2.

As a closing comment, the estimates of the magnetically induced displacement in aqueous media, provided above, concern only a single magnetic particle. For colloidal and particulate dispersions, the body force effects are enhanced by inter-particle interactions through the incompressible fluid and depend on the volume fraction of the particles, particularly around 1% and above [37, 41, 42]. The cooperative effect on the sedimentation of particulate dispersion has been reported for volume fractions as low as 0.01% [43]. Therefore, the above comparisons between the magnetic and other types of particle motion in microscale should be considered only as rough estimates [44].

---

laminar flow of the maximum linear velocity $v = 1$ cm/s, as depicted in the sketch above (the vertical curve in the upper panel indicates ~1 cm range of a nonvanishing magnetic gradient of maximum magnitude of 100 T/m). Inspired by Oberteuffer [45].

References

[1] Purcell, E. M. (1985). Electricity and magnetism. Berkeley Physics Course—Volume 2. McGraw-Hill, New York.

[2] Feynman, R. P., Leighton, R. B. and Sands, M. (1977). Lectures on physics: Mainly electromagnetism and matter. Addison-Wesley, Reading, MA.

[3] Becker, R. (1982). Electromagnetic fields and interactions. Dover Publications, Inc., New York.

[4] Stratton, J. A. (1941). Electromagnetic theory. McGraw-Hill, New York.

[5] Cullity, B. D. (1972). Introduction to magnetic materials. Addison-Wesley, Reading, MA.

[6] Jiles, D. (1998). Introduction to magnetism and magnetic materials. 2nd ed. Chapman and Hall, London.

[7] Jacubovics, J. P. (1994). Magnetism and magnetic materials. Cambridge University Press, Cambridge.

[8] Simon, M. D. and Geim, A. K. (2000). Diamagnetic levitation: Flying frogs and floating magnets (invited). Appl. Phys. A *87*, 6200–4.

[9] Bozorth, R. M. (1947). Magnetism. Rev. Mod. Phys. *19*, 29–86.

[10] Bleaney, B. I. and Bleaney, B. (1991). Electricity and magnetism, Vol. 1. Oxford University Press, Oxford.

[11] Weast, R. C. (1986). CRC handbook of chemistry and physics. 67 ed. CRC Press, Inc., Boca Raton, FL.

[12] Pauling, L. and Coryell, C. D. (1936). The magnetic properties and structure of hemoglobin, oxyhemoglobin and carbonmonoxyhemoglobin. Proc. Natl. Acad. Sci. USA *22*, 210–6.

[13] Fabry, M. E. and San George, R. C. (1983). Effect of magnetic susceptibility on nuclear magnetic resonance signals arising from red cells: A warning. Biochemistry *22*, 4119–25.

[14] Spees, W. M., Yablonskiy, D. A., Oswood, M. C. and Ackerman, J. J. (2001). Water proton MR properties of human blood at 1.5 Tesla: Magnetic susceptibility, T(1), T(2), T*(2), and non-Lorentzian signal behavior. Magn. Reson. Med. *45*, 533–42.

[15] Svoboda, J. (2000). Separation of red blood cells by magnetic means. J. Magn. Magn. Mater. *45*, 533–42.

[16] Moore, L. R., Milliron, S., Williams, P. S., Chalmers, J. J., Margel, S. and Zborowski, M. (2004). Control of magnetophoretic mobility by susceptibility-modified solutions as evaluated by cell tracking velocimetry and continuous magnetic sorting. Anal. Chem. *76*, 3899–907.

[17] Plyavin, Y. and Blum, E. (1983). Magnetic parameters of blood cells and high gradient paramagnetic and diamagnetic phoresis. Magnetohydrodynamics *19*, 349–59.

[18] Russell, A. P., Evans, C. H. and Westcott, V. C. (1987). Measurement of the susceptibility of paramagnetically labeled cells with paramagnetic solutions. Anal. Biochem. *164*, 181–9.

[19] Evans, C. H. (1990). Biochemistry of the lanthanides. Plenum Press, New York.

[20] Weinmann, H. J., Brasch, R. C., Press, W. R. and Wesbey, G. E. (1984). Characteristics of gadolinium-DTPA complex: A potential NMR contrast agent. Am. J. Roentgenol. *142*, 619–24.

[21] Zborowski, M., Malchesky, P. S., Jan, T. F. and Hall, G. S. (1992). Quantitative separation of bacteria in saline solution using lanthanide Er(III) and a magnetic field. J. Gen. Microbiol. *138*, 63–8.

[22] Harrison, P. M., Andrews, S. C., Artymiuk, P. J., Ford, G. C., Lawson, D. M., Smith, J. M. A., Treffry, A. and White, J. L. (1990). Ferritin. In: Iron Transport and Storage (Ponka, P., Schulman, H. M. and Woodworth, R. C., eds.). CRC Press, Inc., Boca Raton, FL, pp. 81–101.

[23] Gider, S., Awschalom, D. D., Douglas, T., Mann, S. and Chaparala, M. (1995). Classical and quantum magnetic phenomena in natural and artificial ferritin proteins. Science *268*, 77–80.

[24] Odette, L. L., McCloskey, M. A. and Young, S. H. (1984). Ferritin conjugates as specific magnetic labels. Implications for cell separation. Biophys. J. *45*, 1219–22.

[25] Zborowski, M., Fuh, C. B., Green, R., Sun, L. and Chalmers, J. J. (1995). Analytical magnetapheresis of ferritin-labeled lymphocytes. Anal. Chem. *67*, 3702–12.

[26] Savicki, J. P., Lang, G. and Ikeda-Saito, M. (1984). Magnetic susceptibility of oxy- and carbonmonoxyhemoglobins. Proc. Natl. Acad. Sci. USA *81*, 5417–9.

[27] Cerdonio, M., Congiu-Castellano, A., Calabrese, L., Morante, S., Pispisa, B. and Vitale, S. (1978). Room-temperature magnetic properties of oxy- and corbonmonoxyhemoglobin. Proc. Natl. Acad. Sci. USA *75*, 4916–19.

[28] Cerdonio, M., Morante, S., Torresani, D., Vitale, S., DeYoung, A. and Noble, R. W. (1985). Reexamination of the evidence for paramagnetism in oxy- and carbonmonoxyhemoglobins. Proc. Natl. Acad. Sci. USA *82*, 102–3.

[29] Weisskoff, R. M. and Kiihne, S. (1992). MRI susceptometry: Image-based measurement of absolute susceptibility of MR contrast agents and human blood. Magn. Reson. Med. *24*, 375–83.

[30] Graham, M. D. (1984). Comparison of volume and surface mechanisms for magnetic filtration of blood cells. J. Physique, Colloque C1 *45 (Suppl.)*, 779–84.

[31] Coryell, C., Stitt, F. and Pauling, L. (1937). The magnetic properties and structure of ferrihemoglobin (methemoglobin) and some of its compounds. J. Am. Chem. Soc. *59*, 633–42.
[32] Carles, P., Huang, Z., Carbone, G. and Rosenblatt, C. (2006). Rayleigh-Taylor instability for immiscible fluids of arbitrary viscosities: A magnetic levitation investigation and theoretical model. Phys. Rev. Lett. *96*, 104501–4.
[33] Mahajan, M. P., Zhang, S., Tsige, M., Taylor, P. L. and Rosenblatt, C. (1999). Stability of magnetically levitated liquid bridges of arbitrary volume subjected to axial and lateral gravity. J. Colloid. Interface. Sci. *213*, 592–5.
[34] Rosensweig, R. E. (1985). Ferrohydrodynamics. Cambridge University Press, Cambridge, MA.
[35] Giddings, J. C. (1993). Field-flow fractionation: analysis of macromolecular, colloidal, and particulate materials. Science *260*, 1456–65.
[36] Dhont, J. K. G. (1996). An introduction to dynamics of colloids. In: Published in studies in interface science (Moebius, D. and Miller, R., eds.), Vol. 2. Elsevier, Amsterdam.
[37] Russel, W. B., Saville, D. A. and Schowalter, W. R. (1989). Colloidal dispersions. In: Published in Cambridge monographs on mechanics and applied mathematics (Batchelor, G. K., ed.). Press Syndicate of the University of Cambridge, Cambridge.
[38] Chatterjee, J., Haik, Y. and Chen, C.-J. (2003). Size dependent magnetic properties of iron oxide nanoparticles. J. Magn. Magn. Mater. *257*, 113–8.
[39] Gupta, A. K. and Gupta, M. (2005). Synthesis and surface engineering of iron oxide nanoparticles for biomedical applications. Biomaterials *26*, 3995–4021.
[40] Vassiliou, J., Mehrotra, V., Russell, M. W., Giannelis, E. P., McMichael, R. D., Shull, R. D. and Ziolo, R. F. (1993). Magnetic and optical properties of gamma-Fe_2O_3 nanocrystals. J. Appl. Phys. *73*, 5109–16.
[41] Happel, J. and Brenner, H. (1965). Low Reynolds number hydrodynamics. Prentice-Hall, Inc., Englewood Cliffs, NJ.
[42] Pozrikidis, C. (1992). Boundary integral and singularity methods for linearized viscous flow. Published in Cambridge texts in applied mathematics. Cambridge Universtiy Press, Cambridge.
[43] Brodkey, R. S. (1995). The phenomena of fluid motions. Dover Publications, Inc., Mineola, NY.
[44] Mikkelsen, C., Hansen, M. F. and Bruus, H. (2005). Theoretical comparison of magnetic and hydrodynamic interactions between magnetically tagged particles in microfluidic systems. J. Magn. Magn. Mater. *293*, 578–83.
[45] Oberteuffer, J. A. (1973). High gradient magnetic separation. IEEE Trans. Magn. MAG-*9*, 303–6.

Laboratory Techniques in Biochemistry and Molecular Biology, Volume 32
Magnetic Cell Separation
M. Zborowski and J. J. Chalmers (Editors)

CHAPTER 2

Magnetic formulary

Maciej Zborowski

Department of Biomedical Engineering, Learner Research Institute, Cleveland Clinic, Cleveland, OH 44195, USA

The behavior of the magnetic field in matter is a topic of numerous textbooks in physics and electrical engineering, and in scientific periodicals, of which only a limited number can be listed here. It covers a vast area of materials with applications in electrical motors, power generation and transmission, information storage, medical imaging, and, increasingly, drug targeting, tumor therapy, and cell separation. The topic is only briefly mentioned for completeness. The interested reader is referred to publications cited in references [1–4].

2.1. Langevin theory of paramagnetism

Perhaps the simplest model of magnetic material based on its microscopic properties is the Langevin theory of paramagnetism [5]. The paramagnetic atoms behave as elementary magnetic dipoles, randomly oriented in space in the absence of the external magnetic field. A natural unit of the atomic magnetic dipole moment is that of the Bohr magneton, associated with the motion of the electron in the ground state of the hydrogen atom:

$$\mu_{\mathrm{B}} = \frac{h}{4\pi}\frac{e}{m_e} \approx 9.274 \times 10^{-24}\,\frac{J}{T} \qquad (2.1)$$

DOI: 10.1016/S0075-7535(06)32002-5

where $h \approx 6.626 \times 10^{-34}$ J·s is the Planck constant, $e \approx 1.602 \times 10^{-19}$ A·s is the elementary charge, and $m_e \approx 9.109 \times 10^{-31}$ kg is the electron mass. In the CGS system of units, $\mu_B \approx 9.274 \times 10^{-21}$ (erg/G).

The Langevin theory of paramagnetism applies to an ideal gas of magnetic dipoles in thermal equilibrium with the environment using methods of classical statistical mechanics. The theory is successful in predicting magnetization and magnetic susceptibility of gases (such as oxygen) and solutions. It provides a convenient, mechanistic model of susceptibility and magnetization saturation, and a suitable point of departure for discussing nonlinear phenomena (ferromagnetism) [5] and the behavior of magnetic colloids [6]. It predicts that with the increasing intensity of the applied field, the initially random orientation of the magnetic dipoles becomes increasingly ordered in the direction of the applied field. Based on the arguments of the statistical mechanics, the average magnetic moment of an ensemble of the magnetic dipoles, $\langle\mu\rangle$, is a function of the applied field and temperature:

$$\frac{\langle\mu\rangle}{\mu_A} = \coth\frac{\mu_A B_0}{kT} - \frac{kT}{\mu_A B_0} \tag{2.2}$$

where μ_A is the magnetic moment of the single dipole. The molar magnetization, M_m, of an Avogadro number, N_A, of the magnetic dipoles is

$$M_m = N_A\langle\mu\rangle \tag{2.3}$$

The dependence of M_m/M_s on $\mu_A B_0/kT$ is illustrated in Fig. 2.1 (where M_s is the saturation magnetization). The function on the right-hand-side of Eq. (2.2) is called the Langevin function, $L(x) \equiv \coth x - 1/x$. In the limit of $x << 1$, $L(x) \approx x/3$, and therefore

$$\frac{\langle\mu\rangle}{\mu_A} \approx \frac{\mu_A B_0}{3kT}, \frac{\mu_A B_0}{kT} << 1 (\text{paramagnetism}) \tag{2.4}$$

$$M_m \approx \frac{N_A \mu_A^2 B_0}{3kT}$$

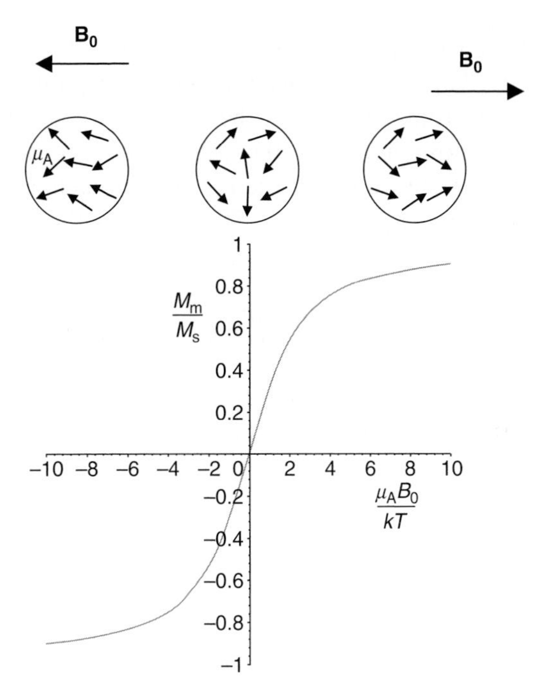

Fig. 2.1. Orientation of magnetic moments of an ideal gas in response to the applied magnetic field $\mathbf{B}_0$ (top) and the corresponding magnetization of the volume of gas, M, as a function of B_0 and temperature, T, as described by the Langevin function.

The molar susceptibility of the substance is the ratio of the molar magnetization and the applied magnetic field, H (Chapter 3):

$$\chi_{\mathrm{m}} = \frac{M_{\mathrm{m}}}{H} = \frac{N_{\mathrm{A}}\langle\mu\rangle}{H} = \frac{N_{\mathrm{A}}\mu_0\mu_{\mathrm{A}}^2}{3kT} \tag{2.5}$$

where the use was made of the relationship $B_0 = \mu_0 H$ (by definition). Note that the magnetization of paramagnetic materials is a linear function of the applied field, and is inversely proportional to temperature, a relationship known as the Curie law [5]. The use of Eq. (2.5) to calculate molar magnetic susceptibility of substances of known molecular dipole moments expressed in Bohr magnetons is illustrated in Chapter 1. Certain paramagnetic substances undergo conversion to ferromagnetic state below a characteristic temperature T_C. In those instances the calculation of the magnetic susceptibility from the atomic magnetic dipole moment is performed using a modified version of Eq. (2.5), known as the Curie–Weiss equation [5]:

$$\chi_m = \frac{N_A \mu_0 \mu_A^2}{3k(T - T_C)} \tag{2.6}$$

where T_C is the Curie temperature. In cases where the assumptions of the theories of Langevin, Curie and Weiss do not strictly apply, the atomic magnetic moment, μ_A, is substituted by the so-called effective magnetic moment, μ_{eff}.

$$\chi_m = \frac{N_A \mu_0 \mu_{eff}^2}{3k(T - T_C)} \tag{2.7}$$

2.2. *Paramagnetic substances: lanthanide solutions*

Lanthanide solutions are strongly paramagnetic because of the high magnetic atomic moment of lanthanide ions. Lanthanides are a class of elements that are paramagnetic due to unpaired inner 4f electrons [5]. These are sufficiently screened so that their magnetic properties are not lost on ionization or binding. Lanthanides almost always form ionic rather than covalent bonds [7], and the predominant valence number is 3+. Lanthanides chelated to diethylenetriaminepentaacetic acid (DTPA) have very high stability constants, ensuring there is no interaction with physiological ligands when administered parenterally for MRI studies. Their presence is detected by a decrease in T1 and T2 relaxation times compared to surrounding structures [8].

Gadolinium has the fourth highest magnetic moment (7.95 Bohr magnetons) of the 15 lanthanides [5]. Correspondingly, the tabulated molar susceptibility of Gd^{3+} in CGS units is 0.02739 in room temperature (in agreement with Eq. (2.7)). The high magnetic moment, plus the extreme stability of the Gd–DTPA complex, makes it an ideal choice for MRI. Commercial preparations of chelated gadolinium, such as Magnevist™ (Berlex Labs, Richmond, CA) and Optimark™ (Mallinckrodt Inc., St. Louis, MO), have undergone clinical trials verifying their lack of toxicity and rapid renal clearance rates. Both have trivalent gadolinium, Gd^{3+}, at a concentration of 0.5 M, allowing significant dilution in most of applications, which is fortuitous in bringing the osmolality nearer the plasma value of 285 mOsm. The strictly paramagnetic behavior of lanthanide solutions is important for studies of magnetic nanoparticle binding to synthetic microspheres that serve as models of the magnetic cell labeling, as discussed in Chapter 8. It is also important for calibration of the instrument used for measuring cell magnetophoresis by cell tracking velocimetry (CTV). The paramagnetic solutions of erbium have been evaluated for use as cell magnetization reagents for magnetic cell separation [9, 10]. The application of Eq. (2.7) to calculation of the magnetic susceptibility of the gadolinium chloride, $GdCl_3$, is illustrated in the example in Chapter 1.

2.3. Paramagnetic substances: hemoglobin and its derivatives

Another example of paramagnetic species important in the context of magnetic cell separation is hemoglobin, the oxygen carrying protein in the red blood cells. The paramagnetic contribution comes from the heme group but not from the globin part, and only when the hemoglobin is dissociated from the oxygen molecule (deoxyhemoglobin) [11–13]. The effective magnetic moment of the deoxy heme group is 5.46 Bohr magnetons, and the total paramagnetic contribution of the heme increases to four times that value because of the presence of four heme groups in the hemoglobin molecule. With the binding of the oxygen molecule, the electronic structure of the heme group changes so that its magnetic dipole moment vanishes. The paramagnetic forces acting on the deoxygenated erythrocytes are sufficiently high to observe their motion in the magnetic field [14].

The presence of heme in the erythrocyte gives rise to another paramagnetic species important for the magnetic cell separation and, potentially, disease diagnosis. Malaria infection results in the parasites lodging in the erythrocytes and feeding on hemoglobin [15] resulting in free heme release that is highly toxic to the parasite. Rather than being excreted from the erythrocyte, it is sequestered in the cell by conversion to an insoluble form known as hemozoin, or malarial pigment [16]. The hemozoin appears as characteristic brown crystals in the digestive vacuole of the parasite [17]. The number and the size of the hemozoin crystals in the erythrocyte depends on the stage of the parasite development, with the least amount of the hemozoin detected in the ring stage and the highest amount in the schizont stage. Because of its importance to parasite survival inside the erythrocyte, hemozoin has been the subject of intensive physicochemical and crystallographic studies. The spectroscopic and crystallographic analyses indicate that the hemozoin crystal structure is identical to that of a synthetic biomineral, β-hematin [18], which is the hematin dimer (ferriprotoporphyrin-IX)$_2$ in which the propionate side chain on one hematin is coordinated to the iron center of the other [19]. Hemozoin is better described as a biomineral rather than a polymer and attains lengths on the order of 1 μm, which makes it clearly visible under a visible light microscope [20]. Its molecular structure and composition is more complex than that of another biogenic magnetic particle, the magnetosome found in prokaryotes (such as *Magnetospirillum magnetotacticum*) that is typically a single domain crystal ($\sim$100 nm) of magnetite (Fe_3O_4) or greigite (Fe_3S_4) [21, 22]. Although the hemozoin structure is known, the mechanism of its formation remains uncertain [23, 24]. The molecular structure of β-hematin has been resolved at the single atom level [25], and its magnetic properties have been determined using electron paramagnetic resonance (EPR) and Mössbauer spectroscopy [19]. It has been shown that the Fe atom exists in a high-spin Fe(III) state, $S = 5/2$. The hemozoin heme electron configuration with five unpaired electrons corresponds to a ferriheme that is known to be a part of another high-spin hemoglobin species, methemoglobin [26]. Methemoglobin-rich erythrocytes have been shown to possess higher magnetophoretic mobility than low-spin, oxygenated hemoglobin erythrocytes ($S = 0$) [14].

Attempts at using magnetic properties of malaria pigment, or hemozoin for concentration and capture of malaria infected cells, date back to 1946 when it was shown that positioning an electromagnet, generating field of 0.5 tesla (T) next to a tube containing a suspension of infected blood causes enrichment of the parasitized erythrocyte fraction from 0.17% to over 24% in the course of 6 to 12 h [27]. Improved results were reported in the late 1970s using the technique of high-gradient magnetic separation (HGMS, discussed in Chapter 4) [28]. Commercially developed HGMS columns have been used to synchronize or enrich in vitro *P. falciparum* cultures from blood samples [29] and murine malaria parasite *P. berghei* ookinetes (that invade the mosquito midgut wall) for further in vitro studies [30]. Those studies identified early the presence of the paramagnetic hemozoin in the infected erythrocyte as the cause of the erythrocyte attraction by the magnetic field.

2.4. *Superparamagnetic particles and ferrofluids*

The Langevin theory of paramagnetism provides a convenient framework to describe properties of magnetic colloidal particles. The same treatment as described above has been extended to freely suspended, small, hard magnets (of the size of up to 10 nm) [6]. The small size of the particle ensures that they stay suspended in solution, and therefore they form colloidal suspensions (do not precipitate from suspension, and therefore are treated as a "magnetic gas" in a manner analogous to van't Hoff's treatment of colloidal solutions [31]). Because of the similarity of the dependence of M on B_0 to that of the paramagnetic

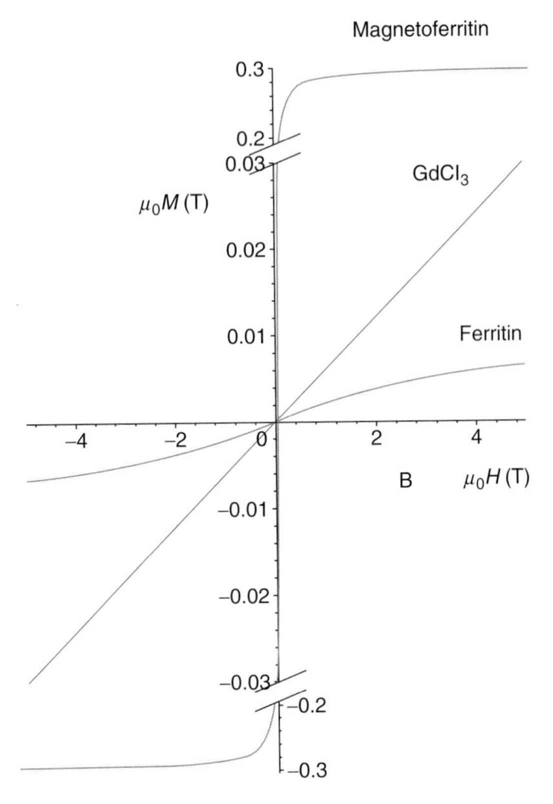

Fig. 2.2. Magnetization curves of paramagnetic (gadolinium chloride, $GdCl_3$) and superparamagnetic (ferritin and magnetoferritin) substances calculated from their reported effective magnetic moments, at room temperature.

molecules, such small magnetic particles are known as superparamagnetic particles [32–34], Fig. 2.2. The magnetic moment of the superparamagnetic particles is significantly higher than that of the paramagnetic molecules, μ_A, and can reach the order of 10,000 to 100,000 Bohr magnetons [35]. The combined colloidal and magnetic properties of such suspensions cause their unique behavior in the magnetic field as magnetic fluids, known as ferrofluids [36, 37].

The discovery of ferrofluids had important theoretical consequences for the science of magnetostatics (such as revision of the Earnshaw theorem [38]) and brought about important engineering applications, such as liquid magnetic seals. It had also important applications to biology, because it led to the development of bioferrofluids [37], and an improved understanding of the magnetic material properties at the nanoscale. The magnetic colloids became an important reagent used in the biology laboratory and in the clinic for cell separation [39].

The prototypical superparamagnetic particles are ferritin [40, 41] and magnetoferritin [42, 43]. Free iron in aqueous solutions of physiological pH 7.4 oxidizes to the ferric state (Fe^{3+}) by forming iron hydroxide and iron chloride which are poorly soluble in water and precipitate. In the course of evolution, organisms developed efficient mechanisms of iron transport and storage in the form of bound iron in metalloproteins, such as transferrin and ferritin [44]. There are two ferric ions in the transferrin molecule (molecular weight of 84,000) but as many as 4500 ferric ions in the ferritin molecule (molecular weight of 480,000 and higher). Ferritin has a complex structure consisting of a spherical polypeptide shell (apoferritin, diameter 12 nm) and a mineral core of hydrated iron oxide ferrihydrite (diameter approximately 8 nm). The reduction of ferric ion to the ferrous state renders the iron soluble and allows the release of free iron from ferritin. Synthetic ferritins with reconstituted cores involving iron and other metals are obtained. In particular, high magnetic moment ferritin has been synthesized with predominantly magnetite and maghemite core (magnetoferritin) [40].

The magnetic moment of the native ferritins is reported to be in the range of 250 to 400 Bohr magnetons [45]. In order to calculate the corresponding value of the magnetic susceptibility, one needs to know the temperature dependence of the magnetic properties of ferritin [46]. The superparamagnetism of ferritin is limited to the range of temperatures between the so-called blocking temperature at the lower limit, and the Néel temperature at the upper limit. Below the blocking temperature, the ferritin is ferrimagnetic.

Above the Néel temperature, the ferritin loses its superparamagnetic properties and behaves as a paramagnetic substance [47–49]. In calculating the magnetic susceptibility of the superparamagnetic ferritin from its effective magnetic moment, one substitutes blocking temperature, T_b, for the Curie temperature, T_c, in Eq. (2.7). The blocking temperature of the native ferritin is $T_b = 15$ K. Taking $\mu_{eff} = 300\mu_B$ for the magnetic moment of ferritin with fully saturated iron core (4500 iron atoms), one obtains $\chi_m = 2150 \times 10^{-6}$ from Eq. (2.7). The susceptibility of ferritin with the half saturated iron core is four times smaller, 537.5×10^{-6}. The molar susceptibility of magnetoferritin with $\mu_{eff} = 10{,}000\mu_B$ and the blocking temperature of $T_b = 30$ K calculated from Eq. (2.7) is 0.2, or $200{,}000 \times 10^{-6}$.

A comparison of room temperature magnetization curves of the paramagnetic gadolinium chloride, $GdCl_3$, and superparamagnetic ferritin and magnetoferritin, calculated from their respective effective dipole moments, is shown in Fig. 2.2. Note differences in the saturating field values: approximately 0.2 T for magnetoferritin and 2 T for native ferritin. The gadolinium chloride does not saturate until well above 10 T (deviation from linearity starting at approximately 50 T).

The cationized form of the native ferritin has been shown to enhance the magnetophoresis of the white blood cells [50, 51]. Magnetoferritin conjugated to antibodies was used to demonstrate specific leukocyte capture in the magnetic field [52]. The biological synthesis of ferritin makes it a prototypical bionanoparticle for magnetic labeling of cells.

2.5. Ferromagnetism and magnetic properties of iron

In solids, the close vicinity of the atomic magnetic dipoles gives rise in certain instances to exchange interactions that lock the orientation of the dipoles over a significant volume of the solid [5]. This results in a permanent magnetization of the solid, which itself becomes a source of an external magnetic field, with distinct North and South poles. The volume of the solid with the atomic dipoles locked in the same direction in space is known as a magnetic domain [53]. The earliest known example of such a solid is a mineral, lodestone (magnetite), that consists chiefly of a mixture of iron oxides at different oxidation states, 2+ and 3+, with the chemical formula designated as Fe_3O_4. Pure metallic iron is ferromagnetic up to its Curie temperature of 770 °C, above which it becomes paramagnetic.

The rise of spontaneous magnetization of ferromagnetic materials affects the shape of their magnetization curve, in that the magnetization depends on the starting point and the direction of change of the applied field. In other words, the final magnetization at a given applied field depends on whether the applied field value is approached from above or from below of its set value. It also depends on the initial magnetization of the ferromagnetic material. In summary, unlike the paramagnetic (and diamagnetic) materials, the magnetization of the ferromagnetic materials is not uniquely determined by the applied magnetic field, but also depends on the magnetization history of the material. The nonuniqueness of magnetization of ferromagnetic material gives rise to hysteresis of the magnetization curve, such as the one illustrated for pure iron in Fig. 2.3. Note the high saturation magnetization of pure iron, $M_s = 2.17$ T, a low saturating field, 0.015 mT, and low coercivity, 0.004 mT. The corresponding effective magnetic dipole moment of an iron atom in the elemental iron sample is 2.2 Bohr magnetons [5]. When retraced between the arbitrarily selected initial and final values of the applied field, one obtains a family of hysteresis loops, Fig. 2.4. A single curve that is representative of a given ferromagnetic material is obtained when traced between extreme values of the applied field, at which the magnetization reaches saturation [54].

The second quadrant of the magnetization hysteresis curve is important in determining the properties of the permanent magnets [5]. A desirable feature of a permanent magnet is its high remnant magnetization, M_r, and a high coercive force, H_c. A plot of the

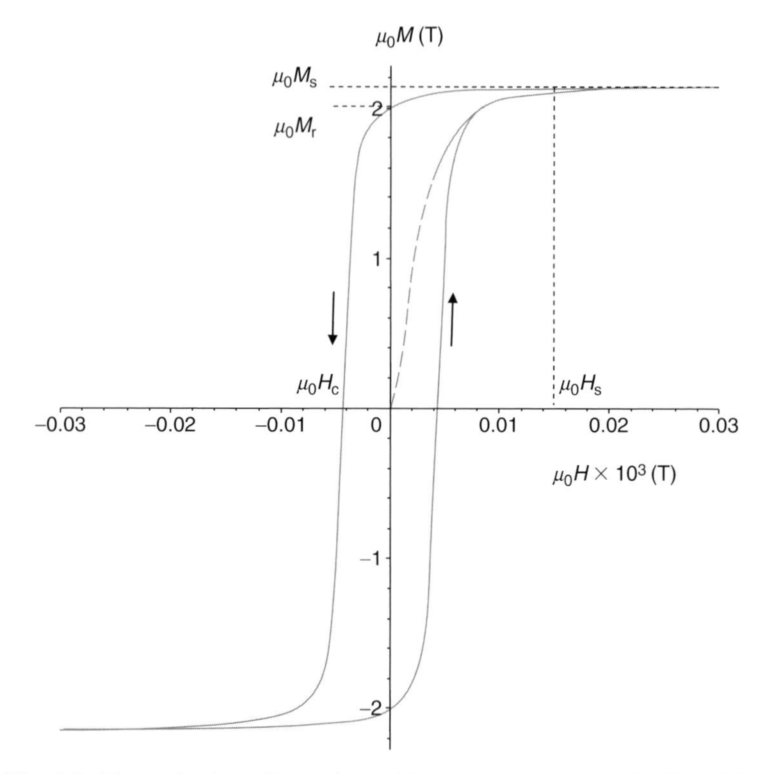

Fig. 2.3. Magnetization of pure iron. M_s – saturation magnetization, M_r – remnant magnetization, H_c – coercive force, H_s – saturating field. Note difference in scale between axes. The broken line shows initial magnetization from demagnetized state, the solid lines represent the hysteresis loop (iron purified in hydrogen, adapted from Bozorth).

energy product, $\mu_0 MH$, for the second quadrant of the magnetization hysteresis loop (Fig. 2.5) is used to determine the maximum energy product, a figure of merit when comparing different permanent magnet materials [3, 55]. The rapid progress in the development of permanent materials over the past 50 years is best illustrated by a nearly exponential growth of the available maximum energy product, from under 10 MGOe (mega-gauss-oersted,

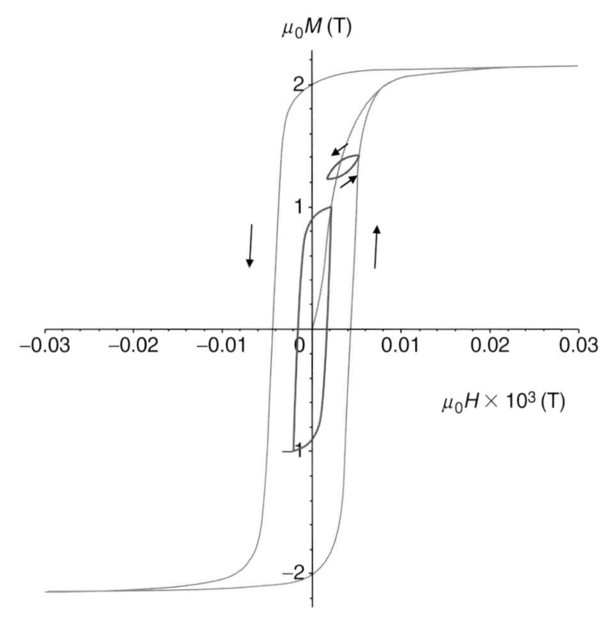

Fig. 2.4. An example of minor hysteresis loops. The nonlinear magnetization of an iron core electromagnet poses a special set of problems in applications requiring fine control of the magnetic field using the applied current, such as magnetic field sweep through zero (as can be appreciated from the above plots).

or 10^6 G·Oe) for Alnico magnets in the 1950s to over 50 MGOe for the newest, commercially available neodymium–iron–boron (NdFeB) magnets (Fig. 2.6). The strong permanent magnets are used as a source of the magnetic fields in all the current, commercially available magnetic cell separators.

The presence of oxides and impurities (such as carbon) significantly affect the ferromagnetic properties of iron compounds, in particular, by turning them into soft or hard magnetic materials (of low- or high-coercive force, respectively, as illustrated in Fig. 2.7). The hardest known magnetic materials are the iron and iron oxide particles whose size is just below the size of a single magnetic domain, or approximately 10 nm. Such particles have

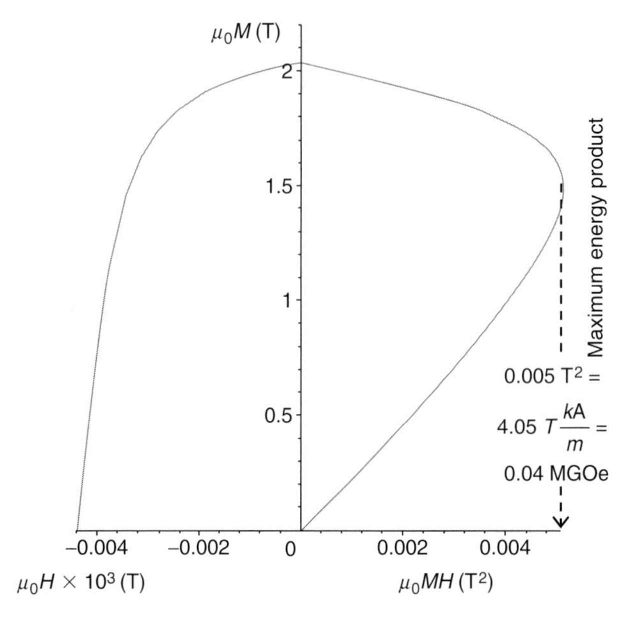

Fig. 2.5. Second quadrant of the magnetic hysteresis loop (the demagnetization curve, left) and the corresponding plot of the magnetic energy product, $\mu_0 MH$ (on the right) showing the maximum energy product of 0.04 MGOe (soft iron).

been developed for use in magnetic storage media (widely used in computer hard drives) and as components of magnetic particles for cell separation. In addition to iron, cobalt and manganese also exhibit ferromagnetic properties, and have been used in metallic form or in combination with iron, as alloys, for preparation of superparamagnetic particles [56, 57]. The preparation of magnetic particles for cell separation has to take into account not only the magnetic properties of the particle, but also its biological activity, including toxicity and biocompatibility [58, 59]. The details of the magnetic particle synthesis and characterization, and toxicity determination, are described in Chapter 6.

The area enclosed by the hysteresis loop on the $M - B$ plot is equal to work expended on the material magnetization, to increase the boundaries of magnetic domains aligned with the field at

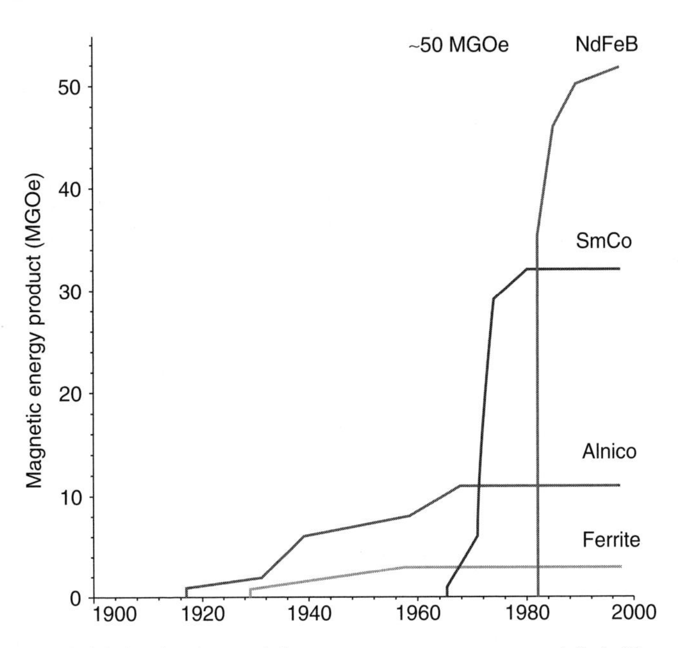

Fig. 2.6. Rapid development of the permanent magnet materials is illustrated by the increase in their magnetic energy product (adapted from Dexter, Inc. website, http://www.dextermag.com). SmCo: samarium-cobalt, NdFeB: neodymium–iron–boron.

the cost of contracting boundaries of the magnetic domains that are not, and to realign the atomic dipole moments. The work is dissipated in the form of heat. The large hysteresis of hard magnets, such as ferromagnetic micro- and nanoparticles, makes them an ideal material for application to local thermotherapy using oscillating magnetic fields [60].

The presence and role of iron in human physiology has an influence on the magnetic properties of human tissues [61]. The most obvious example is the role of iron in oxygen transport in blood. The iron of heme group in hemoglobin exists in the 2+ oxidation state (ferrous). The electronic structure of the chemical bonds between iron, the four nitrogen atoms of the heme

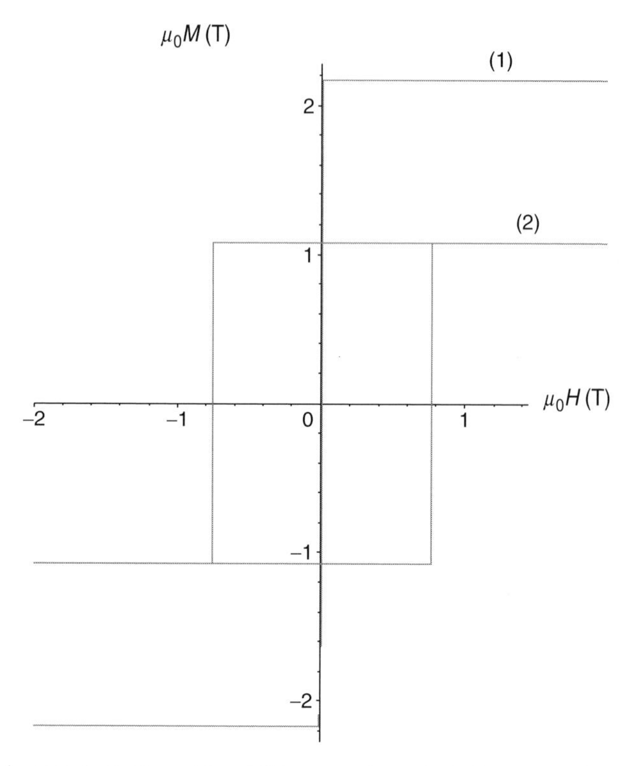

Fig. 2.7. Soft (1) and hard (2) magnetic material hysteresis loops.

porphyrin ring and the nitrogen atom of histidine in the globin molecule, depends on binding of the oxygen molecule. In oxyhemoglobin (the oxygen–hemoglobin complex) the chemical bonds are covalent and the magnetic moment is zero. In deoxyhemoglobin, the chemical bonds are ionic and the magnetic moment is 5.46 Bohr magnetons. The changes of magnetic moment of oxygen–hemoglobin complexes with binding of oxygen affect the local magnetic field and the NMR signal from the water hydrogen atoms. The effect has proven useful in mapping the local tissue oxygenation by MRI to tissue function and became the basis of functional MRI (fMRI for short) [13].

2.6. Diamagnetism

The diamagnetic properties of water or, for that matter, all substances, are related to the orbital spin of the electrons [62]. In classical electrodynamics, the orbital electron spin is associated with the circular flow of electric current, which produces a magnetic dipole moment. The reason why water or other substances do not exhibit a permanent dipole moment is that the effect of orbital spins cancels out when averaged over the electrons in the atom or a molecule. With the application of an external magnetic field, however, the field interaction with the moving electrons affects their motion in such a way that there arises a net magnetic moment of an atom that is directed against the applied field. The opposition of the induced magnetic dipole to the applied field is known in classical electrodynamics as the Lenz rule, which is a direct consequence of the energy conservation law [63]. The diamagnetic moment is orders of magnitude smaller than the paramagnetic or ferromagnetic dipole moments for fields under 1 T, and typically is neglected when considering forces in the magnetic field. However, the effect does not saturate and in principle, for sufficiently high magnetic fields, may produce sufficiently high forces to counteract gravity (diamagnetic levitation) [62, 64, 65]. The relationship between the diamagnetic moment, the applied field, and the electron orbital motion can be approximated by the motion of a classical charge in the magnetic field, resulting in the following formula

$$\Delta\mu = -\frac{e^2 r^2}{4m_e} B_0 \tag{2.8}$$

where $\Delta\mu$ is the diamagnetic dipole moment, B_0 is the applied field, $e = 1.602 \times 10^{-19}$ A·sec is the elementary charge of the electron, r is the radius of electron orbit, and $m_e = 9.11 \times 10^{-31}$ kg is the electron rest mass. For the electron in its lowest orbit $r = r_0 = 0.53 \times 10^{-10}$ m, or the Bohr radius. The resulting mass susceptibility of water is estimated by observing that there are approximately two electrons per unit atomic mass, and therefore approximately $N_A/2$ electrons in one gram of water (or any other substance, where $N_A \approx 6.02 \times 10^{23}$ is the Avogadro number). Therefore, specific susceptibility is

$$\chi_g = \frac{M_g}{H} \approx \frac{N_A \Delta\mu}{2H} = -\frac{N_A e^2 r_0^2}{8m_e}\frac{B_0}{H} = -\frac{N_A \mu_0 e^2 r_0^2}{8m_e} \tag{2.9}$$

where M_g is the magnetization of water and $B_0 = \mu_0 H$. By substituting numerical values for various constants entering Eq. (2.9), one obtains

$$\begin{aligned}\chi_g &\approx -\frac{6.02 \times 10^{23}\left[\frac{1}{g}\right] \times 4\pi \times 10^{-7}\left[\frac{T \cdot m}{A}\right] \times 1.602^2 \times 10^{-38}[A^2 \cdot s^2] \times 0.53^2 \times 10^{-20}[m^2]}{8 \times 9.11 \times 10^{-31}[kg]} \\ &= -7.48 \times 10^{-12}\left[\frac{T \cdot A \cdot m^3 \cdot s^2}{10^{-3} kg^2}\right] = -7.48 \times 10^{-9}\left[\frac{N \cdot m^2 \cdot s^2}{kg^2}\right] \\ &= -7.48 \times 10^{-9}\left[\frac{m^3}{kg}\right]\end{aligned} \tag{2.10}$$

By multiplying both sides of Eq. (2.10) by the density of water at 273 K, $\rho = 1000\,kg/m^3$, one obtains for the volume susceptibility of water

$$\chi = \chi_g \rho \approx -7.48 \times 10^{-6} \tag{2.11}$$

Considering the simplicity of the model used to describe the diamagnetic effect in matter [62], limited to a classical charge moving in the magnetic field, the calculated value is very close to the measured value of $\chi = -9.04 \times 10^{-6}$ (Eq. 1.9).

2.7. Magnetic field vectors

The static magnetic field is fully characterized by two vectors, **H** and **B**, the magnetic field strength and the magnetic field intensity [62, 66, 67]. The vector **B** is also called the magnetic induction or the magnetic flux intensity in the electrical engineering literature [1, 68], often depending on the context. The vector **H** describes the strength of the magnetic field source, such as an electric line current or a permanent magnet, and the vector **B** describes the resulting effect on the surrounding space, such as the force acting on a neighboring current-carrying line or a compass needle (a magnetic dipole). Thus the vector **H** is analogous to a "driving force" (a cause) and the vector **B** the coupled "flux density" (an effect). The analogies are found in the description of other vector fields (such as temperature and heat flow in thermodynamics, or pressure and fluid flow in fluid mechanics [63, 69]).

In free space (a vacuum), the magnetic flux density is directly proportional to the field strength:

$$\mathbf{B}_0 = \mu_0 \mathbf{H} \tag{2.12}$$

where μ_0 is the magnetic permeability of the free space, a numerical constant fixed by the definition of the electric current unit ($\mu_0 = 4\pi \times 10^{-7}$ T·m/A) (for relationship between selected units, see Appendix A). In this text the vector quantities will be indicated by bold face. The vector notation will be used to simplify typography. The relationship between vector notation and component notation is summarized in Appendix B.

2.8. Magnetic field magnitude

The field magnitude, occasionally referred to as field strength, is represented by the length of the field vector. Inside a long, thin solenoid, the magnetic field strength equals the current per unit length of the solenoid, $H = I \cdot n$, where I is the current flowing in the solenoid wire in amperes, and n is the number of windings (single layer) per unit length of the solenoid [63, 70]. As an illustration of how much current is required to produce the magnetic field, consider the following example:

Example Magnetic field inside a long, thin solenoid.
Electric current, $I = 1$ A
Number of windings per unit length of solenoid, $n = 1{,}000$/m
Magnetic field strength, $H = I \cdot n = 1{,}000$ A/m
Magnetic field intensity (flux density), $B_0 = \mu_0 \cdot H = 0.00126$ T $= 12.6$ G

The field inside a long solenoid represents the case of a highly homogeneous field, whose intensity is independent of spatial coordinates. It also represents the case of a highly efficient use of electric current to produce a magnetic field, because the entire field (or almost all of it, in practical cases) is confined inside the solenoid, but none (or almost none) strays outside the solenoid. Because of its homogeneity, however, the field inside the solenoid is not useful for magnetic separation applications, as discussed below.

2.9. Magnetic field sources

One notes considerable power losses involved in generating magnetic field by electric currents alone. Assuming that the electrical resistivity of the wire making the solenoid is such that it takes the voltage of a car battery (12 V) to produce a current of 1 A, and that we consider a solenoid made of 10 turns of a wire per 1 cm length (that is, the resistivity is 12 ohm/cm of length of the solenoid), one obtains the power loss of 12 W/cm, or a significant 1.2 kW/m of the length of the solenoid (the power loss per unit length of the solenoid increases with the square of the electric current intensity times the resistivity per unit length). Here we assume that one has a suitable source of the electromotive force capable of generating the necessary voltage of 12 kV in order to maintain current of 1 A per meter length of the solelnoid. When not applied to do useful work, all the power loss is converted to heat, which could be quite considerable at 1.2 kW/m. One also notes that the magnetic field intensity thus produced inside the solenoid, 12.6 G as calculated in the example above, is only about 20-fold higher than that of the Earth at sea level.

In order to achieve the field intensity comparable to that used in modern magnetic resonance imaging (MRI) scanners, such as 1.26 T, one would have to use current of 1000 A, with the concomitant power loss of 1.2 MW/m. Apart from obvious problems of the power and heat management, one is confronted with dielectric breakdown of air ($\sim$10 kV/m, much lower than the needed voltage drop along the solenoid of 1.2 MV/m). The considerable engineering challenges involved in building large solenoids for generation of strong magnetic fields in large volumes were successfully solved by Francis Bitter [70]. Small volume solenoids require significantly less power but are limited by finite heat dissipation rate, which limits their useful maximum field. In recent years, the technology involved in building superconducting magnets capable of generating fields in excess of 1 T inside the room temperature bore has progressed to a point where one can contemplate purchase of such a laboratory superconducting magnet without the need of liquid helium for refrigeration (substituted by a refrigeration unit running on household current) [71]. Still, the technology is out of reach for a typical biology or clinical laboratory interested in the magnetic cell separation.

The above example shows clear advantages of using permanent magnets rather than electric currents in applications that do not require large volumes, such as in cell separation. The permanent magnet technology advanced rapidly during the last 50 years to a point where it is routine to achieve fields of 1 T in volumes useful for cell separation, for a price that fits the budget of a typical research laboratory [39].

2.10. Field gradients

An ideal solenoid produces high fields but no field gradients. There are other current configurations that generate lower magnetic fields but are capable of producing significant field gradients, of interest to cell separation applications. The extreme case is that of a single current-carrying loop, where the field extends into the entire space surrounding the loop but quickly decreases with distance (inversely to the cube of the distance from the center of the loop at large distances from the loop). The field of the current loop is spatially highly inhomogeneous and therefore significant for magnetic separations; however, the useful fields are limited only to the immediate vicinity of the loop because of the fast field decay with distance. An intermediate case is the field of a long, current-carrying wire whose field decreases with distance at a slower rate as unity over the distance (Figs. 2.8–2.10). Every field geometry that can be produced by an arrangement

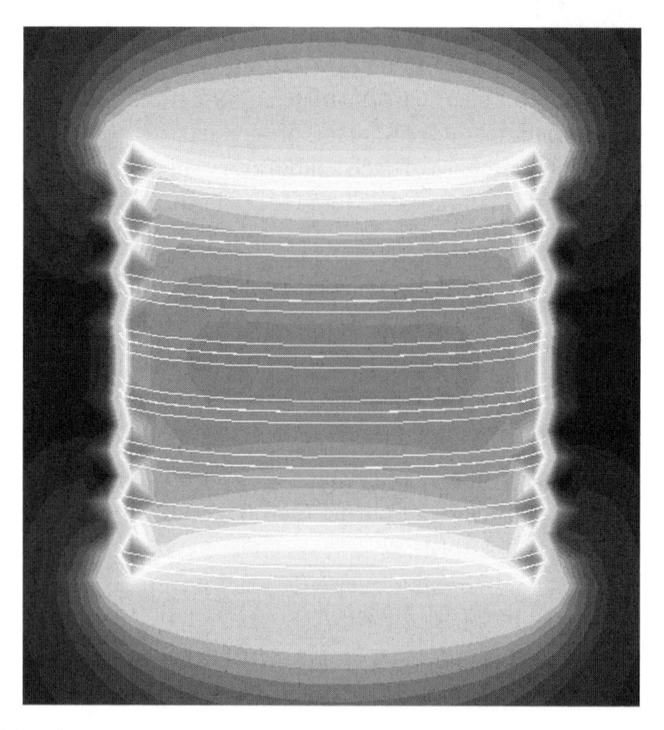

Fig. 2.8. Direct current magnetic field sources, in the order of increasing field inhomogeneity and field gradients that they produce. Solenoid: Note uniformly high field inside the solenoid, $B_0 \approx$ const., as indicated by uniform color. Note also field gradient at the rim of the solenoid, as indicated by color gradation with distance. (See Color Insert.)

Fig. 2.9. Direct current magnetic field sources, in the order of increasing field inhomogeneity and field gradients that they produce. Current line: Note decrease of field intensity with distance from the wire, $B_0 \propto 1/r$, as indicated by change of color intensity from light to dark. (See Color Insert.)

of electric currents can also be produced by an arrangement of permanent magnets. The selection of one field source over another is dictated by economy and ease of use.

2.11. Magnetic field lines

The magnetic fields $B_0 = \mu_0 H$ of a solenoid, line current, and the current loop are illustrated in Figs. 2.8–2.10, respectively. There, the field of a permanent magnet is shown in Fig. 2.11. There, the field is visualized using the magnetic field lines, Fig. 2.12, defined as curves that are tangent everywhere to the direction of the local field **H**. On a plane, magnetic field lines are defined by the equation:

$$\frac{\mathrm{d}y}{\mathrm{d}x} = \frac{H_y}{H_x} \tag{2.13}$$

(see Fig. 2.13 for illustration). The resulting differential equation describes a family of curves, $y = y(x)$, that is a visual representation

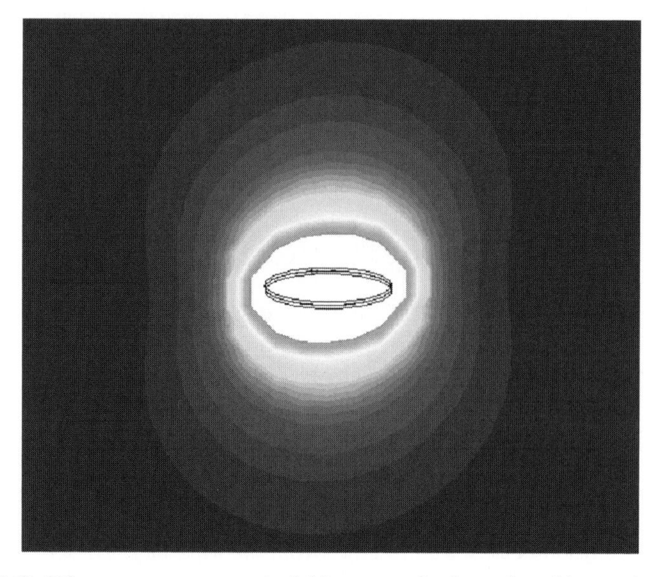

Fig. 2.10. Direct current magnetic field sources, in the order of increasing field inhomogeneity and field gradients that they produce. Current ring: Note decrease of field intensity with distance from the ring, that is much faster than that seen for the current line. The far field of the ring current is equivalent to the field of a magnetic dipole, $B_0 \propto 1/r^3$. (See Color Insert.)

of the magnetic field. Each curve is determined by the properties of the magnetic dipole or the electric current so that the boundary conditions determine the distribution of the field lines H in the entire surrounding space. The fundamental property of the static magnetic field, shared with other vector fields (such as electrostatic and incompressible fluid flow) is that the magnetic flux associated with a fixed number of magnetic field lines is conserved [68, 72] (Fig. 2.14). What follows is that the flux density (the magnetic field intensity) is high where the magnetic line density is high (measured in the plane perpendicular to the direction of the magnetic flux lines) and low where the magnetic line density is low. The high visual impact of the field line representation of the electromagnetic fields was appreciated by the

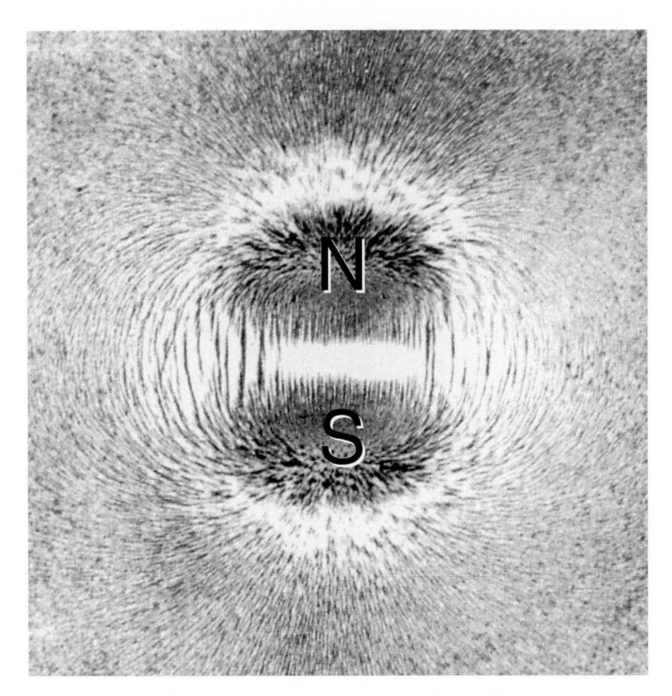

Fig. 2.11. Field of a permanent magnet visualized by a photograph of iron filings. The direction of iron filings follows the field lines (compare with Fig. 2.12).

early work of Faraday and Maxwell and continues to bear fruit in innumerable physics and electrical engineering applications [73]. Examples of magnetostatic fields used for cell separations are shown in the next chapters.

2.12. Magnetic field in matter

The introduction of matter in the magnetic field modifies the field fluxes. The magnetic properties of matter are defined by magnetic polarization, **M**, which describes how much the magnetic flux density

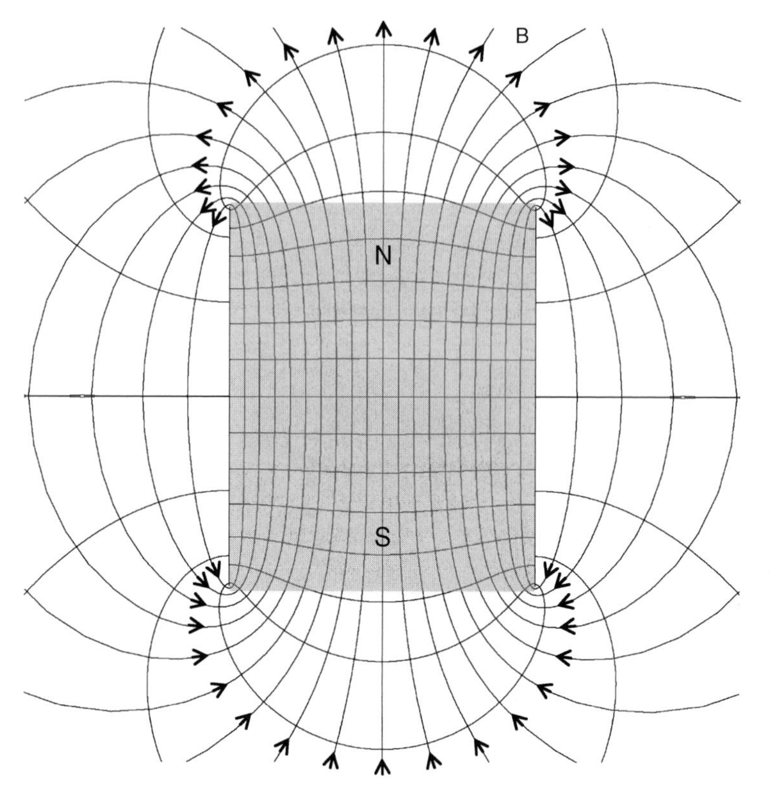

Fig. 2.12. Field of a permanent magnet (represented by a shaded area), **B** field lines are marked by arrowheads (2-D approximation). Note similarities with Fig. 2.11. Also shown are magnetostatic equipotential lines.

changes with the introduction of matter:

$$\mathbf{M} = \frac{\mathbf{B}}{\mu_0} - \mathbf{H} \tag{2.14}$$

where **B** is the magnetic flux density in matter. The vector **M** is also called the magnetization. In isotropic media the magnetization vector is parallel to the vector **H**:

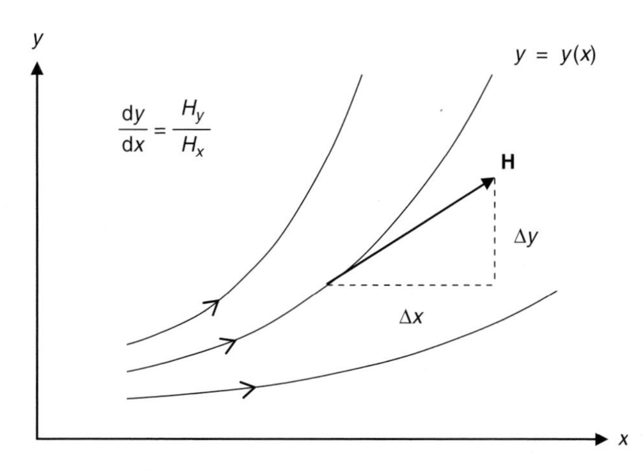

Fig. 2.13. Magnetic field lines $y = y(x)$ are tangent everywhere to the magnetic field vector **H**.

$$\mathbf{M} = \chi \mathbf{H} \tag{2.15}$$

where χ is the magnetic susceptibility, a material characteristic of the medium. Thus, the effect of magnetic susceptibility on the local magnetic flux is determined by combining Eqs. (2.14) and (2.15) to obtain:

$$\mathbf{B} = \mu_0(\mathbf{M} + \mathbf{H}) = \mu_0(\chi \mathbf{H} + \mathbf{H}) = \mu_0(\chi + 1)\mathbf{H} \equiv \mu_0 \mu_r H \equiv \mu_{\mathrm{m}} \mathbf{H} \tag{2.16}$$

or

$$\mathbf{B} = \mu_r \mathbf{B}_0 \tag{2.17}$$

where

$$\mu_r \equiv \chi + 1 \tag{2.18}$$

$$\mu_{\mathrm{m}} \equiv \mu_0 \mu_r$$

are defined as the relative and the absolute magnetic permeabilities of the medium, respectively [67], and $\mathbf{B}_0 = \mu_0 \mathbf{H}$. For cells, tissues and the vast majority of the biological material, and, indeed, water the absolute value of the magnetic susceptibility is very small as compared

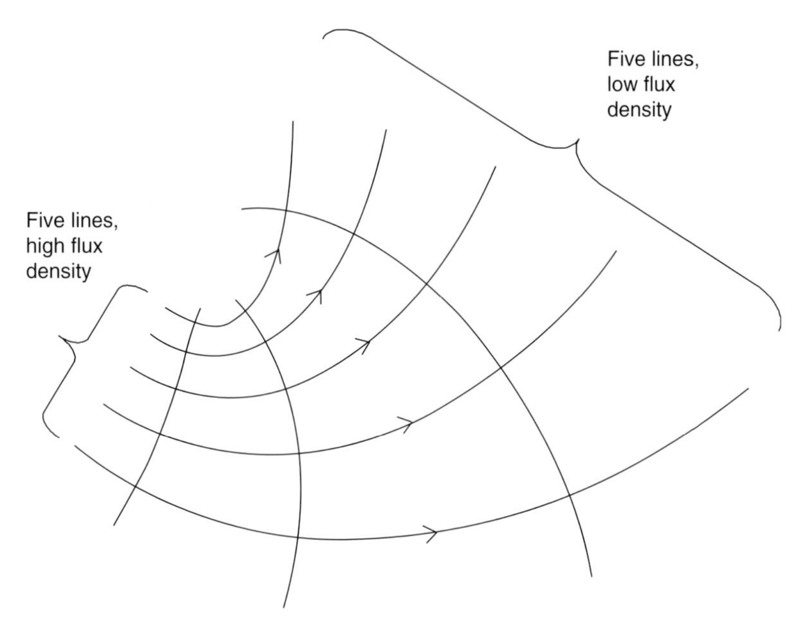

Fig. 2.14. Density of the magnetic field lines (marked by arrowheads) is directly proportional to the magnitude of the magnetic field strength, *H*. Or, the field flux enclosed between two field lines is constant. (In three dimensions, the field enclosed by filed isosurfaces, or tubes, is constant). Also shown are equipotential lines (not considered further).

to unity, typically on the order of 10^{-6} [see Eq. (2.11) and the accompanying discussion for magnetic susceptibility of water]. Also, typically the sign of magnetic susceptibility is negative for such materials, which makes them diamagnetic (as is water). At the other extreme are iron and ferromagnetic alloys, for which the magnetic susceptibility can be as high as 10^{-1}. Thus, there is around a five order of magnitude difference in the magnetic susceptibility of cells and that of iron. This explains the strong effect of even a small addition of iron to cells, such as by magnetic nanoparticle binding, on the magnetic properties of the cell–iron complex.

References

[1] Melcher, J. R. (1981). Continuum electromechanics. The MIT Press, Cambridge, MA.
[2] Häfeli, U., Schuett, W., Teller, J. and Zborowski, M. (1997). Scientific and Clinical Applications of Magnetic Carriers. Plenum Press, New York.
[3] Livingstone, J. D. (1996). Driving force: The natural magic of magnets. Harvard University Press, Cambridge, MA.
[4] Whitesides, G. M. and Love, J. C. (2001). The art of building small. Sci. Am. *285*, 38–48.
[5] Bozorth, R. M. (1947). Magnetism. Rev. Mod. Phys. *19*, 29–86.
[6] Weissleder, R., Elizondo, G., Wittenberg, J., Rabito, C. A., Bengele, H. H. and Josephson, L. (1990). Ultrasmall superparamagnetic iron oxide: Characterization of a new class of contrast agents for MR imaging. Radiology *175*, 489–93.
[7] Evans, C. H. (1990). Biochemistry of the lanthanides. Plenum Press, New York.
[8] Brasch, R. C., Weinmann, H. J. and Wesbey, G. E. (1984). Contrast-enhanced NMR imaging: Animal studies using gadolinium-DTPA complex. AJR Am. J. Roentgenol. *142*, 625–30.
[9] Graham, M. D. and Selvin, P. R. (1982). Separation of lanthanide-binding cells. IEEE Trans. Magn. *Mag-18*, 1523–5.
[10] Zborowski, M., Malchesky, P. S., Jan, T. F. and Hall, G. S. (1992). Quantitative separation of bacteria in saline solution using lanthanide Er(III) and a magnetic field. J. Gen. Microbiol. *138*, 63–8.
[11] Pauling, L. and Coryell, C. D. (1936). The magnetic properties and structure of hemoglobin, oxyhemoglobin and carbonmonoxyhemoglobin. Proc. Natl. Acad. Sci. *22*, 210–6.
[12] Savicki, J. P., Lang, G. and Ikeda-Saito, M. (1984). Magnetic susceptibility of oxy- and carbonmonoxyhemoglobins. Proc. Natl. Acad. Sci. USA *81*, 5417–9.
[13] Spees, W. M., Yablonskiy, D. A., Oswood, M. C. and Ackerman, J. J. (2001). Water proton MR properties of human blood at 1.5 Tesla: Magnetic susceptibility, T(1), T(2), T*(2), and non-Lorentzian signal behavior. Magn. Reson. Med. *45*, 533–42.
[14] Zborowski, M., Ostera, G. R., Moore, L. R., Milliron, S., Chalmers, J. J. and Schechter, A. N. (2003). Red blood cell magnetophoresis. Biophys. J. *84*, 2638–45.
[15] Greenwood, B. and Mutabingwa, T. (2002). Malaria in 2002. Nature *415*, 670–2.

[16] Goldberg, D. E., Slater, A. F. G., Cerami, A. and Henderson, G. B. (1990). Hemoglobin degradation in the malaria parasite *Plasmodium falciparum*: An ordered process in a unique organelle. Proc. Natl. Acad. USA *87*, 2931–5.
[17] Goldberg, D. E., Slater, A. F., Beavis, R., Chait, B., Cerami, A. and Henderson, G. B. (1991). Hemoglobin degradation in the human malaria pathogen *Plasmodium falciparum*: A catabolic pathway initiated by a specific aspartic protease. J. Exp. Med. *173*, 961–9.
[18] Egan, T. J. (2002). Physico-chemical aspects of hemozoin (malaria pigment) structure and formation. J. Inorg. Biochem. *91*, 19–26.
[19] Bohle, D. S., Debrunner, P., Jordan, P. A., Madsen, S. K. and Schulz, C. E. (1998). Aggregated heme detoxification byproducts in malarial trophozoites: Beta-hematin and malaria pigment have a single $S = 5/2$ iron environment in the bulk phase as determined by EPR and magnetic Moessbauer spectroscopy. J. Am. Chem. Soc. *120*, 8255–6.
[20] Noland, G. S., Briones, N. and , Sullivan, D. J., Jr. (2003). The shape and size of hemozoin crystals distinguishes diverse Plasmodium species. Mol. Biochem. Parasitol. *130*, 91–9.
[21] Dunin-Borkowski, R. E., McCartney, M. R., Frankel, R. B., Bazylinski, D. A., Posfai, M. and Buseck, P. R. (1998). Magnetic microstructure of magnetotactic bacteria by electron holography. Science *282*, 1868–70.
[22] Bazylinski, D. A. and Frankel, R. B. (2004). Magnetosome formation in prokaryotes. Nat. Rev. Microbiol. *2*, 217–30.
[23] Egan, T. J., Combrinck, J. M., Egan, J., Hearne, G. R., Marques, H. M., Ntenteni, S., Sewell, B. T., Smith, P. J., Taylor, D., van Schalkwyk, D. A. and Walden, J. C. (2002). Fate of haem iron in the malaria parasite *Plasmodium falciparum*. Biochem. J. *365*, 343–7.
[24] Bendrat, K., Berger, B. J. and Cerami, A. (1995). Haem polymerization in malaria. Nature *378*, 138–9.
[25] Pagola, S., Stephens, P. W., Bohle, D. S., Kosar, A. D. and Madsen, S. K. (2000). The structure of malaria pigment beta-haematin. Nature *404*, 307–10.
[26] Coryell, C., Stitt, F. and Pauling, L. (1937). The magnetic properties and structure of ferrihemoglobin (methemoglobin) and some of its compounds. J. Am. Chem. Soc. *59*, 633–42.
[27] Heidelberger, M., Mayer, M. M. and Demarest, C. R. (1946). Studies in human malaria. I. The preparation of vaccines and suspensions containing plasmodia. J. Immunol. *52*, 325–30.
[28] Melville, D., Paul, F. and Roath, S. (1975). Direct magnetic separation of red cells from whole blood. Nature *255*, 706.

[29] Uhlemann, A.-C., Staalsoe, T., Klinkert, M.-O. and Hviid, L. (2000). Analysis of *Plasmodium falciparum*-infected red blood cells. MACS & more *4*, 7–8.

[30] Carter, V., Cable, H. C., Underhill, B. A., Williams, J. and Hurd, H. (2003). Isolation of *Plasmodium berghei* ookinetes in culture using Nycodenz density gradient columns and magnetic isolation. Malar. J. *2*, 35.

[31] Katchalsky, A. and Curran, P. F. (1965). Nonequilibrium thermodynamics in biophysics. Harvard University Press, Cambridge, MA.

[32] St. Pierre, T. G., Jones, D. H. and Dickson, D. P. E. (1987). The behaviour of superparamagnetic small particles in applied magnetic fields: A Mossbauer spectroscopic study of ferritin and haemosiderin. J. Magn. Magn. Mater. *69*, 276–84.

[33] Bean, C. P. and Livingston, J. D. (1959). Superparamagnetism. J. Appl. Phys. *30(Suppl.)*, 120S–9S.

[34] Vassiliou, J., Mehrotra, V., Russell, M. W., Giannelis, E. P., McMichael, R. D., Shull, R. D. and Ziolo, R. F. (1993). Magnetic and optical properties of gamma-Fe_2O_3 nanocrystals. J. Appl. Phys. *73*, 5109–16.

[35] Gider, S., Awschalom, D. D., Douglas, T., Mann, S. and Chaparala, M. (1995). Classical and quantum magnetic phenomena in natural and artificial ferritin proteins. Science *268*, 77–80.

[36] Rosensweig, R. E. (1985). Ferrohydrodynamics. Cambridge University Press, Cambridge, MA.

[37] Liberti, P. A. and Feeley, B. P. (1991). Analytical- and process-scale cell separation with bioreceptor ferrofluids and high gradient magnetic separation. In: Cell Separation Science and Technology (Compala, D. S. and Todd, P., eds.), Vol. 464. ACS Symposium Series, Washington, pp. 268–88.

[38] Rosensweig, R. E. (1966). Buoyancy and stable levitation of a magnetic body immersed in a magnetizable fluid. Nature *216*, 613–4.

[39] Kantor, A. B., Gibbons, I., Miltenyi, S. and Schmitz, J. (1998). Magnetic cell sorting with colloidal superparamagnetic particles. In: Cell separation methods and applications (Recktenwald, D. and Radbruch, A., eds.). Marcel Dekker Inc., New York, pp. 153–73.

[40] Mann, S., Williams, J. M., Treffry, A. and Harrison, P. M. (1987). Reconsitiuted and native iron-cores of bacterioferritin and ferritin. J. Mol. Biol. *198*, 405–16.

[41] St. Pierre, T. G., Bell, S. H., Dickson, D. P. E., Mann, S., Webb, J., Moore, G. R. and Williams, R. J. P. (1986). Mossbauer spectroscopic studies of the cores of human, limpet and bacterial ferritins. Biochim. Biophys. Acta *870*, 127–34.

[42] Meldrum, F. C., Heywood, B. R. and Mann, S. (1992). Magnetoferritin: In vitro synthesis of a novel magnetic protein. Science *257*, 522–3.

[43] Bulte, J. W., Douglas, T., Mann, S., Frankel, R. B., Moskowitz, B. M., Brooks, R. A., Baumgarner, C. D., Vymazal, J. and Frank, J. A. (1994). Magnetoferritin. Biomineralization as a novel molecular approach in the design of iron-oxide-based magnetic resonance contrast agents. Invest. Radiol. *29(Suppl. 2)*, S214–6.

[44] Ponka, P., Schulman, H. M. and Woodworth, R. C. (1990). Iron transport and storage. CRC Press, Boca Raton, FL.

[45] Brem, F., Stamm, G. and Hirt, A. M. (2006). Modeling the magnetic behavior of horse spleen ferritin with a two-phase core structure. J. Appl. Phys. *99*, 123906.

[46] Harris, J. G. E., Grimaldi, J. E., Awschalom, D. D., Chiolero, A. and Loss, D. (1999). Excess spin and the dynamics of antiferromagnetic ferritin. Phys. Rev. B *60*, 3453–6.

[47] Tejada, J., Zhang, X. X., del Barco, E. and Hernandez, J. M. (1997). Macroscopic resonant tunneling of magnetization in ferritin. Phys. Rev. Lett. *79*, 1754–7.

[48] Resnick, D., Gilmore, K. and Idzerda, Y. U. (2004). Modeling of the magnetic behavior of gamma-Fe_2O_3 nanoparticles mineralized in ferritin. J. Appl. Phys. *95*, 7127–9.

[49] Seehra, M. S. and Punnoose, A. (2001). Deviations from the Curie-law variation of magnetic susceptibility in antiferromagnetic nanoparticles. Phys. Rev. B *64*, 1–4.

[50] Zborowski, M., Malcheski, P. S., Savon, S. R., Green, R., Hall, G. S. and Nose, Y. (1991). Modification of ferrography method for analysis of lymphocytes and bacteria. Wear *142*, 135–49.

[51] Zborowski, M., Fuh, C. B., Green, R., Sun, L. and Chalmers, J. J. (1995). Analytical magnetapheresis of ferritin-labeled lymphocytes. Anal. Chem. *67*, 3702–12.

[52] Zborowski, M., Fuh, C. B., Green, R., Baldwin, N. J., Reddy, S., Douglas, T., Mann, S. and Chalmers, J. J. (1996). Immunomagnetic isolation of magnetoferritin-labeled cells in a modified ferrograph. Cytometry *24*, 251–9.

[53] Jacubovics, J. P. (1994). Magnetism and magnetic materials. Cambridge University Press, Cambridge.

[54] Furlani, E. P. (2001). Permanent magnet and electromechanical devices. Academic Press, San Diego, CA.

[55] Hatch, G. P. and Stelter, R. E. (2001). Magnetic design considerations for devices and particles used for biological high-gradient magnetic separation (HGMS) systems. J. Magn. Magn. Mater. *225*, 262–76.

[56] Sun, S., Zeng, H., Robinson, D. B., Raoux, S., Rice, P. M., Wang, S. X. and Li, G. (2004). Monodisperse MFe_2O_4 ($M = Fe, Co, Mn$) nanoparticles. J. Am. Chem. Soc. *126*, 273–9.

[57] Hyeon, T., Chung, Y., Park, J., Lee, S. S., Kim, Y.-W. and Park, B. H. (2002). Synthesis of highly crystalline and monodisperse cobalt ferrite nanocrystals. J. Phys. Chem. B *106*, 6831–3.

[58] Harris, L. A., Goff, J. D., Carmichael, A. Y., Riffle, J. S., Harburn, J. J., St. Pierre, T. G. and Saunders, M. (2003). Magnetite nanoparticle dispersions stabilized with triblock copolymers. Chem. Mater. *15*, 1367–77.

[59] Häfeli, U. O. and Pauer, G. J. (1999). In vitro and in vivo toxicity of magnetic microspheres. J. Magn. Magn. Mater. *194*, 76–82.

[60] Jordan, A., Wust, P., Fahling, H., John, W., Hinz, A. and Felix, R. (1993). Inductive heating of ferrimagnetic particles and magnetic fluids: Physical evaluation of their potential for hyperthermia. Int. J. Hyperthermia. *9*, 51–68.

[61] St. Pierre, T. G., Chua-anusorn, W., Webb, J. and Macey, D. (2000). Iron overload diseases: The chemical speciation of non-heme iron deposits in iron loaded mammalian tissues. Hyperfine. Interact. *126*, 75–81.

[62] Purcell, E. M. (1985). Electricity and magnetism. Berkeley physics course. Vol. 2. McGraw-Hill Book Co., New York.

[63] Becker, R. (1982). Electromagnetic fields and interactions. Dover Publications, Inc., New York.

[64] Simon, M. D. and Geim, A. K. (2000). Diamagnetic levitation: Flying frogs and floating magnets (invited). Appl. Phys. A *87*, 6200–4.

[65] Rosenblatt, C., Yager, P. and Schoen, P. E. (1987). Orientation of lipid tubules by a magnetic field. Biophys. J. *52*, 295–301.

[66] Stratton, J. A. (1941). Electromagnetic theory. McGraw-Hill Book Company Inc., New York.

[67] Lorrain, P. and Corson, D. R. (1997). Electromagnetism: Principles and applications. W.H Freeman and Co., New York.

[68] Hammond, P. and Sykulski, J. K. (1995). Engineering electromagnetism: Physical processes and computation. Oxford University Press, Oxford.

[69] Longair, M. S. (1994). Theoretical concepts in physics. Cambridge University Press, Cambridge.

[70] Bitter, F. (2004). Mathematical physics. Dover Publications Inc., Mineola, NY.

[71] Guimaraes, A. P. (2005). From lodestone to supermagnets: Understanding magnetic phenomena. Wiley-VCH Verlag GmbH and Co. KGaA, Weinheim Germany.

[72] Weber, E. (1960). Electromagnetic fields. Theory and applications (Vol. 1, Mapping of fields). John Wiley and Sons Inc., New York.

[73] Maxwell, J. C. (1954). A treatise on electricity and magnetism. Vol. 1. Dover Publications Inc., New York.

Laboratory Techniques in Biochemistry and Molecular Biology, Volume 32
Magnetic Cell Separation
M. Zborowski and J. J. Chalmers (Editors)

CHAPTER 3

Maxwell stress and magnetic force

Maciej Zborowski

Department of Biomedical Engineering, Lerner Research Institute, Cleveland Clinic, Cleveland, OH 44195, USA

3.1. Introduction

A magnetic field exerts a stress on space that results in motion and deformation of a magnetically susceptible, continuous medium exposed to the field. The exact form of the so-called Maxwell stress and its relationship to the magnetic force density is a rather complicated affair and therefore an interested reader is referred to other sources for details [1, 2]. A number of simple cases are illustrated in Appendix C.

In a homogeneous, isotropic, linear medium the Maxwell stress reduces to a scalar quantity that has a pressure-like property of acting with equal force in all directions in space [3, 4]:

$$\frac{1}{2}HB \tag{3.1}$$

where H and B denote the magnitude of the magnetic vectors $\mathbf{H}$ and $\mathbf{B}$. For distinction between H and B fields see Section 2.12. The above quantity is also referred to simply as the magnetic pressure (in matter) [4]. The Maxwell stress provides a convenient means of calculating magnetic force acting on a finite size magnetic particle by enclosing it in a "bubble" that separates the particle from the rest of the field and integrating the stress over the surface of the bubble (the so-called Maxwell surface) [1]. The B field on the Maxwell surface is the

DOI: 10.1016/S0075-7535(06)32003-7

magnitude of the vector sum of the applied field and the induced field of the magnetic particle. Interestingly, the resulting force does not depend on the selection of the Maxwell surface (as long as it wholly encloses the magnetic particle) and therefore the choice is usually dictated by computational convenience. This approach has become accessible to a wider group of users apart from the physics and electrical engineering community because of the availability of computer algebra and numerical solver software packages for personal computers (such as Mathematica, Maple, or Femlab). An illustration of the Maxwell surface and the associated Maxwell stresses for calculating force on a particle for simple cases of axially symmetric particles (a sphere and an ellipsoid of revolution), which are simplified models of a cell shape, are shown in Appendix C.

3.2. Magnetic force density and the body force

It has been shown by Maxwell that the local magnetic force density, **f**, can be calculated directly from the distribution of the magnetic stresses in matter. In the general case, the magnetic force density is equal to the divergence of the Maxwell stress tensor [1, 5]. In the limited case that is of interest to cell separation applications, that formula can be considerably simplified by assuming that the physical properties of the cell and the suspending fluid are regular (i.e., they are homogenous and isotropic), and that the field does not vary strongly with the distance characteristic of cell dimensions. In that case, the divergence of the Maxwell stress tensor is nearly equal to the ratio of the total stress over volume enclosed by the Maxwell surface, and in the limit of very small volumes [3]:

$$\mathbf{f} = \nabla\left(\frac{1}{2}HB\right) \tag{3.2}$$

where ∇ denotes a gradient operator (see Appendix C for the derivation of the above formula). Again, the formula is noteworthy for its resemblance to a similar expression in fluid dynamics linking the body force and the gradient of pressure. The extensive analysis of magnetic body forces in applications to the mechanics of continuous media is described in detail in texts on magnetohydrodynamics and ferrohydrodynamics [3–5].

Taking into account that the magnetic field in matter, B, is a sum of the imposed magnetic field, $B_0 = \mu_0 H$, and the magnetization of matter, $\mu_0 M$, one obtains

$$\frac{1}{2}HB = \frac{1}{2}H\mu_0(M+H) = \frac{1}{2}H\mu_0 M + \frac{1}{2}H\mu_0 H = \frac{1}{2}MB_0 + \frac{1}{2}HB_0 \tag{3.3}$$

The term $(1/2)HB_0$ is the Maxwell stress in free space. Of historical interest, the term contributed to the hypothesis of an electromagnetic ether, that is, a hypothetical medium that is a seat of the electromagnetic field. This term drops out from the expression for the magnetic body force when the Eq. (3.3) is inserted into Eq. (3.2). Consequently, one obtains

$$\mathbf{f} = \nabla\left(\frac{1}{2}HB\right) = \nabla\left(\frac{1}{2}MB_0\right) \tag{3.4}$$

The term $(1/2)MB_0$ is often used to characterize permanent magnet materials in the technical literature, where it is referred to as the "energy product," as discussed in Chapter 2.

Considering that the magnetization of matter, M, is directly proportional to its magnetic susceptibility, χ, as discussed in Chapter 2, one further obtains

$$\mathbf{f} = \nabla\left(\frac{1}{2}MB_0\right) = \nabla\left(\frac{1}{2}\chi HB_0\right) \tag{3.5}$$

In the case of linearly polarizable magnetic media, such as paramagnetic and diamagnetic materials, for which the magnetic susceptibility χ is independent of the applied magnetic field, one obtains

$$\mathbf{f} = \chi\nabla\left(\frac{1}{2}HB_0\right) \tag{3.6}$$

The last formula has a desirable feature of separating variables that describe properties of matter (χ) from those that describe the properties of the imposed field, $\nabla(1/2HB_0)$. This considerably simplifies engineering approaches to magnetic cell separation [6]. In particular, the term $(1/2)HB_0$ is also described as the "external magnetic energy product" in the permanent magnet materials literature [7]. Note that the direction of force on the paramagnetic material, $\chi > 0$, is in the direction of Maxwell stress gradient, whereas that for the diamagnetic material, $\chi < 0$, is opposite to the direction of the Maxwell stress gradient.

3.3. Magnetic force acting on a small, magnetically susceptible particle (diamagnetic and paramagnetic materials)

The relationship between the magnetic force density, **f**, and the total magnetic force, **F**, acting on a small, homogeneous particle is equal to **F** divided by the particle volume, V:

$$\mathbf{f} = \frac{\mathbf{F}}{V} \tag{3.7}$$

(here "small particle" means that the field does not vary significantly over distances characteristic of particle dimensions). By combining Eqs. (3.6) and (3.7) one obtains the following expression for the magnetic force acting on a small paramagnetic or diamagnetic particle of susceptibility, χ:

$$\mathbf{F} = \chi V \nabla\left(\frac{1}{2}HB_0\right) \tag{3.8}$$

In one dimension the above formula simplifies to

$$F_x = \chi V \frac{\mathrm{d}}{\mathrm{d}x}\left(\frac{1}{2}HB_0\right) \tag{3.9}$$

Considering that $B_0 = \mu_0 H$ (Chapter 2), one obtains the following alternative expressions for the force on a test (induced) magnetic dipole:

$$F_{\mathrm{x}} = \chi V \frac{\mathrm{d}}{\mathrm{d}x}\left(\frac{B_0{}^2}{2\mu_0}\right) \tag{3.10}$$

Alternatively

$$F_{\mathrm{x}} = \frac{1}{2}\mu_0 \chi V \frac{\mathrm{d}H^2}{\mathrm{d}x} \tag{3.11}$$

Considering that $\mathrm{d}H^2/\mathrm{d}x = 2H(\mathrm{d}H/\mathrm{d}x)$, one obtains

$$F_{\mathrm{x}} = \mu_0 \chi V H \frac{\mathrm{d}H}{\mathrm{d}x} \tag{3.12}$$

By similar arguments applied to B_0, and reverting to the formula $B_0 = \mu_0 H$, one obtains

$$F_x = \frac{\chi V}{2\mu_0} \frac{\mathrm{d}B_0{}^2}{\mathrm{d}x} \tag{3.13}$$

and

$$F_x = \chi V H \frac{\mathrm{d}B_0}{\mathrm{d}x} \tag{3.14}$$

All forms of the formula for the magnetic force, presented above, are used in the literature [8–16]. Selection of the particular form depends on context and the ease of use. For instance, Eqs. (3.8)–(3.11) are often encountered in applications to continuous, linear magnetic media and in the magnetohydrodynamics. There, the term $B_0{}^2/2\mu_0$ is often referred to as "the imposed magnetic pressure" [4]. Equations (3.12) and (3.14) are encountered when dealing with discrete particles [12, 14] and magnetic levitation [17]. Another convention encountered in the literature is the use of the term "ponderomotive" in application to the magnetic force acting on massive magnetic dipoles [12, 18, 19].

3.4. Magnetic force on a small particle acting as a permanent magnetic dipole (ferromagnetic materials)

The formula for force acting on a permanent magnetic dipole moment can be obtained directly from Eq. (3.14) by considering the magnetic susceptibility, χ, entering the equation as an "effective" magnetic susceptibility, χ_{eff}, as a ratio of the magnetization saturation, M_{s}, to the imposed magnetic field H:

$$M_{\mathrm{s}} = \chi_{\mathrm{eff}} H \tag{3.15}$$

In that case, from Eq. (3.14)

$$F_x = \chi_{\mathrm{eff}} V H \frac{\mathrm{d}B_0}{\mathrm{d}x} = V M_{\mathrm{s}} \frac{\mathrm{d}B_0}{\mathrm{d}x} \tag{3.16}$$

3.5. Potential energy of an elementary magnetic dipole

The potential energy of a magnetic dipole, U, associated with a small, homogeneous particle in a magnetic field represents the amount of mechanical energy stored in the system that is available for performing useful work, such as rotation and a spatial displacement of the particle against the reaction forces of the viscous medium [1, 5]. It is equal to the difference between the total energy of the system, $-\boldsymbol{\mu}\cdot\mathbf{B}_0$, and the internal energy lost to dissipation, such as heat of magnetization, Q

$$U = -\boldsymbol{\mu}\cdot\mathbf{B}_0 - Q \tag{3.17}$$

where $\boldsymbol{\mu}$ is the magnetic dipole moment and $\mathbf{B}_0$ is the applied magnetic field (measured in the absence of the magnetic dipole, that is in free space). The force acting on the dipole is equal to the negative gradient of its potential energy:

$$\mathbf{F} = -\nabla U \tag{3.18}$$

By combining the above two equations, one obtains

$$\mathbf{F} = \nabla(\boldsymbol{\mu}\cdot\mathbf{B}_0 + Q) = (\boldsymbol{\mu}\cdot\nabla)\mathbf{B}_0 + (\mathbf{B}_0\cdot\nabla)\boldsymbol{\mu} + \nabla Q \tag{3.19}$$

where the quantities in parentheses on the right-hand side (RHS) represent gradient in the direction of the magnetic moment, $(\boldsymbol{\mu}\cdot\nabla)$, and gradient in the direction of the magnetic field, $(\mathbf{B_0}\cdot\nabla)$, also shown in component notation in Appendix B. Now, the second term on the RHS depends on whether or not the magnetic dipole moment is a function of the imposed magnetic field, $\mathbf{B_0}$, that is, whether it represents an induced or a permanent dipole, respectively. If it is an induced dipole, then the term $(\mathbf{B_0}\cdot\nabla)\boldsymbol{\mu}$ represents the change of magnetization with the change of dipole position in the direction of field $\mathbf{B_0}$, corresponding to the change in the internal energy that is dissipated as the heat of magnetization. Therefore

$$(\mathbf{B_0}\cdot\nabla)\boldsymbol{\mu} + \nabla Q = 0 \quad \text{induced dipole} \tag{3.20}$$

In the case of the permanent dipole, however, there is no change in the magnetic dipole moment with position, and therefore no heat dissipation related to change in the magnetization

$$(\mathbf{B}\cdot\nabla)\boldsymbol{\mu} = \nabla Q = 0 \quad \text{permanent dipole} \tag{3.21}$$

As a result, by inserting either Eq. (3.20) or Eq. (3.21) into Eq. (3.19), one obtains the same formula for the force acting on an induced magnetic dipole (representing paramagnetic and diamagnetic substances) and a permanent dipole (representing ferromagnetic substance), respectively:

$$\mathbf{F} = (\boldsymbol{\mu}\cdot\nabla)\mathbf{B_0} \quad \text{induced and permanent dipoles} \tag{3.22}$$

For a freely suspended magnetic dipole (free to align with the magnetic field $\mathbf{B_0}$) the above formula reduces to

$$\mathbf{F} = \mu\nabla B_0, \mu \equiv |\boldsymbol{\mu}|, B_0 \equiv |\mathbf{B_0}| \quad \text{induced and permanent dipoles} \tag{3.23}$$

where μ and B_0 are magnitudes of vectors $\boldsymbol{\mu}$ and $\mathbf{B_0}$, as indicated. For a permanent dipole whose magnetic moment is independent of position, one may move the symbol μ for the magnetic moment inside the argument of the gradient operator:

$$\mathbf{F} = \nabla(\mu B_0) \quad \text{permanent dipole only} \tag{3.24}$$

The above operation is not permissible for the induced dipole. Additional formulas showing coordinate notation are provided in Appendix B.

The above formula leads to the same expression of force as that derived from the Maxwell stress, Eqs. (3.4) and (3.7), or Eq. (3.16), because the particle magnetization, **M**, is the ratio of the magnetic moment, $\boldsymbol{\mu}$, to volume, V, of the particle:

$$\mathbf{M} = \frac{\boldsymbol{\mu}}{V} \tag{3.25}$$

This, and the comparison of Eqs. (3.4), (3.7), (3.24), and (3.25), shows that the local Maxwell stress in regular (homogeneous, isotropic, linear) magnetic media is equal to half the absolute value of energy density of a freely suspended, magnetic dipole, $\mu B_0/V$. It follows that for such media the surfaces of constant magnetic pressure (magnetic pressure isosurfaces) coincide with the surfaces of the constant magnetic field energy density, $(1/2)HB_0 = const.$

The thermodynamics involved in the energy conversion of the magnetic system into useful work and the associated magnetization energy dissipation is treated in detail in textbooks on the electromechanical mechanics of continuous media [5].

3.6. *Magnetic field-induced particle motion*

There is no experimental evidence of elementary magnetic charges but only elementary magnetic dipoles [20] (however, see discussion in Rosensweig on the importance of the magnetic charge concept to engineering applications [3]). This causes considerable problems in visualizing the magnetic field effects on motion of small magnetic particles. Yet, such visualization could be very important for developing intuitive understanding of the magnetic field effects during cell separation. The difficulties arise from the fact that, historically, the practical applications of magnetic fields focused on power generation, electric motor development, and elementary particle physics, for which the notion of the magnetic field line provided a sufficiently convenient visual representation of the field interactions with electric currents. In electrostatics, electric field lines are force lines that act on an electrical test charge (point-like, mass-less electrical charges) [1]. They also represent trajectories of such particles (any inertial effects are avoided by assuming zero mass of the test charges).

In contrast, magnetic field lines do not represent force lines acting on magnetic particles because such particles behave like pairs of opposite charges. Rather, they represent local orientation of freely rotating but fixed in space magnetic dipoles (as illustrated in Fig. 3.1). In other words, the magnetic field lines are lines tangent to the elementary magnetic dipoles whose degree of freedom is limited to rotation (no translation). When the restriction on displacement is removed and the magnetic dipoles are allowed to move, their trajectories do not follow the magnetic field lines (Fig. 3.2). One notes that the elementary dipole trajectories may take any direction relative to the magnetic field lines, from nearly parallel to nearly perpendicular.

3.7. Magnetic pathlines

The need for a more intuitive representation of the magnetic field effect on magnetic cell motion is evident from publications describing the operation of magnetic separators [10, 21, 22]. Indeed, it appears that a visually intuitive representation of the magnetic field for magnetic separation applications would be highly useful considering the multidisciplinary character of magnetic separation research and development. Our work suggests that such a representation could be provided by the Maxwell stress isolines (or magnetic pressure isobars) and magnetic pressure gradients, because they directly represent magnetic body forces, as described in preceding sections. Therefore, this notion is further elaborated below for a

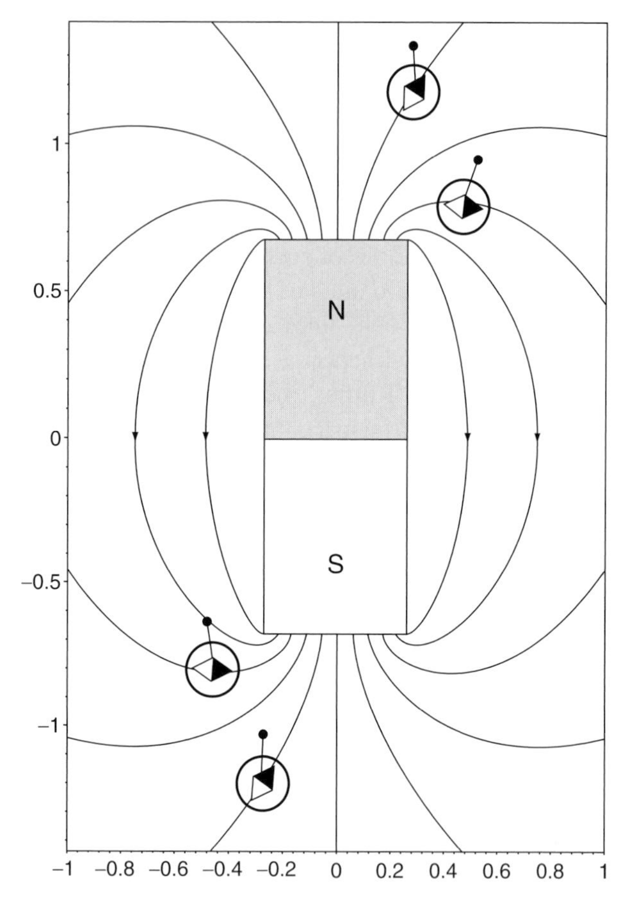

Fig. 3.1. Field of a bar magnet showing field lines and orientation of elementary magnetic dipoles, represented by compass needles. The dipoles are free to rotate but not to translate, as indicated ideographically by pins. Note that the north poles of the compass needles (marked in black) point to the south pole of the magnet. This is different from the convention used to describe polarity of the geomagnetic field (kept for historical reasons) where the north pole of the compass needle points to the (magnetic) North Pole of the Earth (thus, the geomagnetic North Pole is a south pole in physical terms) [27].

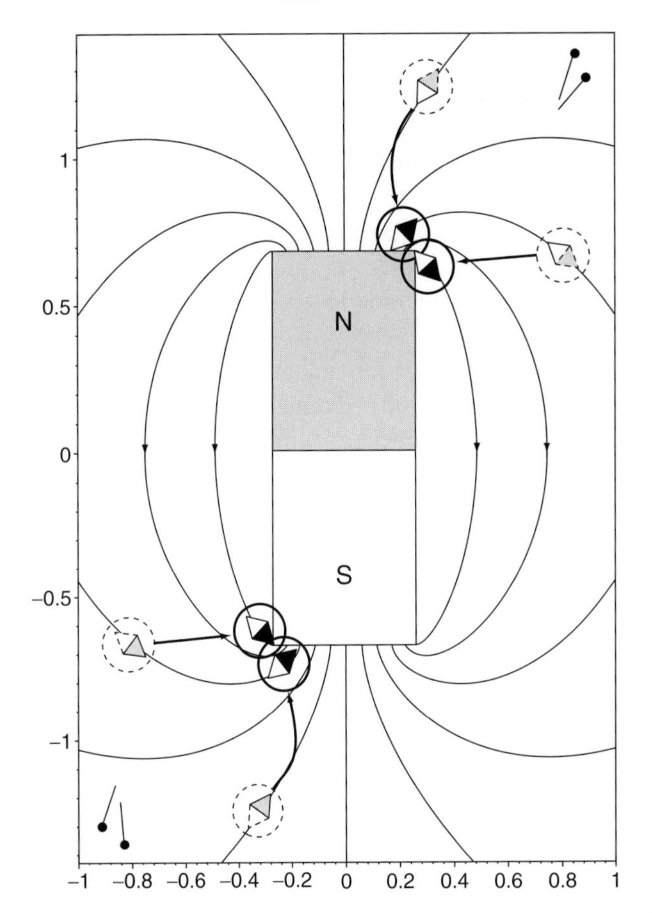

Fig. 3.2. As in Fig. 3.1, except that the restriction on translational motion of the compass needles has been removed, as indicated by the pins discarded at the borders of the graph. Note that the trajectories of the compass needles (thick arrows) are not related to the magnetic field lines but rather follow magnetic pressure gradient, as discussed in the text.

magnetic test particle that is sufficiently small, which can be approximated by a single (elementary) magnetic dipole and whose mass is sufficiently small that inertial effects can be neglected. The small size of the test particle ensures that it can accurately trace even the most rapid changes of the field with position, and that its own field does not interfere with the applied field.

A suitable test particle is a particle whose material property (its magnetic susceptibility) does not depend on the imposed magnetic field, and therefore is paramagnetic or diamagnetic in nature. The paramagnetic particle represents a more practical choice because paramagnetic effects are much stronger than diamagnetic effects and therefore one is aided by the intuitive understanding of its expected behavior from observation of macroscopic (ferromagnetic) objects (paramagnetic and ferromagnetic objects are attracted by the magnet). The choice of a diamagnetic test particle is perhaps more appealing on theoretical grounds because the magnetic body force points in the direction of decreasing magnetic pressure, in analogy to the fluid body force pointing in the direction of decreasing hydraulic pressure, and because there are no magnetization saturation effects in diamagnetism. A particle with a permanent magnetic moment is not suitable as a test particle because such moment is not a material property of the particle (in a sense that its susceptibility, or magnetic permeability, is) and the moment typically depends on the size and the magnetization history of the particle. On the other hand, all magnetic nanoparticles will be saturated at typical magnetic field strengths used for cell separation. One then treats such a particle as a special case, for which the magnetic susceptibility is not independent of the applied field but rather is a higher order function of the applied field [11].

Consequently, the magnetic pathline of a small, homogeneous magnetic particle is defined as a curve that is everywhere tangent to the direction of the local force field **F** acting on the elementary magnetic dipole. On a plane, the magnetic pathlines are defined by the equation:

$$\frac{\mathrm{d}y}{\mathrm{d}x} = \frac{F_y}{F_x} \tag{3.26}$$

The resulting differential equation describes a family of curves, $y = y(x)$, that is a visual representation of the magnetic force.

Unlike the magnetic field lines, however, magnetic pathline density does not represent the magnitude of the local magnetic force (magnetic force is not an analytical function of spatial coordinates in a mathematical sense that the field strength is) [23–25]. An alternative method of visualizing the local force magnitude is the length of the particle trajectory traced for a fixed interval of time under the influence of the imposed magnetic field. Therefore, for a set of randomly selected initial particle coordinates, the magnetic force field can be visualized by the magnetic path lines that are traced in equal periods of time, Δt, by solving the set of equations:

$$\frac{\mathrm{d}v_y}{\mathrm{d}v_x} = \frac{F_y}{F_x} = \frac{\left(\nabla\left(\frac{1}{2}HB_0\right)\right)_y}{\left(\nabla\left(\frac{1}{2}HB_0\right)\right)_x}, \Delta t = const \qquad (3.27)$$

where v_x, v_y are the particle velocity components (see Fig. 3.3 for illustration). By tracing particle trajectories moving in a quasistatic motion, that is in a manner by which the magnetic force is balanced everywhere by the external reaction forces, one obtains

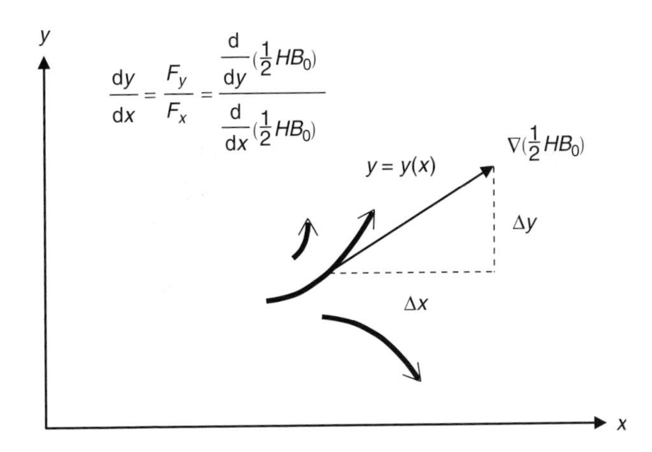

Fig. 3.3. Magnetic path lines and Maxwell stress isolines: magnetic pathlines $y = y(x)$ are tangent everywhere to the magnetic pressure gradient (Maxwell stress), $\nabla(\frac{1}{2}HB_0)$, and their lengths correspond to the length of the elementary dipole trajectories traced in equal lengths of time.

the desirable, graphic representation of the effect of the magnetic field that is compatible with the field description by Maxwell stresses, $(1/2)HB_0$. Moreover, by taking a snapshot of such motions over the entire area of interest, one is also provided with a visual cue of the magnitude of the effect by which the short trajectories represent weak effects and long trajectories represent strong magnetic effects. Since the local concentration of trajectories depends on the local density of starting points, one also assumes uniform distribution of starting points. An overall effect is that of a strong magnetic field in the area with the concentrated, long trajectories and of a weak magnetic field in the area of short or diffuse, point-like trajectories. This approach is similar to the representation of particle motion in fluid flow by the method of pathlines [26]. A comparison of field representations by field lines and magnetic pathlines is provided in Fig. 3.4 and discussed further in Chapter 4.

3.8. *Cell deposition contour surfaces*

For regular magnetic media that approximate magnetic cell separation conditions, the plane of constant magnetic pressure (constant magnetic energy density) approximates a surface formed by the accumulation of magnetic particles on the magnet, by virtue of Eqs. (3.6), (3.17), and (3.18). Those equations show that a small magnetic particle in quasistatic motion travels so that its trajectory is perpendicular to the surface of constant magnetic energy. Consequently, for a rain of equally sized, noninteracting, rigid spheres falling on the surface of a magnet, the spheres distribute so that their potential energy is minimized. Given the reaction forces from the growing layers of rigid spheres that arrive first at the magnet surface, the magnetic deposition layers grow uniformly in the direction of magnetic path lines and thus conform to the shape of the constant magnetic energy U (or the magnetic pressure isosurfaces). Obviously, this is a highly idealized picture of the real process of magnetic cell separation because of the viscoelastic properties of

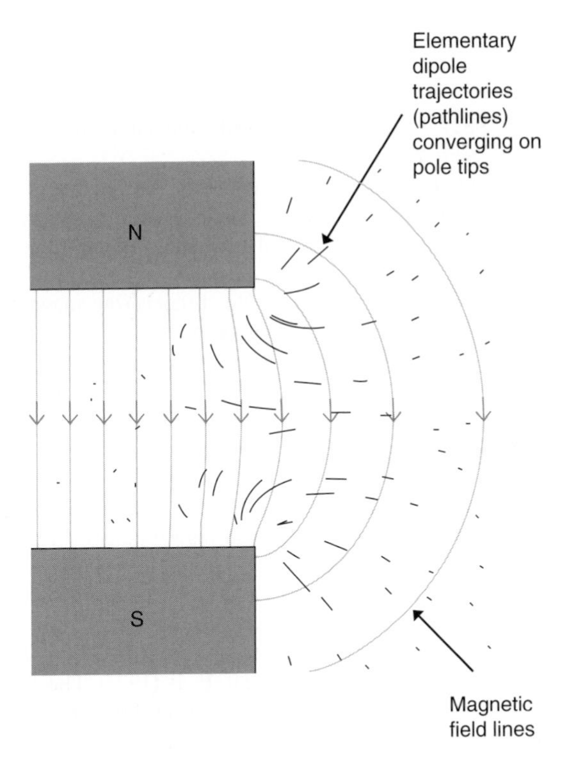

Fig. 3.4. Magnetic field lines (thin, with arrowheads) and the associated pathlines (thick). The length of the pathline correlates with the velocity of the magnetic dipoles. Note the arbitrary direction of pathlines relative to field lines, and their dynamic, visual impact.

cells that are far from those of rigid spheres, and because of cell–cell interactions comprising long-range interaction forces mediated by the incompressible aqueous medium and short-range adhesion forces that cause cells to adhere to neighboring cells or to foreign surfaces. Nevertheless, the constant magnetic energy contour surfaces provide a rapid and relatively simple way of visualizing magnetic cell capture distribution in the complex field of magnetic separators, as illustrated in Chapter 4.

References

[1] Becker, R. (1982). Electromagnetic fields and interactions. Dover Publications Inc., New York.
[2] Stratton, J. A. (1941). Electromagnetic theory. McGraw-Hill Book Company Inc., New York.
[3] Rosensweig, R. E. (1985). Ferrohydrodynamics. Cambridge University Press, Cambridge, MA.
[4] Sutton, G. W. and Sherman, A. (2006). Engineering magnetohydrodynamics. Dover Publications Inc, Mineola, NY.
[5] Melcher, J. R. (1981). Continuum electromechanics. MIT Press, Cambridge, MA.
[6] Zborowski, M., Moore, L. R., Williams, P. S. and Chalmers, J. J. (2002). Separations based on magnetophoretic mobility. Sep. Sci. Technol. *37*, 3611–33.
[7] Hatch, G. P. and Stelter, R. E. (2001). Magnetic design considerations for devices and particles used for biological high-gradient magnetic separation (HGMS) systems. J. Magn. Magn. Mater. *225*, 262–76.
[8] Zimmels, Y. (1990). Magnetic forces and magnetic diffusion in magnetizable dispersions. J. Magn. Magn. Mater. *83*, 439–41.
[9] Watson, J. H. P. (1973). Magnetic filtration. J. Appl. Phys. *44*, 4209–13.
[10] Lawson, W. F., Simons, W. H. and Treat, R. P. (1977). The dynamics of a particle attracted by a magnetized wire. J. Appl. Phys. *48*, 3213–24.
[11] Henjes, K. (1994). The traction force in magnetic separators. Meas. Sci. Technol. *5*, 1105–8.
[12] Furlani, E. P. (2001). Permanent magnet and electromechanical devices. Academic Press, San Diego, CA.
[13] Smolkin, M. R. and Smolkin, R. D. (2006). Calculation and analysis of the magnetic force acting on a particle in the magnetic field of separator. Analysis of the equations used in the magnetic methods of separation. IEEE Trans. Magn. *42*, 3682–93.
[14] Jones, T. B. (1995). Electromechanics of particles. Cambridge University Press, Cambridge.
[15] Slepian, J. (1950). Electromagnetic ponderomotive forces within material bodies. Physics *36*, 485–97.
[16] Zborowski, M. (1997). Physics of magnetic cell sorting. In: Scientific and Clinical Applications of Magnetic Carriers (Haefeli, U., Schuett, W., Teller, J. and Zborowski, M., eds.). Plenum Press, New York, pp. 205–31.
[17] Simon, M. D. and Geim, A. K. (2000). Diamagnetic levitation: Flying frogs and floating magnets (invited). Appl. Phys. A *87*, 6200–4.

[18] Tchikov, V., Winoto-Morbach, S., Kabelitz, D., Kroenke, M. and Schuetze, S. (2001). Adhesion of immunomagnetic particles targeted to antigens and cytokine receptors on tumor cells determined by magnetophoresis. J. Magn. Magn. Mater. *225*, 285–93.

[19] Winoto-Morbach, S., Tchikov, V. and Mueller-Ruchholtz, W. (1994). Magnetophoresis: I. Detection of magnetically labeled cells. J. Clin. Lab. Anal. *8*, 400–6.

[20] Kolm, H. H., Villa, F. and Odian, A. (1971). Search for magnetic monopoles. Phys. Rev. D *4*, 1285–96.

[21] Tibbe, A. G. J., de Grooth, B. G., Greve, J., Liberti, P. A., Dolan, G. J. and Terstappen, L. W. M. M. (1999). Optical tracking and detection of immunomagnetically selected and aligned cells. Nat. Biotechnol. *17*, 1210–13.

[22] Tibbe, A. G. J., de Grooth, B. G., Greve, J., Dolan, G. J., Rao, C. and Terstappen, L. W. M. M. (2002). Magnetic field design for selecting and aligning immunomagnetic labeled cells. Cytometry *47*, 163–72.

[23] Kellogg, O. D. (1954). Foundations of potential theory. Dover Publications, New York.

[24] Weber, E. (1960). Electromagnetic Fields. Theory and Applications. Vol. 1. Mapping of fields. John Wiley and Sons, Inc., New York.

[25] Dettman, J. W. (1984). Applied complex variable. Dover Publications Inc., New York.

[26] Brodkey, R. S. (1995). The phenomena of fluid motions Dover Publications Inc., Mineola, NY.

[27] Livingstone, J. D. (1996). Driving force: The natural magic of magnets. Harvard University Press, Cambridge, MA.

Laboratory Techniques in Biochemistry and Molecular Biology, Volume 32
Magnetic Cell Separation
M. Zborowski and J. J. Chalmers (Editors)

CHAPTER 4

Basic magnetic field configurations

Maciej Zborowski

Department of Biomedical Engineering, Lerner Research Institute, Cleveland Clinic, Cleveland, OH 44195, USA

4.1. Introduction

High magnetic permeability ferromagnetic wires concentrate the applied magnetic field at the interface with a low magnetic permeability medium such as air or water (Fig. 4.1) [1–4]. The resulting local magnetic field gradients are the source of magnetic force on magnetically susceptible material, such as magnetic microparticles and magnetically labeled cells, that concentrates the material on the wire surface. The magnitude of the magnetic force depends on the radius of curvature of the interface—the smaller the radius of curvature, the higher the gradient and, therefore, the higher the force exerted on a magnetic particle such as a magnetized cell [5]. This property has been the basis of a great many designs of magnetic separators, starting from the mining and processing industry [6, 7] to the magnetic cell separators [8–11].

4.2. Infinite cylinder

The magnetic field distribution around a single wire is conveniently approximated by a field around an infinite cylinder (Fig. 4.1). The type of magnetic separators using the matrices of high-permeability

DOI: 10.1016/S0075-7535(06)32004-9

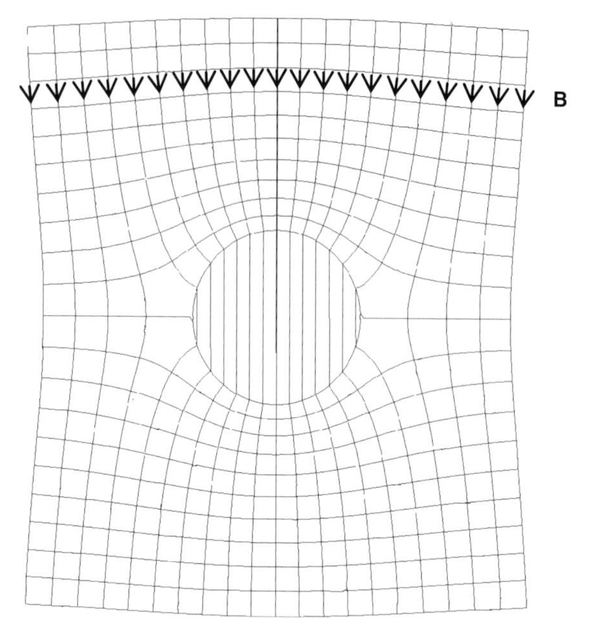

Fig. 4.1. Distortion of an applied magnetic field (assumed uniform at infinity) by a long, ferromagnetic wire (approximated by a cylinder, represented by its circular cross section). Note concentration of field lines inside the wire, and a high level of distortion at the wire surface, indicating region of a high field gradients, and a high magnetic force acting on the magnetically susceptible media.

wires or microspheres (discussed below) is typically referred to as a high gradient magnetic separator (HGMS).

The distribution of magnetic pathlines shows that the influence of the wire on the surrounding magnetic particles decays strongly with distance from the surface of the wire (Fig. 4.2). The published accounts of experimental and theoretical work on magnetic particle trajectories in the vicinity of a ferromagnetic wire show that for most practical applications the useful force range is limited to approximately three wire diameters from the surface [12]. The range of the effective field gradient determines "the cross section for magnetic capture" around the magnetic wire and this cross section on average corresponds to the volume of three wire diameters around the wire

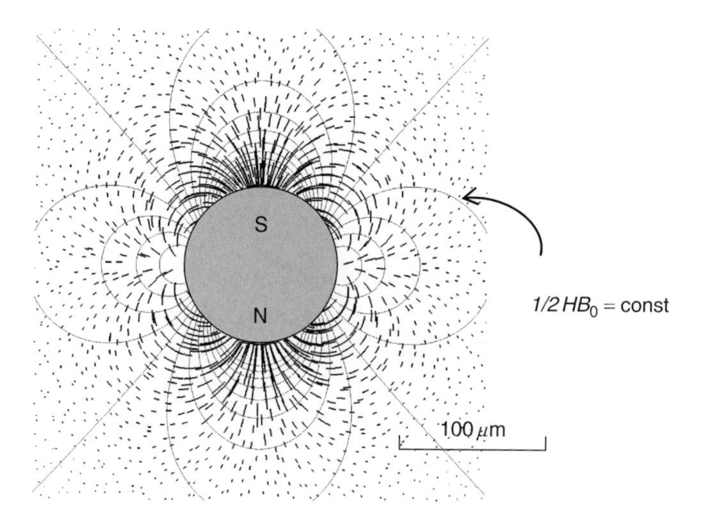

Fig. 4.2. Magnetic pathlines (thick) and magnetic pressure contours (represented by magnetic pressure isobars, thin). The external sources of the magnetic field are as in Fig. 4.1. Note direction of the pathlines perpendicular to the magnetic pressure isobars. Paramagnetic particles are attracted to the filament in the direction of the high magnetic pressure (N–S axis), depopulating the areas of the low magnetic pressure (orthogonal to the N–S axis). An efficient cell capture requires 3D filament packing in the form of filter matrices.

(more about the cross section for capture in Chapter 5). This, in turn, dictates the most efficient wire configuration for matrices and filter beds inserted between pole pieces of a magnet [7].

The deposition of paramagnetic particles on a ferromagnetic wire matrix is illustrated in Figs. 4.3 and 4.4. The wire matrices are used in industrial magnetic separators from which they were adapted for building early versions of the magnetic cell separators. Note unequal distribution of magnetic material on the wire that is a function of the relative orientation of the applied magnetic field and the wire axes. The buildup of the magnetically trapped material on the wire over time closes the interstitial spaces of the wire matrix and thus determines the separation capacity of the magnetic filter. The direct experimental observation confirms the characteristic, bipolar distribution of

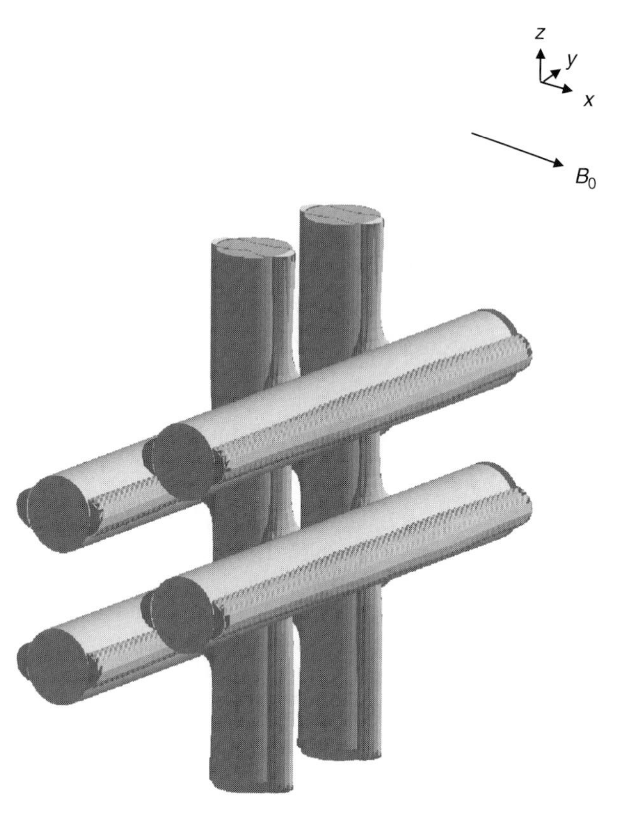

Fig. 4.3. Accretion of cellular deposit on a wire mesh, as simulated by magnetic pressure isobar surfaces. Note uneven accumulation of the deposit (rendered in light gray) on the wire surface, in the direction of the field.

the magnetic particulate deposit (Fig. 4.5) [7]. The efficiency of the magnetic filters has been calculated from first principles using models of particle motion in the vicinity of a ferromagnetic wire [4]. The accumulation of magnetic particles on single wires has been observed microscopically and used for analysis of magnetic migration velocity [1, 13]. The motion of the magnetic particles and magnetically susceptible cells (erythrocytes) induced by gradients and created by

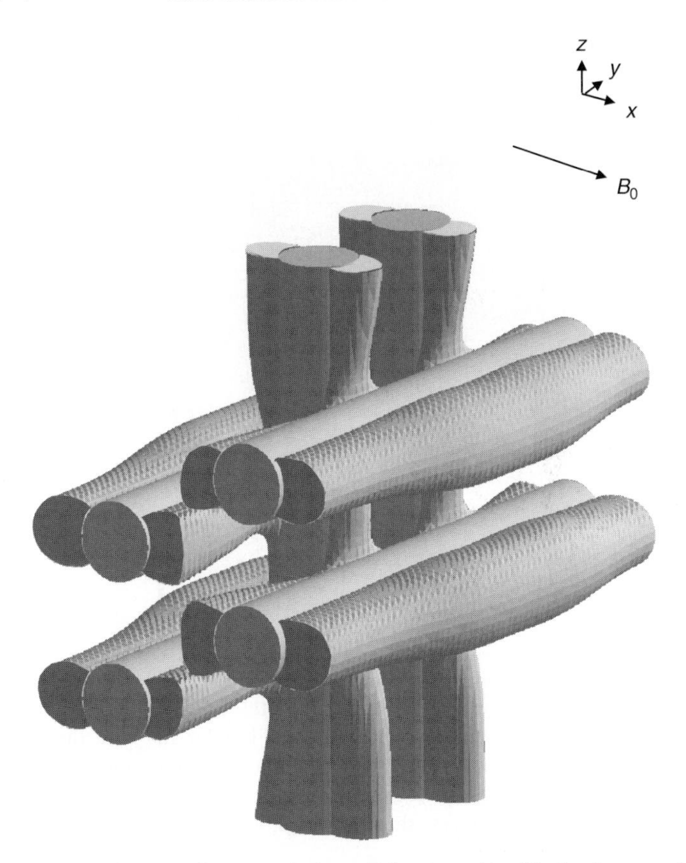

Fig. 4.4. Note increased accumulation of the deposit with the increased cell load, and closing of the open spaces suggesting loss of fluid contact between different parts of the matrix, necessary for elution of the nonmagnetic material, ultimately leading to the flow blockage, nonspecific entrapment, and loss of separation specificity.

the presence of high-permeability wires in the magnetic field has been used to study properties of particles and cells [14]. Single-wire magnetic sorters have been built and evaluated for diamagnetic and paramagnetic particle and cell sorting [15].

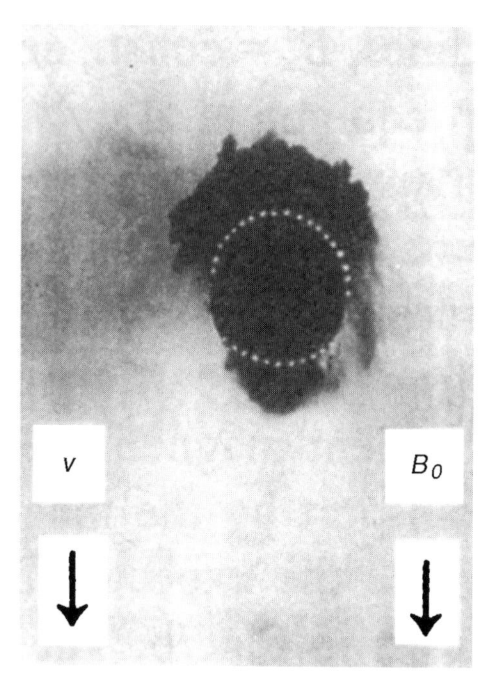

Fig. 4.5. Microphotograph of the magnetic particle deposition on a ferromagnetic wire showing the characteristic, bipolar shape of the captured particle mass (adapted from [7]). The microphotograph provides experimental support for the validity of cell deposition simulation by Maxwell stress isosurfaces, shown in Figs. 4.3 and 4.4.

4.3. Sphere and a stack of spheres

Design of an efficient magnetic cell capture device requires maximizing the ratio of the surface area that is the source of high field gradient to the cell suspension volume. This is provided by a stack of equally sized, ferromagnetic beads. The close packing of equally sized spheres is characterized by a relatively high void volume. Here, the void volume is defined as unity minus the fraction of volume filled by the spheres. It is independent of the sphere radius and is not less than 25%. The interstitial space forms a highly regular system of interconnected chambers that serve as channels for the flow of cell suspension. By adjusting the size of the spheres so that the effective magnetic gradient and the corresponding magnetic capture cross section cover the entire volume of the chambers formed by the neighboring spheres, one obtains a highly efficient magnetic cell capture system.

The field around a single ferromagnetic bead in the plane of a magnetic field vector has similar, polar characteristics to the field of the ferromagnetic wire (discussed above, Fig. 4.1), except that the field gradients are steeper. This implies stronger forces at the bead surface and shorter range of the effective field gradient (with other parameters held constant compared to the ferromagnetic wire system). The Maxwell stress distribution on the surface of a ferromagnetic sphere inside a stack of equal-sized spheres is illustrated in Fig. 4.6. One notes the concentration of magnetic stress at the points of contact between the neighboring spheres, which suggests that these are the sites of increased magnetic cell deposition.

The accumulation of the magnetic material, such as magnetized cells, in the interstitial spaces of stacked ferromagnetic spheres is further illustrated in Fig. 4.7. One notes accretion of the cellular deposit at the points of contact between the spheres, as suggested by the false color rendition in Fig. 4.6. With an increasing time of separation, the volume of the cell deposit grows from the points of sphere contacts into the interstitial space, without interrupting the fluid path between different layers of spheres. This provides efficient means for evacuating the fluid with the remaining, nonmagnetized cell suspension from the column filled with the spheres, and thus an efficient way of separation of magnetic cells from the nonmagnetic cells. Eventually, the magnetic deposit grows far enough into the interstitial spaces to block the fluid connection between different regions of the column and thus limits its cell separation capacity (Figs. 4.7–4.9).

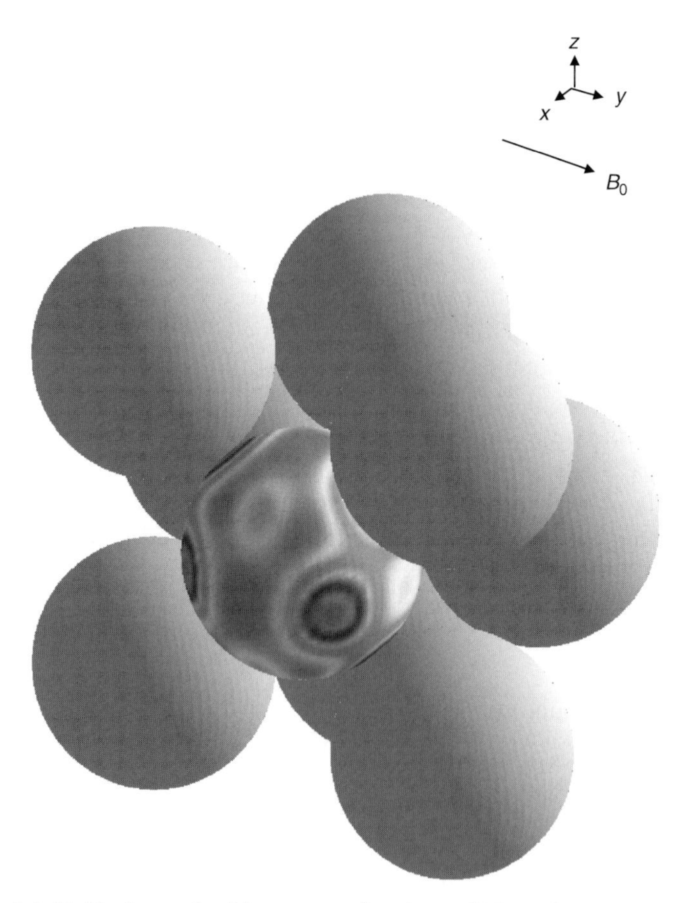

Fig. 4.6. Field of a stack of ferromagnetic spheres. False color representation of the magnetic pressure (Maxwell stress) distribution at the surface of a sphere (violet—high, red—low). The nearest neighbors are rendered in gray. Only 8 of 12 nearest neighbors are shown for clarity. Note magnetic stress concentration at points of contact between spheres. (See Color Insert.)

4.4. Interpolar gap

The fringing field of an interpolar gap is a source of intense magnetic gradients that have been found convenient for cell separation and

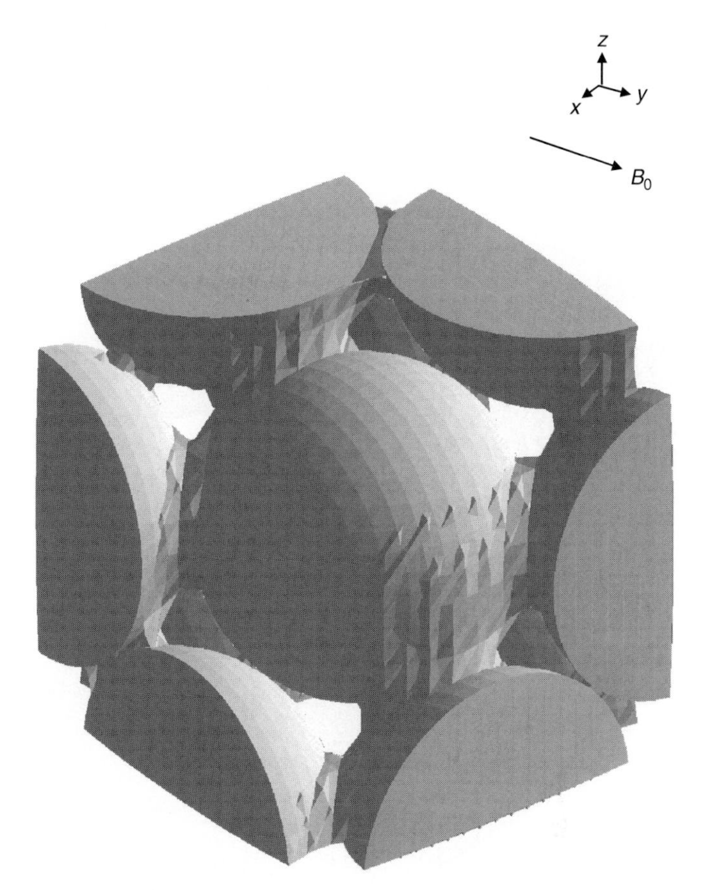

Fig. 4.7. Accretion of cellular deposit inside a stack of ferromagnetic spheres, as simulated by magnetic pressure (Maxwell stress) isosurfaces. Note accumulation of the deposit adjacent to points of contact between the beads.

analysis (Fig. 4.10.). Because of the ability to produce high fields and gradients in the vicinity of an interpolar gap, such fields are used in recording heads in audio and video tape recorders [16]. As an analytical tool, such fields have been first used in the machine industry for analysis of wear particles in lubricating oils [17]. When

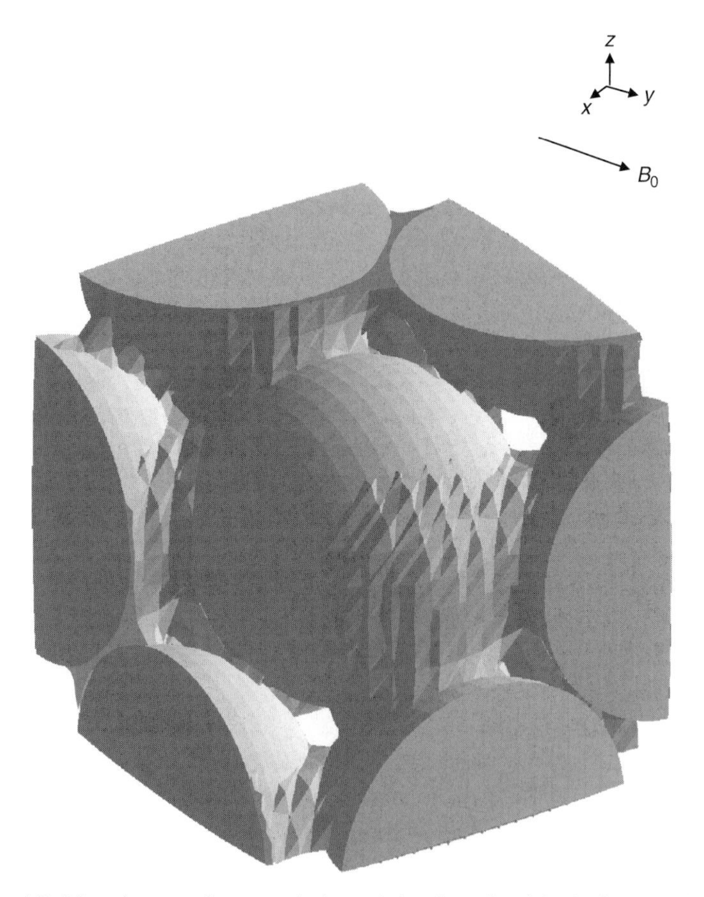

Fig. 4.8. Note increased accumulation of the deposit with the increased cell load. Note open spaces suggesting uninterrupted fluid contact between different parts of the packed sphere column, necessary for elution of the nonmagnetic material.

applied to a thin film of cell suspension pumped over the interpolar gap, the fringing field provides a convenient means of capturing magnetized cells on a thin substrate, suitable for subsequent microscopic analysis in a manner similar to blood smear analysis [18, 19]. Because the cross section for cell capture scales with the width of

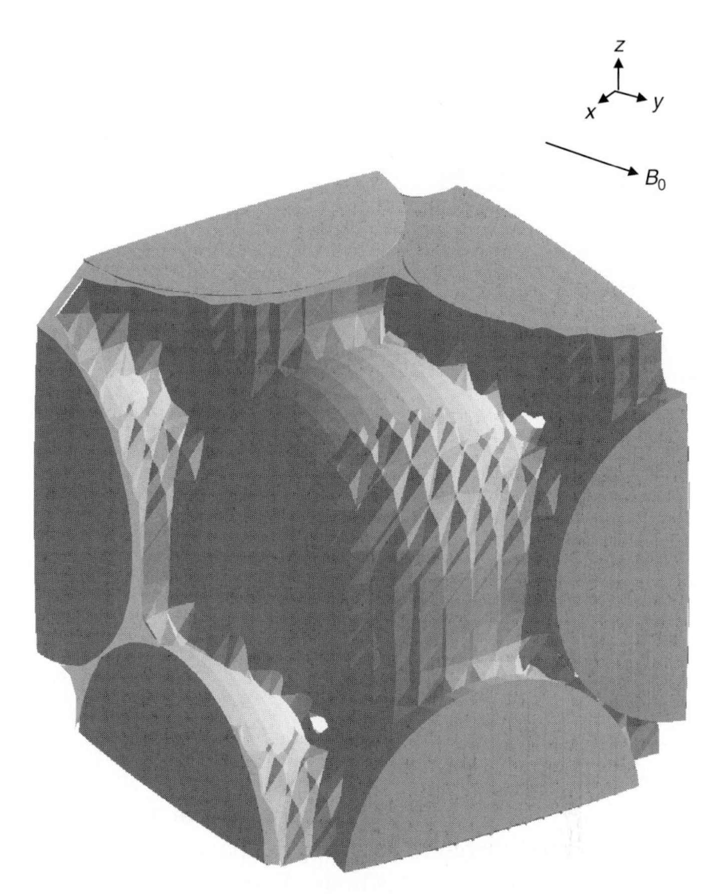

Fig. 4.9. Note further increase of cellular accumulation with the increase of the cell load. Also note blockage of open spaces suggesting overloading of the packed bed of the ferromagnetic spheres with the excessive load by the magnetized cells.

the interpolar gap (assuming that the magnetic field at the pole piece surface is kept constant), the fringing fields were used for debulking large volumes of cell mixtures (Isolex, Baxter Healthcare Corporation).

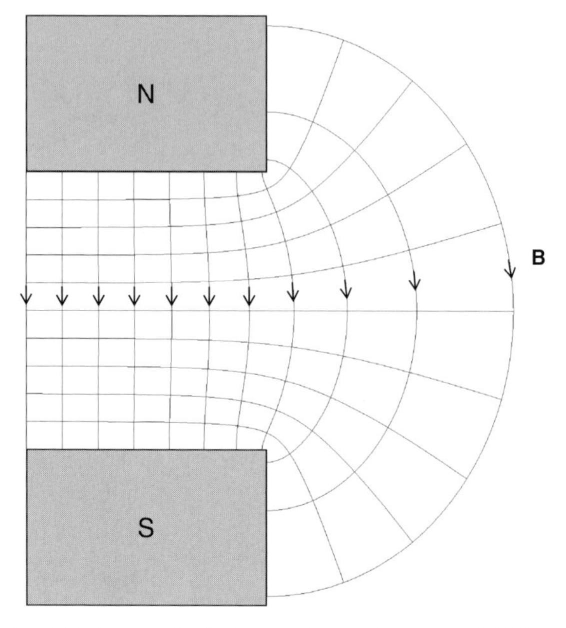

Fig. 4.10. Magnetic field of an interpolar gap. Field lines.

One of the advantages of the fringing field of an interpolar gap as compared to that of a bar magnet is that the field source can be situated remotely from the cell separation area and therefore can be made as bulky (and therefore as strong) as possible for the weakly magnetic cell separation applications. This type of design, where the source of the magnetomotive force and the separation zone are spatially separated, is quite typical and found in various forms of magnetic cell separators. Such designs rely on an efficient conduction of magnetic field lines from their source (typically, a permanent magnet) to the separation zone using high-permeability steel alloys and suitably shaped pole and flux return pieces, and can be manufactured quite inexpensively. On the other hand, the use of permanent magnets directly at the separation zone, when combined with a clever configuration of the component magnetizations, can lead to high local fields, higher than permitted by the saturation magnetization of pure iron (2.2 T), and high gradients (Halbach configurations) [20]. The type of magnetic separator that does not require inserts of high-permeability material (such as wire or microsphere matrices) is referred to as an open gradient magnetic separator (OGMS). Typically, the HGMS configurations, discussed above, generate higher local magnetic field gradients and therefore are used in combination with submicron magnetic particles as the cell labeling reagent (because of their low magnetic moment). In comparison, the OGMS configurations generate lower magnetic gradients and

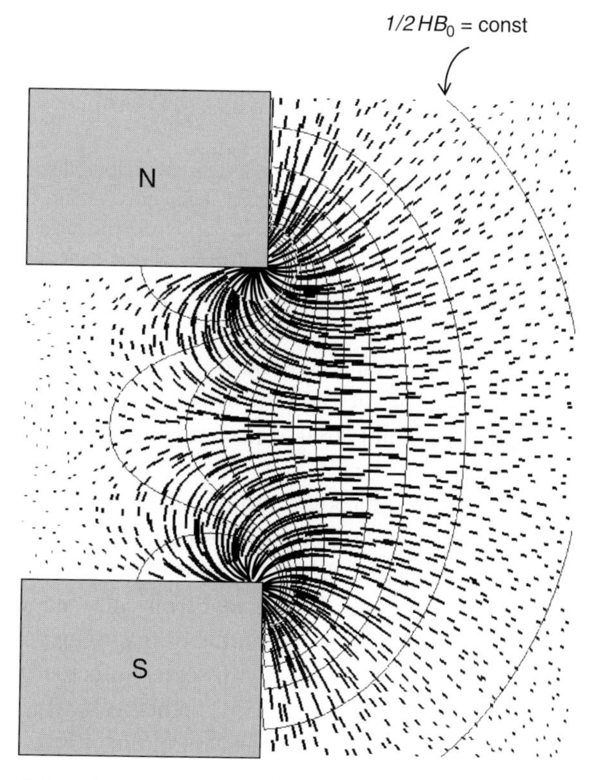

Fig. 4.11. Interpolar gap—magnetic pathlines. Note highly localized field effects, limited to the vicinity of the surface of the interpolar gap but not deep between the pole pieces or away from them.

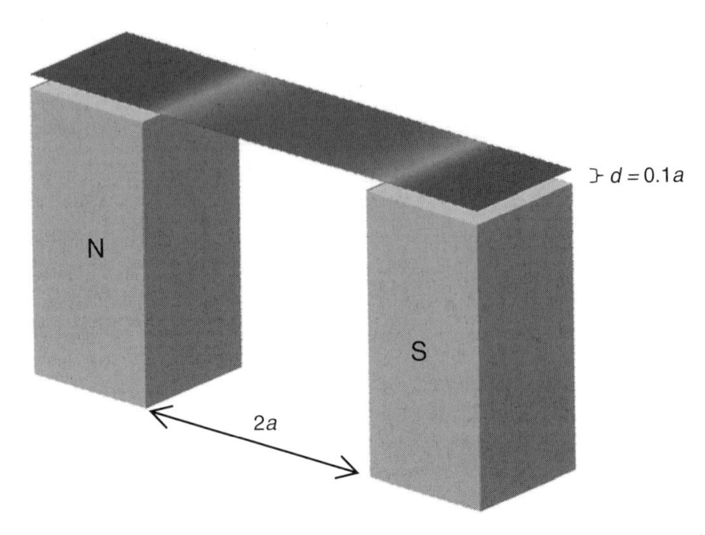

Fig. 4.12. Maxwell stress from an interpolar gap at various distances of cell deposition substrate (gray scale representation, light gray—high, dark gray—low). The deposition surface is at a distance of $d = 0.1a$ from the surface of the pole pieces. Note highly intense and localized stresses at the pole tips with the deposition substrate close to the magnet surface.

require use of micron-sized magnetic particles (of high magnetic moment). The advantage of the OGMS operation is that it does not require high-permeability matrix inserts and typically gives more uniform gradients.

In the vicinity of an interpolar gap, the field gradients are localized at the tips of the pole pieces (Fig. 4.10.). This is an area of intense Maxwell stress (Fig. 4.11) that allows capture of even the most weakly magnetic material, such as intrinsically magnetic cells, not modified by the attachment of extraneous magnetic particles but only by the pathological process (erythrocytes infected by malaria parasites, discussed in Chapter 12) [18]. Further away from the gap, the field becomes more regular, albeit less intense (Figs. 4.12 and 4.13), suitable for capture of a more magnetic material (such as cells modified by the attachment of extraneous magnetic nanoparticles). With the increasing load of the magnetized cells in the mixture

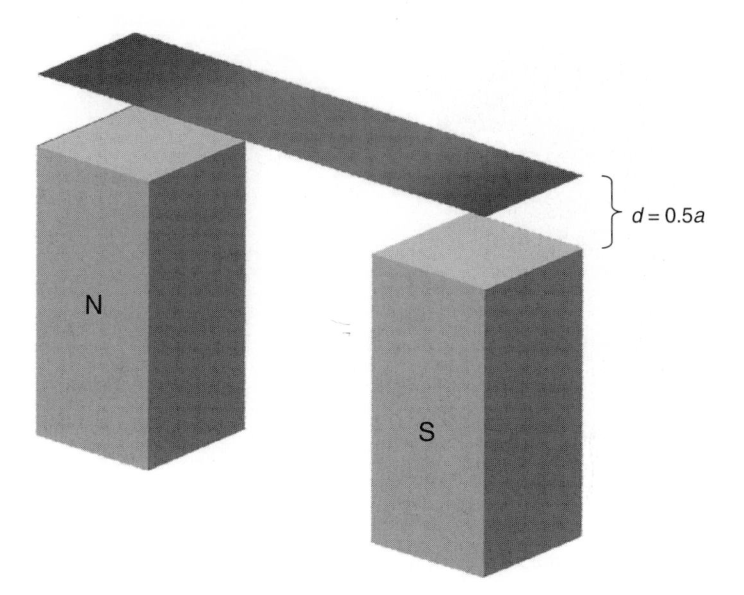

Fig. 4.13. The deposition surface distance of $d = 0.5a$. Note less intense and more diffuse stresses with the substrate moved away from the surface of the magnet. The cellular deposit forms in the areas of the most intense Maxwell stresses.

(or the increasing time of the separation, assuming a constant supply of the cell mixture), the cells initially are trapped in the vicinity of the pole tips to become more evenly distributed between the pole tips with the passage of time (Figs. 4.14 and 4.15).

4.5. Isodynamic field

The so-called "isodynamic field" produces nearly constant magnetic force on linearly polarizable particles, $\mathbf{F} \approx \mathbf{const}$ (compare with equations in Chapter 3) [21–23]. In keeping with the vector notation (Appendix B), the use of the bold type here denotes constant force vector, that is its constant magnitude and direction. The isodynamic field was first introduced for magnetic fractionation of dry

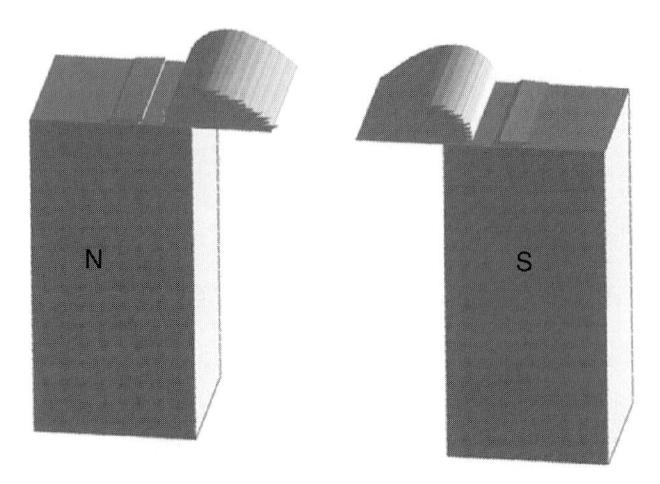

Fig. 4.14. Accretion of cellular deposit at the interpolar gap, for increasing load of the magnetized cells in the cell suspension exposed to the field. Note initial localized accumulation of cellular deposit (rendered in light gray) at the pole tips.

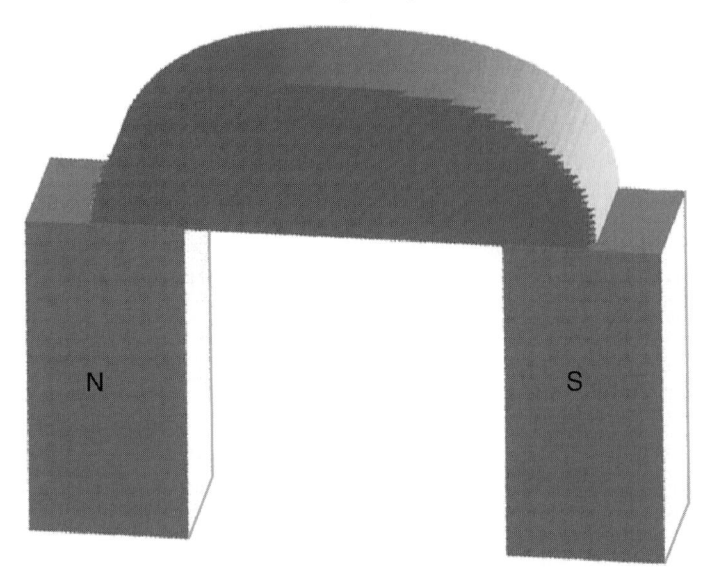

Fig. 4.15. Note broad distribution of the cellular deposit across the interpolar gap surface with the increasing number of the trapped cells.

particulate matter using a vibrating chute, in application to separation of paramagnetic from diamagnetic minerals (in particular, enrichment of diamond) [21]. It was later adopted for field-induced cell velocity determination [24, 25]. The advantage of the isodynamic field over other types of magnetic fields for measuring field effects on the cell velocity is that, within its domain, the cell velocity does not depend on cell position but only on the properties of the cell. This is important for weak magnetic effects, which can only be determined using statistical analysis on large numbers of cells, allowing discrimination between "control" (nonmagnetic) and "test" (magnetic) cell samples, involving from a few hundred to a few thousand cell trajectory measurements. The analysis provides quantitative information about the cell (such as its magnetic

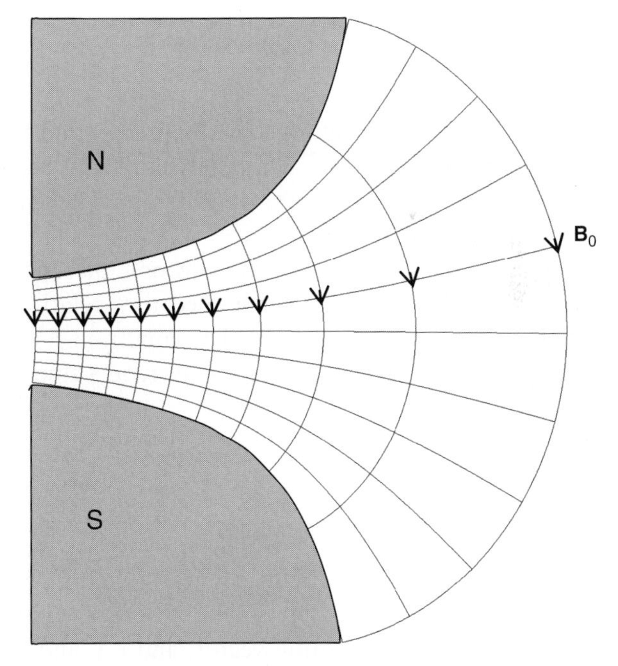

Fig. 4.16. The so-called "isodynamic field." Field lines.

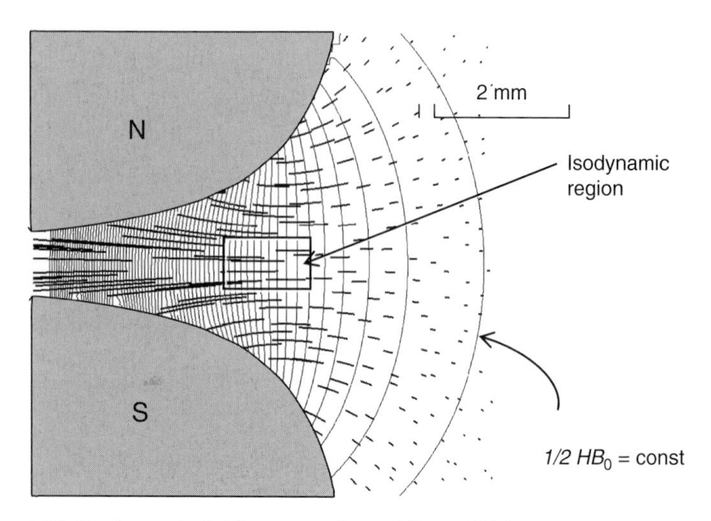

Fig. 4.17. Isodynamic field—magnetic pathlines (thick). Note highly regular pathline distribution in the locally (nearly-)isodynamic field region (parallel, same length).

antibody-binding capacity, discussed in Chapter 8) and therefore is a contribution to cytometry. The isodynamic magnetic field also enables cell fractionation based on differential cell magnetophoresis, as discussed in Chapter 5.

The isodynamic field is generated by two pole pieces of hyperbolic cross section (Fig. 4.16) and is limited to a relatively small fraction of the gap volume (Fig. 4.17). The condition $F = const$ is typically maintained to within 1% of its average value over the isodynamic field domain [24, 25].

4.6. *Quadrupole field*

An ideal quadrupole field is strictly a linear function of spatial coordinates, $\mathbf{B} = \mathbf{B}(\mathbf{r}) = \alpha\mathbf{i}x + \beta\mathbf{j}y + \gamma\mathbf{k}z$, where α, β, and γ are constants; $\mathbf{r} = [x,y,z]$ is the position vector; and $\mathbf{i}$, $\mathbf{j}$, and $\mathbf{k}$ are the unit vectors along the Cartesian coordinates axes $0x$, $0y$ and $0z$,

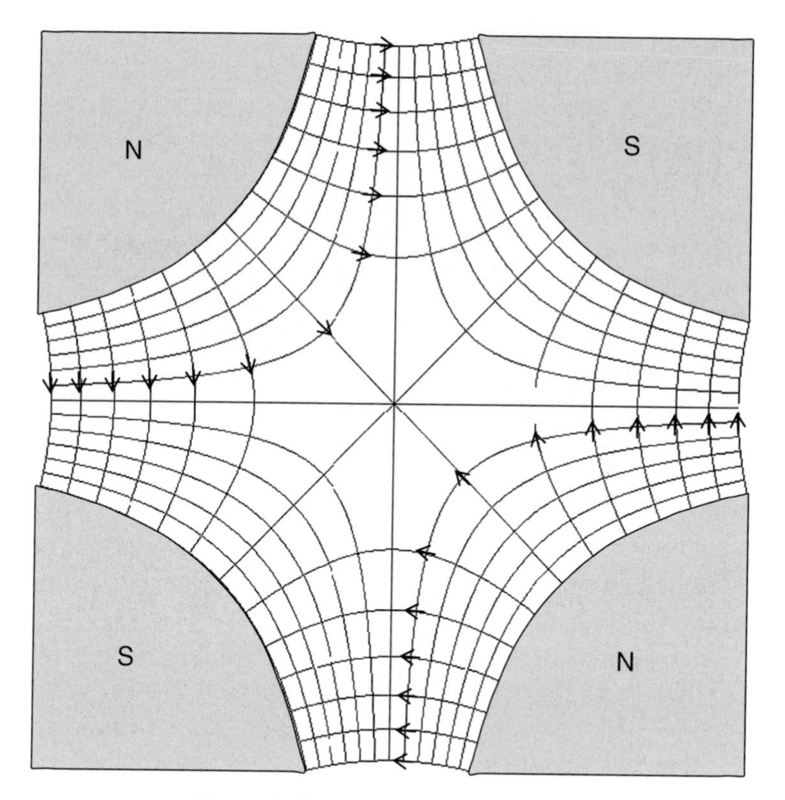

Fig. 4.18. Quadrupole field. Field lines.

respectively. Quadrupole fields play an important role in mass spectroscopy and in ion traps [26–28].

A two-dimensional quadrupole field is produced by four pole pieces having right hyperbolic cross sections (Fig. 4.18) (here $\gamma = 0$, $\alpha = \beta = B_0/r_0$, where B_0 is the field at a characteristic radial distance r_0). A simpler design for applications with permanent magnets is known as a Halbach configuration [20].

The great appeal of quadrupole fields lies in their efficient use of the magnetomotive forces (from electric currents or permanent magnet bricks) with minimal loss due to stray fields. They are

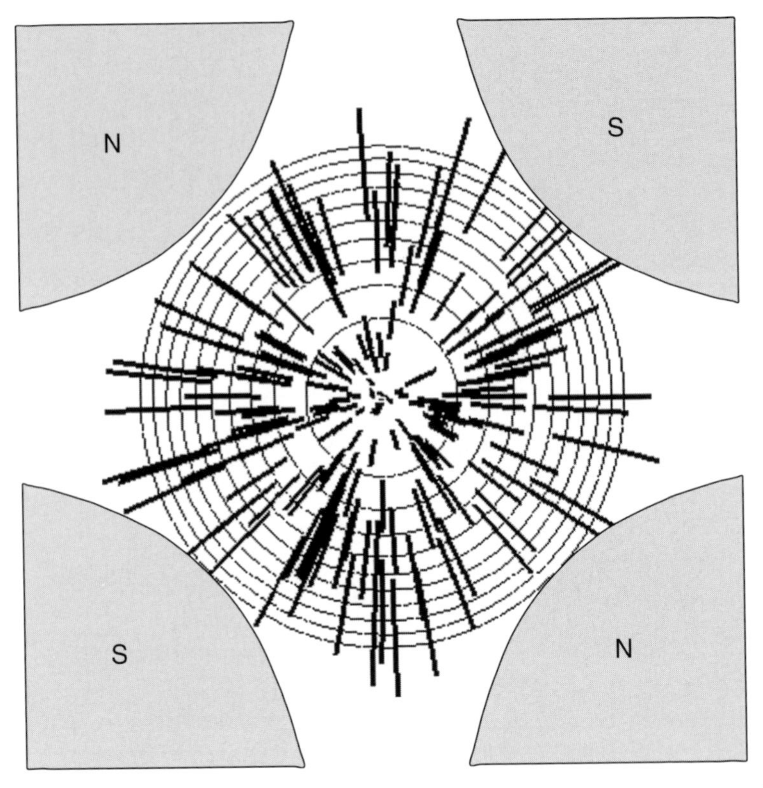

Fig. 4.19. Magnetic pathlines. Note centrifugal (away from the center) distribution of the pathlines.

amenable to compact design capable of producing high fields (at or even above the limit of the saturation magnetization of the iron pole pieces used for its construction) and high gradients (as the field at the bore center is equal to zero). In one particular cell separation application, they are capable of producing an average field of 1.2 T and the field gradient in excess of 0.15 T/mm or 150 T/m in a volume of 3 cm^3 [29]. In an ideal case, when acting on paramagnetic particles, they produce a strictly centrifugal force field (Figs. 4.19 and 4.20).

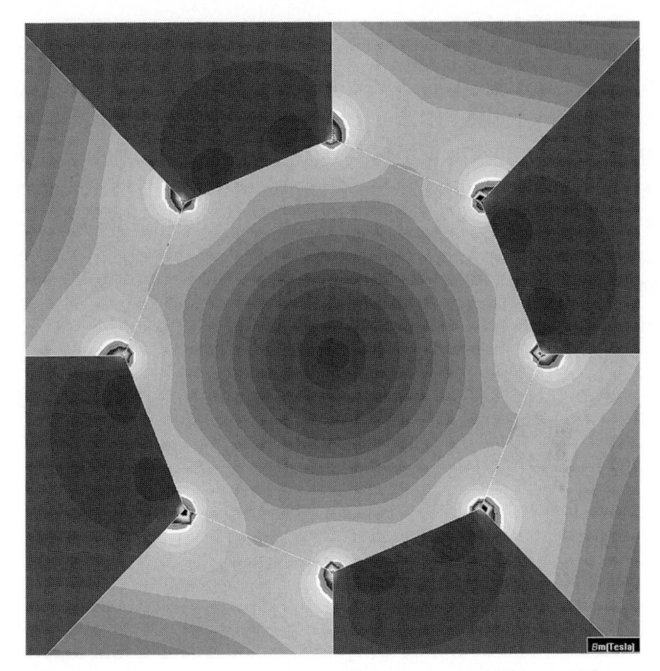

Fig. 4.20. Quadrupole field of Halbach permanent magnet configuration, false color representation of B field (from dark gray—low field to light gray—high field). Note axial symmetry of the field, and a linear increase in B with distance from center, corresponding to the centrifugal distribution of the magnetic pathlines in Fig. 4.19.

A large-scale, high-volume superconducting quadrupole magnet assembly was developed and tested for application to cleaning of coal from sulfur impurities [28]. A permanent magnet-based quadrupole magnet assembly has been applied to continuous cell sorting by magnetophoresis (additional details are provided in Chapter 12).

4.7. Other magnet configurations

There are many other magnet configurations used in application to cell separation. The literature on the subject grows steadily with the increasing interest in application of magnetic fields to separation and manipulation of cells [30, 31]. Particularly, interesting are the recent advances at the extremes of the technology scale of high field and small channel dimensions. High field technology is capable of producing sustained field intensities of up to 40 T and open field gradients of 1000 T/m [32]. The advances in miniaturization of the separation channels using a microelectromechanical systems (MEMS) approach, in combination with innovative approaches to magnetic field generation, provide opportunities for large-scale manufacturing of inexpensive magnetic cell separators [33, 34]. Electromagnet arrays also provide means for transporting and separating magnetic particles using sequential switching of the magnetic field [35].

References

[1] Friedlaender, F. J., Takayasu, M., Rettig, J. B. and Ketzner, C. P. (1978). Particle flow and collection process in single wire HGMS studies. IEEE Trans. Magn. *MAG-14*, 1158–64.

[2] Takayasu, M., Kelland, D. R. and Minervini, J. V. (2000). Feasibility of direct magnetic separation of white cells and plasma from whole blood. IEEE Trans. Appl. Supercond. *10*, 927–30.

[3] Lawson, W. F., Simons, W. H. and Treat, R. P. (1977). The dynamics of a particle attracted by a magnetized wire. J. Appl. Phys. *48*, 3213–24.

[4] Watson, J. H. P. (1973). Magnetic filtration. J. Appl. Phys. *44*, 4209–13.

[5] Oberteuffer, J. A. (1973). High gradient magnetic separation. IEEE Trans. Magn. *Mag-9*, 303–6.

[6] Kolm, H., Oberteuffer, J. and Kelland, D. (1975). High-gradient magnetic separation. Sci. Am. *233*, 47–54.

[7] Watson, J. H. P. (1989). High gradient magnetic separation. In: Solid-Liquid Separation (Svarovsky, L., ed.). Butterworth, London, pp. 661–70.

[8] Thomas, T. E., Richards, A. J., Roath, O. S., Watson, J. H. P., Smith, R. J. S. and Lansdorp, P. M. (1993). Positive selection of human blood cells using improved high gradient magnetic separation filters. J. Hematother. *2*, 297–303.

[9] Molday, R. S. and MacKenzie, D. (1982). Immunospecific ferromagnetic iron-dextran reagents for the labeling and magnetic separation of cells. J. Immunol. Methods *52*, 353–67.

[10] Roath, S., Richards, A. J., Smith, R. J. S. and Watson, J. H. P. (1990). High-gradient magnetic separation in blood and bone marrow processing. J. Magn. Magn. Mater. *85*, 285–9.
[11] Miltenyi, S., Müller, W., Weichel, W. and Radbruch, A. (1990). High gradient magnetic cell separation with MACS. Cytometry *11*, 231–8.
[12] Oberteuffer, J. A. (1974). Magnetic separation: A review of principles, devices, and applications. IEEE Trans. Magn. *Mag-10*, 223–38.
[13] Takayasu, M., Gerber, R. and Friedlaender, F. J. (1983). Magnetic separation of submicron particles. IEEE Trans. Magn. *MAG-19*, 2112–4.
[14] Plyavin, Y. and Blum, E. (1983). Magnetic parameters of blood cells and high gradient paramagnetic and diamagnetic phoresis. Magnetohydrodynamics *19*, 349–59.
[15] Takayasu, M., Maxwell, E. and Kelland, D. R. (1984). Continuous selective HGMS in the repulsive force mode. IEEE Trans. *MAG-20*, 1186–8.
[16] Livingstone, J. D. (1996). Driving Force: The natural magic of magnets. Harvard University Press, Cambridge, MA.
[17] Evans, C. H. and Tew, W. P. (1981). Isolation of biological materials by use of erbium (III)—induced magnetic susceptibilities. Science *213*, 653–4.
[18] Zimmerman, P. A., Thomson, J. M., Fujioka, H., Collins, W. E. and Zborowski, M. (2006). Diagnosis of malaria by magnetic deposition microscopy. Am. J. Trop. Med. Hyg. *74*, 568–72.
[19] Zborowski, M., Malcheski, P. S., Savon, S. R., Green, R., Hall, G. S. and Nose, Y. (1991). Modification of ferrography method for analysis of lymphocytes and bacteria. Wear. *142*, 135–49.
[20] Hatch, G. P. and Stelter, R. E. (2001). Magnetic design considerations for devices and particles used for biological high-gradient magnetic separation (HGMS) systems. J. Magn. Magn. Mater. *225*, 262–76.
[21] Frantz, S. G. (1936). Magnetic separation methods and means. Patent No. 2,056,426 (United States).
[22] Smolkin, M. R. and Smolkin, R. D. (2006). Calculation and analysis of the magnetic force acting on a particle in the magnetic field of separator. Analysis of the equations used in the magnetic methods of separation. IEEE Trans. Magn. *42*, 3682–93.
[23] Zborowski, M., Moore, L. R., Williams, P. S. and Chalmers, J. J. (2002). Separations based on magnetophoretic mobility. Sep. Sci. Technol. *37*, 3611–33.
[24] Moore, L. R., Zborowski, M., Nakamura, M., McCloskey, K., Gura, S., Zuberi, M., Margel, S. and Chalmers, J. J. (2000). The use of magnetite-doped polymeric microspheres in calibrating cell tracking velocimetry. J. Biochem. Biophys. Methods *44*, 115–30.

[25] Moore, L. R., Fujioka, H., Williams, P. S., Chalmers, J. J., Grimberg, B., Zimmerman, P. A. and Zborowski, M. (2006). Hemoglobin degradation in malaria-infected erythrocytes determined from live cell magnetophoresis. FASEB J. *20*, 747–9.

[26] Ramsey, N. F. (1990). Molecular Beams, published in Tne International Series of Monographs of Physics. Oxford University Press, Oxford.

[27] Ziock, K.-P. and Little, W. A. (1987). One tesla rare-earth permanent quadrupole magnet for spin separation of metal clusters. Rev. Sci. Instrum. *58*, 557–62.

[28] Doctor, R. D., Panchal, C. B. and Swietlik, C. E. (1986). A model of open-gradient magnetic separation for coal cleaning using a superconducting quadrupole field. AICHE Symp. Ser. *82*, 154–68.

[29] Jing, Y., Moore, L. R., Williams, P. S., Chalmers, J. J., Farag, S. S., Bolwell, B. and Zborowski, M. (2007). Blood progenitor cell separation from clinical leukapheresis product by magnetic nanoparticle binding and magnetophoresis. Biotechnol. Bioeng. *96*, 1139–54.

[30] Ghebremeskel, A. N. and Bose, A. (2000). A flow-through, hybrid magnetic-field-gradient, rotating wall device for magnetic colloidal separations. Sep. Sci. Technol. *35*, 1813–28.

[31] Watson, J. H. P. and Beharrell, P. D. (1997). Magnetic separation using a switchable system of permanent magnets. J. Appl. Phys. *81*, 4260–2.

[32] Takeda, S., Mishima, F., Fujimoto, S., Izumi, Y. and Nishijima, S. (1997). Development of magnetically targeted drug delivery system using superconducting magnet. J. Magn. Magn. Mater. *311*, 367–71.

[33] Ahn, C. H., Allen, M. G., Trimmer, W., Jun, Y.-N. and Erramilli, S. (1996). A fully integrated micromachined magnetic particle separator. J. Microelectromech. Syst. *5*, 151–7.

[34] Nath, P., Moore, L. R., Zborowski, M., Roy, S. and Fleischman, A. J. (2006). A method to obtain uniform magnetic-field energy density gradient distribution using discrete pole pieces for a microelectromechanical-system-based magnetic cell separator. J. Appl. Phys. *99*, 08R905.

[35] Ramadan, Q., Samper, C., Poenar, D. and Yu, C. (2006). Microcoils for transport of magnetic beads. Appl. Phys. Lett. *88*, 032501.

Laboratory Techniques in Biochemistry and Molecular Biology, Volume 32
Magnetic Cell Separation
M. Zborowski and J. J. Chalmers (Editors)

CHAPTER 5

Magnetophoresis

Maciej Zborowski

Department of Biomedical Engineering, Lerner Research Institute, Cleveland Clinic, Cleveland, OH 44195, USA

5.1. Magnetophoretic mobility

Magnetic cell separation is made possible by a magnetic force acting on magnetically susceptible cells. The nature of the force is that of a magnetic dipole interaction with the external applied field. The magnetic dipole moment of a eukaryotic cell in its natural environment, in a tissue, is typically nonexistent in the absence of a magnetic field, and typically small, and negative, even with the application of high magnetic fields. The reason for that is that the major components of cells, such as water, phospholipid bilayers, proteins, and DNA, are diamagnetic [1, 2]. There are notable exceptions to that rule, such as cells exhibiting paramagnetic behavior (deoxygenated erythrocytes [3], and erythrocytes modified by the presence of methemoglobin [4] or hemozoin [5]) and ferromagnetic behavior of certain prokaryotic cells (such as magnetotactic bacteria) [6]; however, the overwhelming majority of current cell separation applications concerns cells that are diamagnetic in nature [4, 7]. Therefore, the diamagnetism of the cell is important for current applications of magnetic cell separation in that the cells do not substantially interact with the magnetic field unless artificially modified (typically, by the attachment of paramagnetic or ferromagnetic particles) [8–12].

DOI: 10.1016/S0075-7535(06)32005-0

It is worth noting that the cell diamagnetic moment does not saturate even in the presence of very high magnetic fields, such as in excess of 50 T. Therefore, in principle, for very high magnetic fields and gradients, diamagnetic forces may eventually approach the typical paramagnetic or ferromagnetic forces acting on a cell labeled with current, commercial paramagnetic labels. The potential of these very large-magnitude diamagnetic forces on tissues in high fields has been dramatically demonstrated by the levitation of tissues or entire organisms over superconducting magnets [13]. This potential has created increasing interest in diamagnetic separations of biological materials in fields exceeding 5 T, with the reported field magnitudes of up to 40 T.

The magnetic force acting on a diamagnetic particle suspended in a diamagnetic, fluid medium is described by (Chapter 3)

$$\mathbf{F} = \Delta\chi V \nabla \left(\frac{B_0{}^2}{2\mu_0} \right) \tag{5.1}$$

where $\Delta\chi = \chi_\mathrm{p} - \chi_\mathrm{f}$, V is the volume of the particle, and B_0 is the applied magnetic field. For a micrometer-sized particle in a viscous, aqueous solution, the drag force, $\mathbf{F}_\mathrm{d}$, of the continuous medium counteracts the effect of the magnetic field to such an extent that the particle reaches its terminal velocity in a matter of microseconds, in which case it is possible to assume

$$\mathbf{F}_\mathrm{d} = \mathbf{F} \tag{5.2}$$

In the limit of a creeping flow, applicable to the motion of a diamagnetic particle in a viscous, diamagnetic media, the Stokes formula for the drag force applies

$$\mathbf{F}_\mathrm{d} = 6\pi\eta R\mathbf{v} \tag{5.3}$$

From a combination of Eqs. (5.1)–(5.3) one obtains the terminal velocity of the particle:

$$\mathbf{v} = \frac{\Delta\chi V}{6\pi\eta R} \nabla \left(\frac{B_0{}^2}{2\mu_0} \right) \tag{5.4}$$

An interesting property of the above expression is that it is in the form of a product of two quantities that are independent of each other, one being a combination of material properties of the particle and the suspending fluid medium, $\Delta\chi V/6\pi\eta R$, the other of the applied magnetic field, $\nabla(B_0^2/2\mu_0)$. This suggests a definition of the magnetophoretic mobility as

$$m \equiv \frac{\Delta\chi V}{6\pi\eta R} \tag{5.5}$$

because it has the desirable property of representing an intrinsic property of the particle in fluid suspension, independent of the applied field. Consequently, this fixes the definition of the force strength, associated with the magnetophoresis as

$$\mathbf{S}_{\mathrm{m}} \equiv \nabla\left(\frac{B_0^2}{2\mu_0}\right) \tag{5.6}$$

which has a familiar form of the magnetic pressure gradient (the Maxwell stress) in free space (in the absence of the magnetic particle and the fluid) [14]. The term "magnetophoresis" (a magnetic field-induced motion) has been proposed in analogy to the term "electrophoresis" (an electric field-induced motion) [15, 16]. The phenomenon has been measured for various field and particle combinations [17–24].

A combination of Eqs. (5.4) and (5.5) leads to the expression for the particle field-induced velocity

$$\mathbf{v} = m\mathbf{S}_{\mathrm{m}} = m\nabla\left(\frac{B_0^2}{2\mu_0}\right) \tag{5.7}$$

Therefore, for a given particle and fluid system (described by m), the particle velocity follows the vector $\nabla(B_0^2/2\mu_0)$. In particular, for a field for which $\nabla(B_0^2/2\mu_0) \approx \mathbf{const}$, then $\mathbf{v} \approx \mathbf{const}$, and the particles move in a nearly uniform motion, that is, in nearly parallel directions and nearly constant velocities. This type of a field has been named an isodynamic field and it has played an important role in separation and particle motion analysis [25–27].

The above formula is reminiscent of the expression for the particle velocity in dielectrophoresis, due to the force exerted by an inhomogeneous electric field, E, on the induced electric dipole

$$\mathbf{v} = m_{\text{DEP}} \nabla E^2 \tag{5.8}$$

where m_{DEP} is the dielectrophoretic mobility [28, 29].

As mentioned above, the intrinsic cell magnetophoresis in the currently available magnetic separators is of limited practical value, such as high-gradient magnetic separator (HGMS) enrichment of malaria-infected erythrocytes and magnetic field-oriented motility of the magnetotactic bacteria [30, 31]. In addition, direction of a diamagnetic cell in an aqueous electrolyte solution, approximating physiological composition of interstitial fluids, is away from the magnetic field source, rather than toward it. However, this creates an opportunity for a highly selective separation process whereby the targeted binding of paramagnetic or ferromagnetic particles makes the cells strongly attracted by the source of the magnetic field, and therefore makes them easily separated from the rest of the cell population [12].

To the lowest-order approximation, the magnetic properties of a cell-label ensemble are a weighed average of the properties of the cell and the magnetic label. Typically, because of the weak diamagnetic properties of the cell, the magnetization of the paramagnetic or ferromagnetic label dominates the magnetic forces. This, in turn, reaches saturation in the magnetic fields of the cell separators. Therefore, for most practical applications, the motion of the cell-label ensemble is described as induced by the force acting on a saturated dipole (Eq. 3.23)

$$\mathbf{F} = \mu \nabla B_0 \tag{5.9}$$

where $\mu = const$ is the magnitude of the magnetic dipole moment (note that the above formula applies only to a freely suspended magnetic dipole). Because the saturated dipole-field interactions are much stronger than the diamagnetic forces acting on the cell itself and the aqueous electrolyte suspension, the magnetic properties of the cell and the fluid can be omitted from the above equation.

Following the same line of reasoning as that used for the derivation of the diamagnetic particle velocity induced by the magnetic field, the velocity of the cell-label ensemble is

$$\mathbf{v} = \frac{\mu}{6\pi\eta R}\nabla B_0 \tag{5.10}$$

Here the motion of the cell-label ensemble follows the lines of ∇B_0, which, interestingly, are identical with the lines of $\nabla(B_0{}^2/2\mu_0)$, which govern the motion of the diamagnetic cells. This identity exists by virtue of the relationship $\nabla B_0{}^2 = 2B_0\nabla B_0$ and that B_0 is a scalar quantity, and therefore does not affect the direction of the vector $B_0\nabla B_0$. The magnitude of the cell-label ensemble velocity is, however, much greater than that of the unlabeled cells, typically by several orders of magnitude, which is more than sufficient to achieve separation. The difference between the motion of an induced magnetic dipole and a saturated magnetic dipole in a magnetic field gradient is illustrated in Figs. 5.1–5.4.

Although the motion of the induced and the saturated dipoles follows the same path, the velocity changes differently with position for the two cases, one being proportional to $|\nabla(B_0{}^2/2\mu_0)|$ and the other to $|\nabla B_0|$, as discussed above. This poses a problem of a proper definition of the magnetophoretic mobility that would be applicable to both cases. Clearly, the magnetophoretic mobility as defined for a linearly inducible dipole, Eq. (5.5), becomes a function of position, or, more generally, of the magnetic field intensity, for a saturated dipole because in that case $\chi_p = \chi_p(H)$. Conversely, if one defines the magnetophoretic mobility as $\mu/6\pi\eta R$ on the basis of Eq. (5.10), then this quantity becomes the function of particle coordinates, or generally, of the magnetic field intensity for an induced dipole because in that case $\mu = MV = \chi VH$.

An ad hoc definition that changes depending of the type of the application (diamagnetic separation in strong fields or paramagnetic or ferromagnetic separations in weak fields) and dictated by a particular circumstance does not appear satisfactory. The review of the published literature does not settle the issue either [16–20].

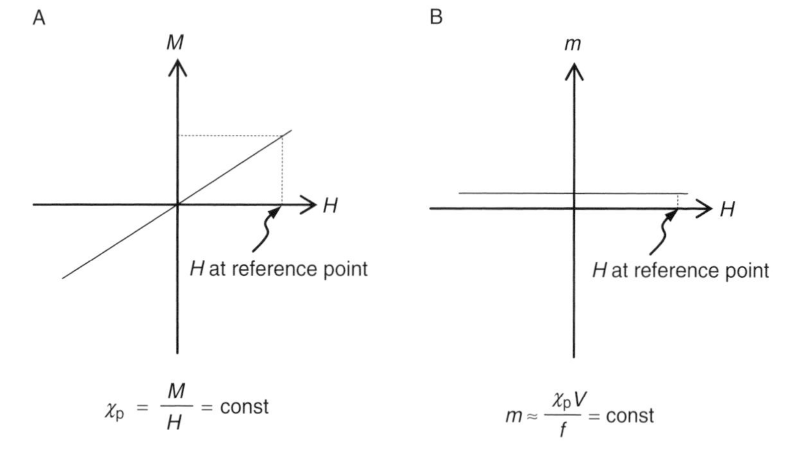

Fig. 5.1. Magnetophoretic mobility is independent of H for paramagnetic and diamagnetic particles. Here $f = 6\pi\eta R$ is particle friction coefficient (Eq. 5.5). (A) Paramagnetic particles, magnetic susceptibility. (B) Paramagnetic particles, magnetophoretic mobility.

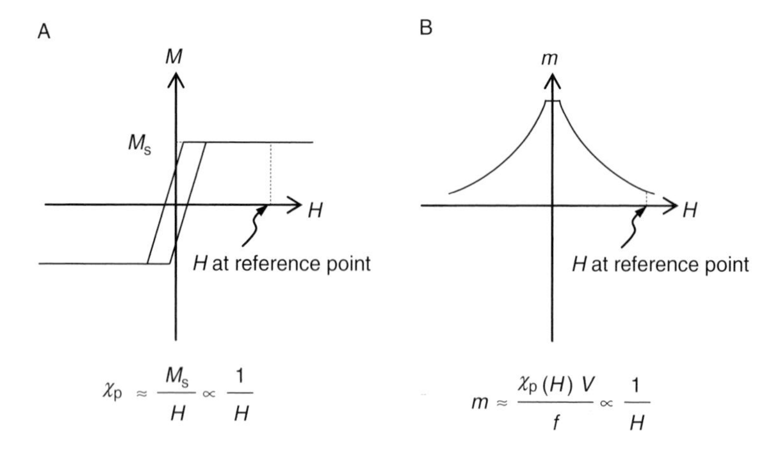

Fig. 5.2. Magnetophoretic mobility depends on H for ferromagnetic and superparamagnetic particles. (A) Ferromagnetic particle, magnetic susceptibility. (B) Ferromagnetic particle, magnetophoretic mobility. Note that the magnetophoretic mobility of ferromagnetic particles greatly exceeds that of the paramagnetic particles in the low H field region (compare with Fig. 5.1).

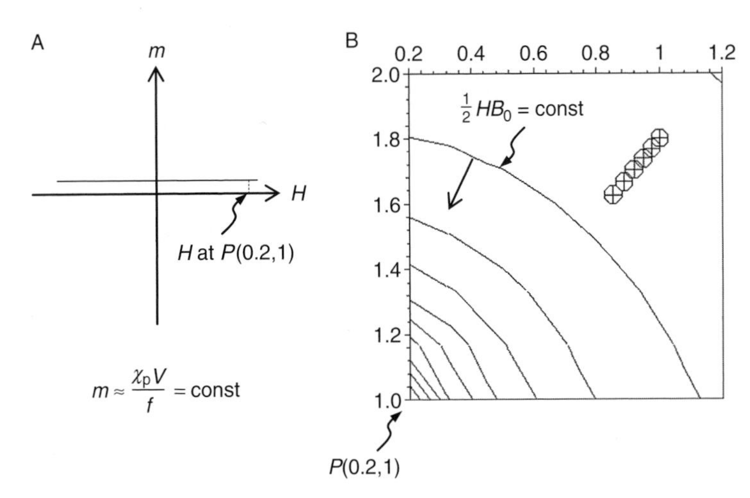

Fig. 5.3. Paramagnetic particles travel smaller distances in equal time intervals under the influence of the magnetic pressure gradient (Maxwell stress), shown in panel B, than the ferromagnetic particles (Fig. 5.4) because of their relatively smaller, field-independent magnetophoretic mobility, *m*, shown in panel A.

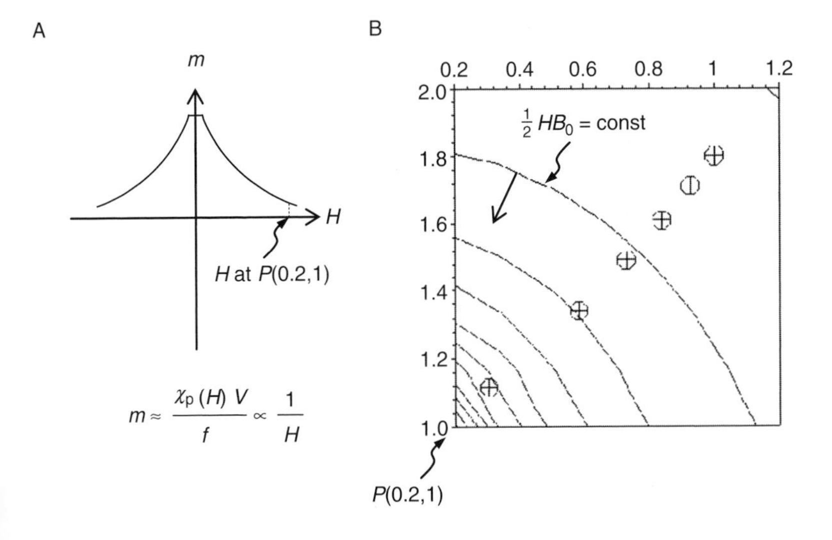

Yet, it is highly desirable to arrive at a definition of the magnetophoretic mobility as a property that is independent of a particular magnetic system and provides means of calculating motion of a cell for any field geometry. The usefulness of such a definition can be compared to that of the electrophoretic mobility in electrophoresis, or sedimentation coefficient in centrifugal separations.

It is the opinion of these authors that the magnetophoretic mobility defined on the basis of Eq. (5.4), as discussed above, is more fundamental than the one suggested by Eq. (5.10). First of all, the diamagnetic properties are exhibited by all matter, whereas paramagnetism and ferromagnetism are limited to only a few, albeit very important, cases [32]. Second, the magnetophoretic mobility as defined in Eq. (5.5) combines particle radius with the material properties of the particle and the fluid media (magnetic susceptibilities and fluid viscosity), unlike the mobility suggested by Eq. (5.10) (magnetic moment is not a material property). Third, there is a precedent in dielectrophoresis in defining the mobility of moving dipoles in viscous media as the ratio of velocity to the field energy gradient, rather than field gradient, as expressed in Eq. (5.8) [28, 29]. Fourth, the extension of the magnetophoretic mobility definition from diamagnetic materials to para- and ferromagnetic materials follows a natural path of admitting the dependence of the magnetic susceptibility on the applied field (Fig. 5.2B), as it is routinely done in the studies of magnetic properties of materials [33].

In conclusion, we propose the following definition of the magnetophoretic mobility that is equally applicable to the linearly

---

Fig. 5.4. Ferromagnetic particle. The initial position of the particle was set the same as in the case of the paramagnetic particle, shown in Fig. 5.3. The saturation magnetization of the ferromagnetic particle was assumed the same as that of the induced magnetization of the paramagnetic particle at the point $P(0.2,1)$ (such a case is more realistic for the superparamagnetic particle than for the paramagnetic particle). Note that the trajectory of the ferromagnetic particle has the same shape as that of the paramagnetic particle, and that it travels a comparatively greater distance than that shown in Fig. 5.3.

polarizable magnetic materials (diamagnetic and paramagnetic) and nonlinear materials, whose magnetic susceptibility is a function of the field (ferromagnetic and superparamagnetic):

$$m \equiv \frac{[\chi_p(H) - \chi_f(H)]V}{6\pi\eta R} \tag{5.11}$$

where we indicate explicitly the dependence of the particle and the fluid susceptibilities on the applied field, $\chi_p = \chi_p(H)$ and $\chi_f = \chi_f(H)$, respectively.

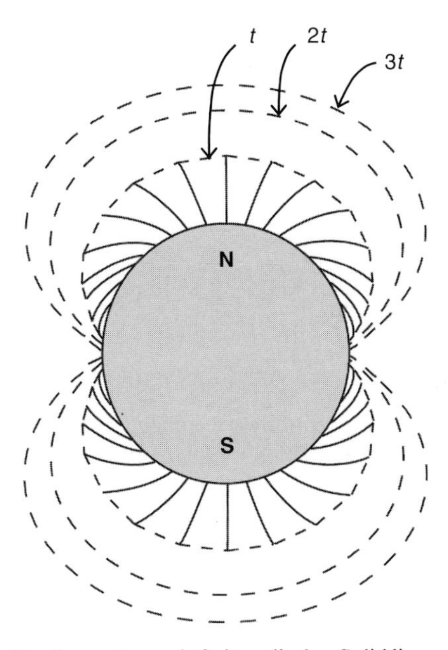

Fig. 5.5. Cross section for capture—infinite cylinder. Solid lines represent magnetic particle trajectories traveled in equal periods of time, ending at the surface of the ferromagnetic cylinder magnetized by an external magnetic field (as in Figs. 4.1 and 4.2). Broken lines represent boundary of the area cleared of the magnetic particles in a given time, *t*. Note diminishing increase in magnetic capture area with the increasing time, as indicated by broken lines marked 2*t* and 3*t*.

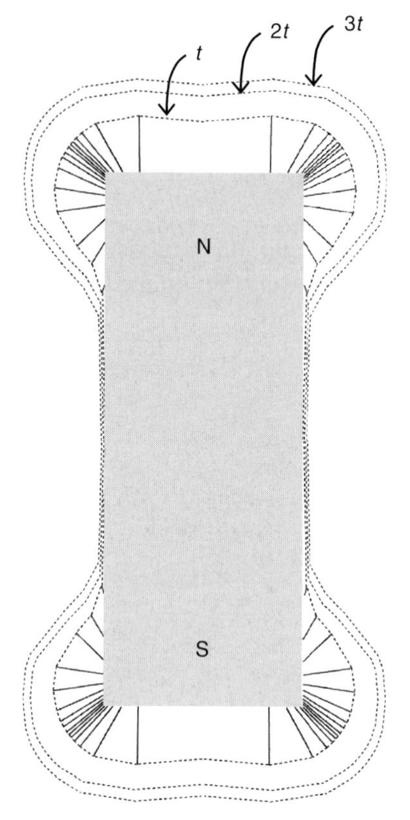

Fig. 5.6. Cross-section for capture—rectangular magnet. See caption for Fig. 5.5 for explanation.

5.1.1. Cross-section for magnetic capture

Cell magnetophoretic mobility, *m*, is the primary determinant of the magnetic separation outcome. There are other parameters, however, that have to be taken into account if the separation is to be accomplished within the constraints imposed by cell biology; such constraints include a maximum separation time and minimum mechanical stresses. A convenient way of summarizing the magnetic

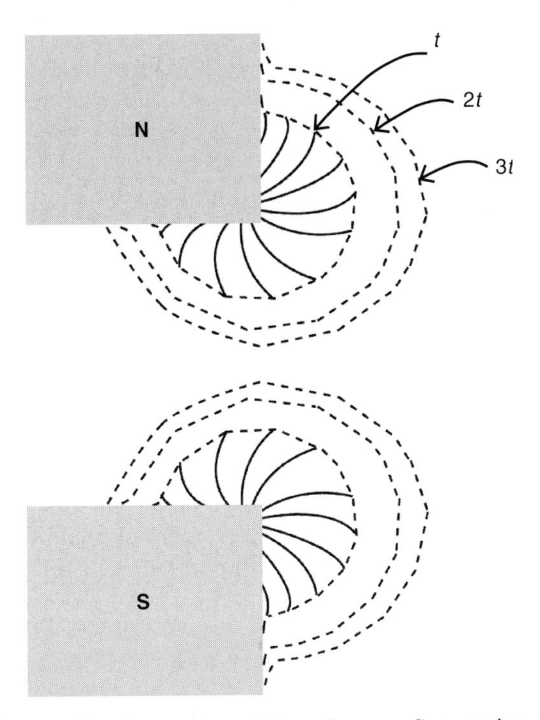

Fig. 5.7. Cross-section for capture—interpolar gap. See caption for Fig. 5.5 for explanation.

separator characteristics is to determine its cross-section for magnetic capture. In simple terms, it describes the area that is cleared of the magnetically mobile cells down to a certain, desirably small fraction of the initial purity. The actual percentage may depend on the application, and it can be perhaps as high as 5% or as low as 0.1%. In geometrical terms, the cross section for capture is an area bound by a curve made of all those points that are the initial positions of the farthest cells captured in a given time t. This applies to a suitable projection of the three-dimensional magnet assembly on a two-dimensional surface that is used to calculate the magnetic capture cross section. Typically, such a surface is normal to the direction of the bulk flow of the cell mixture. For a single wire

HGMS and the bulk flow parallel to the long axis of the wire, the cross section for capture is calculated for a surface normal to the wire. On that surface, the capture cross-section is determined by a family of cell trajectories parametrized by their lengths, l, that are traveled by the equal amounts of time between the point on the curve bounding the cross-section for capture to the surface of the magnet:

$$l = m\left\langle \frac{\mathrm{d}\frac{1}{2}HB_0}{\mathrm{d}l} \right\rangle_l \Delta t, \qquad \Delta t = \mathrm{const} \tag{5.12}$$

In other words, the linear extension of the capture cross section is directly proportional to cell magnetophoretic mobility, m, the average magnitude of the Maxwell stress gradient along the cell trajectory, $\langle \mathrm{d}\frac{1}{2}HB_0/\mathrm{d}l \rangle_l$, and the cell residence time in the magnetic field, Δt. The cross sections are illustrated in Figs. 5.5–5.7 for selected magnet geometries.

References

[1] Torbet, J. (1987). Using magnetic field orientation to study structure and assembly. Trends Biochem. Sci. *12*, 327–30.

[2] Torbet, J. and Dickens, M. J. (1984). Orientation of skeletal muscle actin in strong magnetic fields. FEBS Lett. *173*, 403–6.

[3] Fabry, M. E. and San George, R. C. (1983). Effect of magnetic susceptibility on nuclear magnetic resonance signals arising from red cells: A warning. Biochemistry *22*, 4119–25.

[4] Zborowski, M., Ostera, G. R., Moore, L. R., Milliron, S., Chalmers, J. J. and Schechter, A. N. (2003). Red blood cell magnetophoresis. Biophys. J. *84*, 2638–45.

[5] Bohle, D. S., Debrunner, P., Jordan, P. A., Madsen, S. K. and Schulz, C. E. (1998). Aggregated heme detoxification byproducts in malarial trophozoites: Beta-hematin and malaria pigment have a single S = 5/2 iron environment in the bulk phase as determined by EPR and magnetic Moessbauer spectroscopy. J. Am. Chem. Soc. *120*, 8255–6.

[6] Bazylinski, D. A. and Frankel, R. B. (2004). Magnetosome formation in prokaryotes. Nat. Rev, Microbiol, *2*, 217–30.

[7] Rosenblatt, C., Torres de Araujo, F. F. and Frankel, R. B. (1982). Birefringence determination of magnetic moments of magnetotactic bacteria. Biophys. J. *40*, 83–5.

[8] Ugelstad, J., Stenstad, P., Kilaas, L., Prestvik, W. S., Herje, R., Berge, A. and Hornes, E. (1993). Monodisperse magnetic polymer particles. New biochemical and biomedical applications. Blood Purif. *11*, 347–69.

[9] Miltenyi, S., Müller, W., Weichel, W. and Radbruch, A. (1990). High gradient magnetic cell separation with MACS. Cytometry *11*, 231–8.

[10] Liberti, P. A. and Feeley, B. P. (1991). Analytical- and process-scale cell separation with bioreceptor ferrofluids and high gradient magnetic separation. In: Cell Separation Science and Technology (Compala, D. S. and Todd, P., eds.), Vol. 464. ACS Symposium Series, Washington, pp. 268–88.

[11] Thomas, T. E., Richards, A. J., Roath, O. S., Watson, J. H. P., Smith, R. J. S. and Lansdorp, P. M. (1993). Positive selection of human blood cells using improved high gradient magnetic separation filters. J. Hematother. *2*, 297–303.

[12] Pankhurst, Q. A., Connolly, J., Jones, S. K. and Dobson, J. (2003). Applications of magnetic nanoparticles in biomedicine. DOI:10.1088/0022–3727/36/13/201. J. Phys. D. Appl. Phys. *36*, R167–81.

[13] Simon, M. D. and Geim, A. K. (2000). Diamagnetic levitation: Flying frogs and floating magnets (invited). Appl. Phys. A. *87*, 6200–4.

[14] Sutton, G. W. and Sherman, A. (2006). Engineering magnetohydrodynamics. Dover Publications Inc., Mineola, NY.

[15] Winoto-Morbach, S., Tchikov, V. and Mueller-Ruchholtz, W. (1994). Magnetophoresis: I. Detection of magnetically labeled cells. J. Clin. Lab. Anal. *8*, 400–6.

[16] Leenov, D. and Kolin, A. (1954). Theory of electromagnetophoresis. I. Magnetohydrodynamic forces experienced by spherical and symmetrically oriented cylindrical particles. J. Chem. Phys. *22*, 683–8.

[17] Gill, S. J., Malone, C. P. and Downing, M. (1960). Magnetic susceptibility measurements of single small particles. Rev. Sci. Instrum. *31*, 1299–303.

[18] Wilhelm, C., Gazeau, F. and Bacri, J.-C. (2002). Magnetophoresis and ferromagnetic resonance of magnetically labeled cells. Eur. Biophys. J. *31*, 118–25.

[19] Suwa, M. and Watarai, H. (2001). Magnetophoretic velocimetry of manganese(II) in a single emulsion droplet at the femtomole level. Anal. Chem. *73*, 5214–9.

[20] Watarai, H. and Namba, M. (2002). Capillary magnetophoresis of human blood cells and their magnetophoretic trapping in a flow system. J. Chromatogr. A. *961*, 3–8.

[21] Watarai, H. and Suwa, M. (2004). Magnetophoresis and electromagnetophoresis of microparticles in liquids. Anal. Bioanal. Chem. *378*, 1693–9.

[22] Furlani, E. P. (2007). Magnetophoretic separation of blood cells at the microscale. J. Phys. D: Appl. Phys. *40*, 1313–9.

[23] Tchikov, V., Schuetze, S. and Kroenke, M. (1999). Comparison between immunofluorescence and immunomagnetic techniques of cytometry. J. Magn. Magn. Mater. *194*, 242–7.

[24] McCloskey, K. E., Moore, L. R., Hoyos, M., Rodriguez, A., Chalmers, J. J. and Zborowski, M. (2003). Magnetophoretic cell sorting is a function of antibody binding capacity. Biotechnol. Prog. *19*, 899–907.

[25] Frantz, S. G. (1936). Magnetic separation methods and means. Patent No. 2,056,426 (US).

[26] Smolkin, M. R. and Smolkin, R. D. (2006). Calculation and analysis of the magnetic force acting on a particle in the magnetic field of separator. Analysis of the equations used in the magnetic methods of separation. IEEE Trans. Magn. *42*, 3682–93.

[27] Moore, L. R., Zborowski, M., Nakamura, M., McCloskey, K., Gura, S., Zuberi, M., Margel, S. and Chalmers, J. J. (2000). The use of magnetite-doped polymeric microspheres in calibrating cell tracking velocimetry. J. Biochem. Biophys. Methods. *44*, 115–30.

[28] Pohl, H. A. (1978). Dielectrophoresis: The behavior of neutral matter in nonuniform electric fields. Published in *Cambridge Monographs on Physics*, edited by Woolfson, M. M. and Ziman, J. M. Cambridge University Press, Cambridge.

[29] Markx, G. H. and Pethig, R. (1995). Dielectrophoretic separation of cells: Continuous separation. Biotechnol. Bioeng. *45*, 337–43.

[30] Carter, V., Cable, H. C., Underhill, B. A., Williams, J. and Hurd, H. (2003). Isolation of *Plasmodium berghei* ookinetes in culture using Nycodenz density gradient columns and magnetic isolation. Malar. J. *2*, 35.

[31] Dunin-Borkowski, R. E., McCartney, M. R., Frankel, R. B., Bazylinski, D. A., Posfai, M. and Buseck, P. R. (1998). Magnetic microstructure of magnetotactic bacteria by electron holography. Science *282*, 1868–70.

[32] Purcell, E. M. (1985). Electricity and Magnetism. Berkeley Physics Course-Volume 2. McGraw-Hill Book Co., New York.

[33] Melcher, J. R. (1981). Continuum electromechanics. The MIT Press, Cambridge, MA.

Laboratory Techniques in Biochemistry and Molecular Biology, Volume 32
Magnetic Cell Separation
M. Zborowski and J. J. Chalmers (Editors)

CHAPTER 6

Synthesis and characterization of nano- and micron-sized iron oxide and iron particles for biomedical applications

Shlomo Margel, Tammy Lublin-Tennenbaum, Sigalit Gura, Merav Tsubery, Udi Akiva, Nava Shpaisman, Anna Galperin, Benny Perlstein, Polina Lapido, Yonit Boguslavsky, Jenny Goldshtein, and Ofra Ziv

Department of Chemistry, Ban-Ilan University, Ramat-Gan 52900, Israel

Nano- and micron-sized magnetic particles, because of their spherical shape, high surface area per volume and magnetic properties, have a wide range of biomedical applications, for example specific cell labeling and separation, cell growth, affinity chromatography, diagnostics, specific hemoperfusion, drug delivery, controlled release, contrast agents for MRI, and hyperthermia. Each application requires the design of magnetic particles with specific physicochemical properties. Large number of methods have been described for preparation of various nano-and micron-sized magnetic particles for biomedical applications. Most of the prepared magnetic particles are based on a magnetic iron oxide core and a matrix or shell of a natural or synthetic polymer.

The present manuscript describes the synthesis of unique magnetic particles prepared in the last few years in our laboratory, as

DOI: 10.1016/S0075-7535(06)32006-2

follows: (1) maghemite nanoparticles of ca. 15- to 100-nm diameter with narrow size distribution; (2) air-stable iron nanocrystalline particles; (3) solid and hollow maghemite/polystyrene and silica/maghemite/polystyrene micron-sized composite particles with narrow size distribution; (4) magnetic/nonmagnetic polystyrene/poly (methyl methacrylate) hemispherical composite micron-sized particles with narrow size distribution.

6.1. Introduction

Polymeric nano-and micron-sized particles are considered to be spherical particles in sizes ranging from a few nanometers up to 100 nm and from 100 nm up to $\sim$10 μm, respectively. Nano- and micron-sized particles of narrow size distribution are of special interest. Highly uniform particles are effective for applications such as adsorbents for HPLC, calibration standards, and spacers for liquid crystals [1, 2]. Particles of narrow size distribution are also essential for drug delivery purposes, since heterogeneous colloidal particles are considered to be potential sources of toxicity for injection into humans [3, 4]. Nano-scaled polymeric particles of narrow size distribution are commonly formed by controlled precipitation methods or optimal heterogeneous polymerization techniques such as emulsion or inverse emulsion polymerization methods. Micron-sized particles of narrow size distribution are usually prepared by dispersion polymerization processes, or by a swelling process, where uniform template particles are swollen with an appropriate monomer(s), which then polymerized within the swollen particles. Properties of solid materials undergo drastic changes when their dimensions are reduced to the nanometer size regime. It is important to keep in mind that the smaller the particles are, the larger portion of their constituent atoms is located at the surface. Nanoparticles, particularly below ca. 20 nm, predominantly exhibit surface and interface phenomena that are not observed in bulk materials, for example lower melting and boiling points, lower sintering temperature, and reduced flow resistance. Nano- and micron-scaled particles, because of their spherical shape and high surface area, may provide neat solutions to a variety of problems in materials science, biology, and medicine. The potential use of these particles for applications, such as composite materials, catalysis, three-dimensional structures, photonic uses, and biomedical applications such as specific cell labeling and separation, cell growth, affinity chromatography, diagnostics, specific blood purification by hemoperfusion, drug delivery, and controlled release, has been demonstrated in a few laboratories [5–17]. Each application requires polymeric particles of different optimal physical and chemical properties. The synthesis and use of enormous types of nano- and micron-scaled particles of different surface chemistry, for example a variety of surface functional groups such as hydroxyl, carboxyl, pyridine, amide, aldehyde, and phenyl chloromethyl have already been described [1, 18–22]. In our laboratory, a variety of functional particles of narrow size distribution and different diameters and composition have been prepared and characterized; for example organic and inorganic particles [polyacrolein, polyglutaraldehyde, polymethyl α-(hydroxymethyl)acrylate, polychloromethylstyrene, polyacrylonitrile, polyvinyl α-amino acids, organo-iodide/bromide, cellulose, carbon, titania, and silica]; core-shell composite particles such as polyacrolein or silica coated onto polystyrene particles; hollow particles (silica and titania); nonsymmetrical two-phase particles; and magnetic particles [23–32]. These particles have been designed for various industrial and medical applications, for example enzyme immobilization, oligonucleotide and peptide synthesis, drug delivery, specific cell labeling and separation, medical imaging, biological glues, and flame retardant polymers [33–50].

Of particular interest are particles with magnetic properties, which are usually used for separation of the particles and/or their conjugates from undesired compounds, via magnetic field. These particles due to their magnetic properties have several additional significant applications, for example magnetic recording, magnetic sealing, electromagnetic shielding, contrast agents for MRI, magnetic drug targeting, magnetic cell separation, and magnetic hyperthermia [5–7, 14–19, 51]. In the past few years, extensive efforts to synthesize efficient nano- and micron-scaled particles with magnetic properties have been carried out [1, 14–19, 23–26, 51]. Magnetic iron oxide is intensively investigated for biomedical uses, since this material is nontoxic and biodegradable. The main way to prepare magnetic iron oxide nanoparticles is based on the precipitation of iron salts in aqueous continuous phase in the presence of optimal surfactants. A similar process in the presence of macroporous micron-sized polystyrene (PS) particles leads to the formation of magnetic micron-sized particles composed of magnetic iron oxide entrapped within the porous of the particles [52]. On the other hand, a similar process in the presence of nonporous PS micron-sized particles with special surface properties dispersed in the aqueous phase (i.e., PS particles containing surface hydroxyl groups, or coated with hydrophilic surfactants) leads to the formation of magnetic coating on the surface of these particles [4–7]. Iron particles are also of special interest, since Fe has the highest magnetic moment among the ferromagnetic transition metals. However, Fe is easily oxidized; therefore currently severe efforts are accomplished to prepare and characterize air-stable Fe nano-and micron-sized particles.

Surface modification (without changing the bulk properties) of the magnetic particles is frequently essential for many reasons, for example, changing the surface composition, improving adhesion, stabilization against aggregation, protein immobilization, polymer compatibility, blood and biocompatibility, weathering, and protection [20, 51, 53, 54]. Numerous methods for surface modification of different particles, such as high-energy radiation (e.g., gamma, glow discharge, corona discharge, or photoirradiation) [55], surface adsorption of surfactants and polymers [20, 56, 57], surface-grafted polymerization by methods such as emulsion polymerization or ATRP (atomic transfer radical polymerization) [58–60], and covalent binding of desired ligand onto surface functional groups via different activation methods, have been described in the literature.

The present manuscript describes the synthesis, characterization, surface modification, and biomedical use of unique magnetic particles prepared in the last few years in our laboratory, as follows: (1) maghemite (γ-Fe_2O_3) nanoparticles of ca. 15- to 100-nm diameter with narrow size distribution; (2) air-stable iron nanocrystalline particles; (3) solid and hollow maghemite/polystyrene and silica/maghemite/polystyrene micron-sized composite particles with narrow size distribution; (4) magnetic/nonmagnetic polystyrene/poly(methyl methacrylate) hemispherical composite micron-sized particles with narrow size distribution.

6.2. *Maghemite nanoparticles of narrow size distribution*

6.2.1. *Synthesis and characterization*

Maghemite nanoparticles of sizes ranging from ca. 15 nm up to 100 nm with narrow size distribution were prepared by nucleation followed by controlled growth of maghemite thin films onto porcine gelatin nuclei, according to Fig. 6.1 [3, 23–26]. The nucleation step is based on complexation of Fe^{2+} ions to chelating sites of the porcine gelatin, followed by partial oxidation (up to ~50%) of the chelated Fe^{2+} to Fe^{3+}, so that the water soluble gelatin contains both chelated Fe^{2+} and Fe^{3+} ions. Gelatin nuclei are then formed by adding NaOH aqueous solution up to pH 9.5. The growth of magnetic films onto the gelatin nuclei accomplished by repeating several times the nucleation step. Briefly, nanoparticles of 15-nm average dry diameter were prepared by adding $FeCl_2$ solution (10 mmol/5 ml H_2O) to 80-ml aqueous solution containing 200-mg gelatin, followed by $NaNO_2$ solution (7 mmol/5 ml H_2O). After a reaction time of 10 min, NaOH aqueous solution (1 N) was added up to pH 9.5. This procedure was repeated four times, or more, if larger particles are required. The formed magnetic nanoparticles were then washed from excess reagents using magnetic columns. Fluorescent magnetic nanoparticles were prepared similarly substituting the gelatin with gelatin covalently bonded to a fluorescent dye (e.g., rhodamine). Figure 6.2 demonstrates a TEM picture of magnetic nanoparticles of increased average diameter formed by repeating the thin magnetic

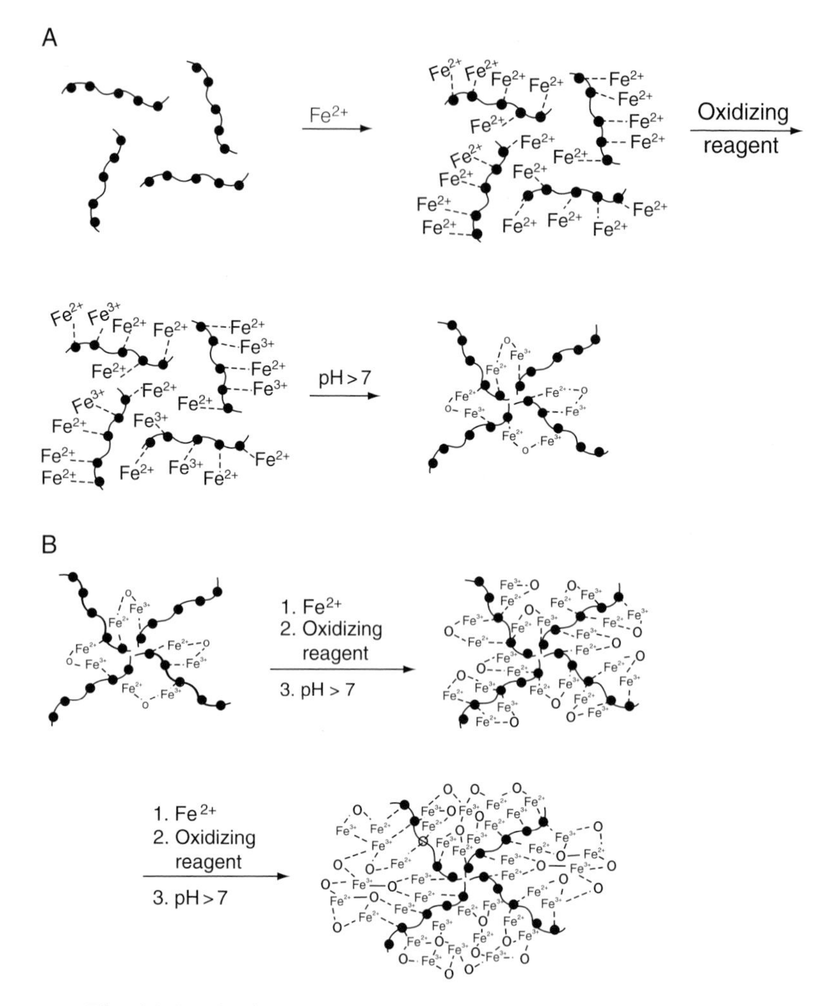

Fig. 6.1. Nucleation (A) and growth (B) of magnetic nanoparticles.

coating process during the growth step: four (A), five (B), six (C), and seven (D) times. The magnetic nanoparticles of 15-nm dry average diameter (prepared as described above) dispersed in water posses one population with hydrodynamic average diameter of

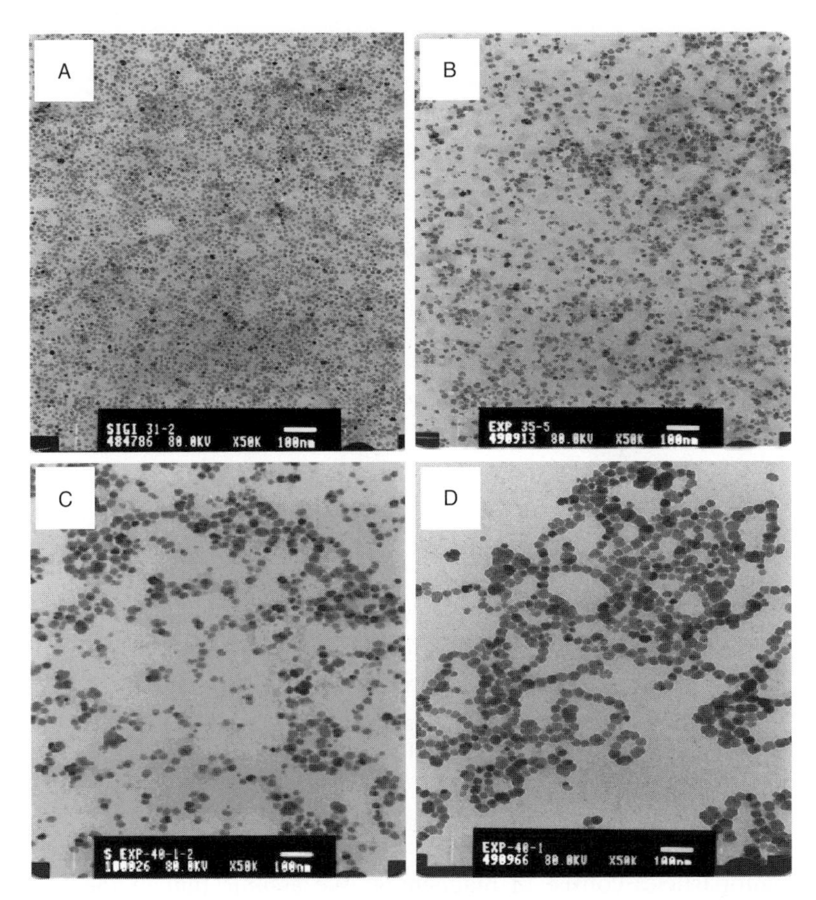

Fig. 6.2. TEM micrographs of magnetic nanoparticles of various sizes.

ca. 100 nm, as shown in Fig. 6.3. High-resolution TEM (HRTEM) picture (Fig. 6.4A) demonstrates crystalline structure with d-spacing of 0.479 nm. Electron diffraction picture (Fig. 6.4B) represents sharp rings indicating the crystalline character of the magnetic nanoparticles. X-ray diffraction (XRD) investigation (Fig. 6.5) shows that the crystalline cores of these nanoparticles consist nearly completely of maghemite (γ-Fe_2O_3). From X-ray line broadening one deduces a mean diameter of the magnetic cores of 15 nm.

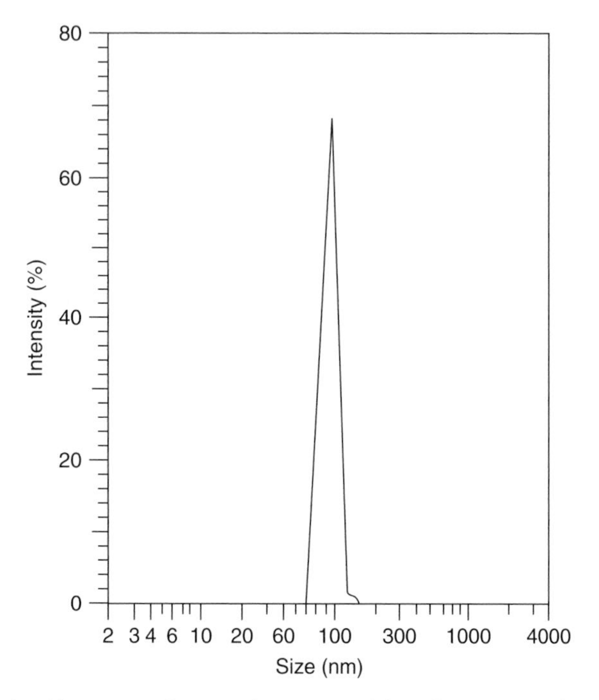

Fig. 6.3. An histogram of magnetic nanoparticles of ca. 20-nm dry diameter dispersed in water.

Mossbauer spectrum (Fig. 6.6) also shows that these magnetic nanoparticles consist of maghemite. We assume that in the first stage magnetite (Fe_3O_4) nanoparticles were produced by this nucleation and growth process. These magnetite nanoparticles were then oxidized to the more thermodynamic stable iron oxide: maghemite. Figure 6.7 represents the hysteresis loop at room temperature of the maghemite nanoparticles of 15-nm dry diameter. This figure shows that the M(H) curve of these nanoparticles does not saturate at 10,000 Oe, and that the obtained magnetic moment at 10,000 Oe is ca. 41 emu g^{-1}. Also, the M(H) curve does not exhibit any coercivity. Both features are typical of superparamagnetic behavior.

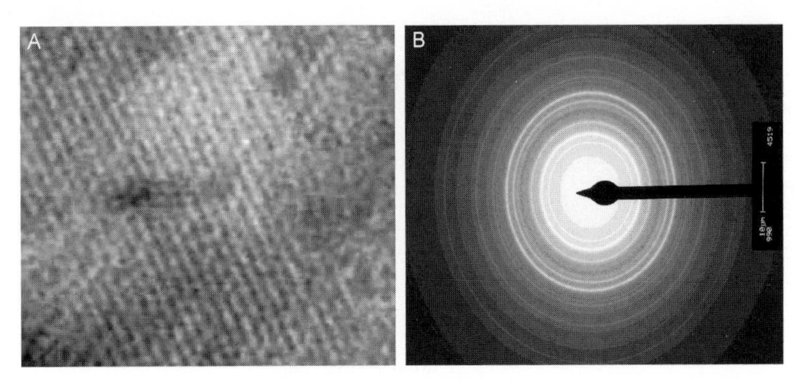

Fig. 6.4. HRTEM (A) and ED patterns (B) of the magnetic nanoparticles.

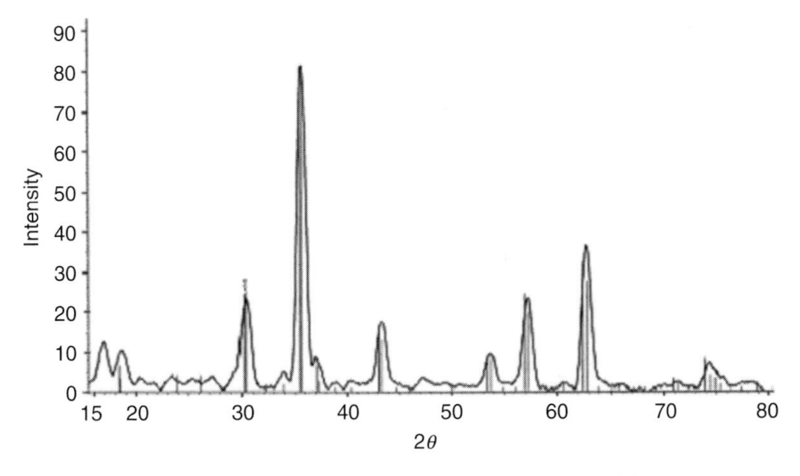

Fig. 6.5. XRD pattern of the magnetic nanoparticles.

6.2.2. Surface modification

Surface modification of the meghemite nanoparticles has been accomplished by two major ways: (A) polymer adsorption, according to Fig. 6.8, and (B) seeded emulsion polymerization.

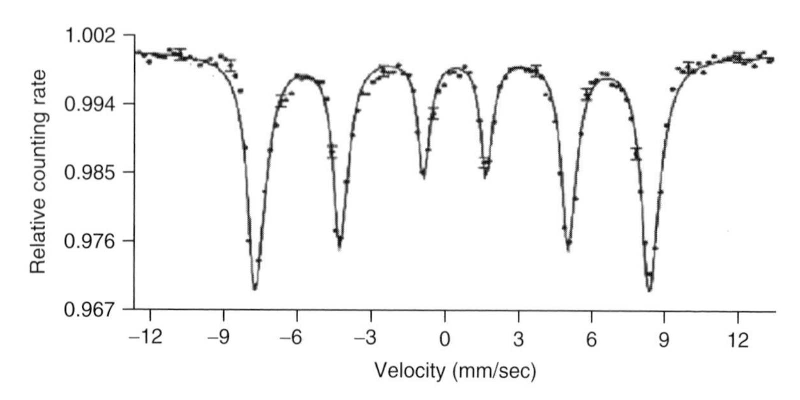

Fig. 6.6. Mossbauer spectrum of the magnetic nanoparticles.

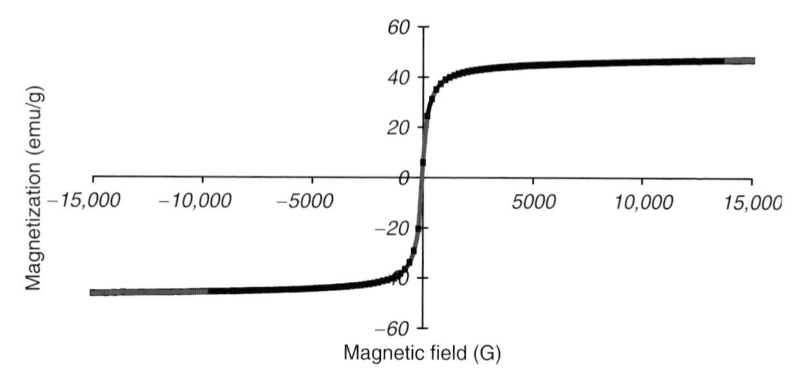

Fig. 6.7. Room temperature magnetization (VSM) loop for the magnetic nanoparticles.

6.2.2.1. Polymer adsorption

Briefly, gelatin coating was performed by shaking the aqueous dispersion of the magnetic nanoparticles of 15-nm dry diameter (2 mg/ml) containing 0.2% gelatin at 85 °C for a few hours. The aqueous dispersion was then cooled to room temperature. The gelatin-coated nanoparticles were then washed by means of magnetic columns. Dextran (MW 48,000) coating was performed similarly substituting the 0.2% gelatin for 1% dextran. Aldehyde groups

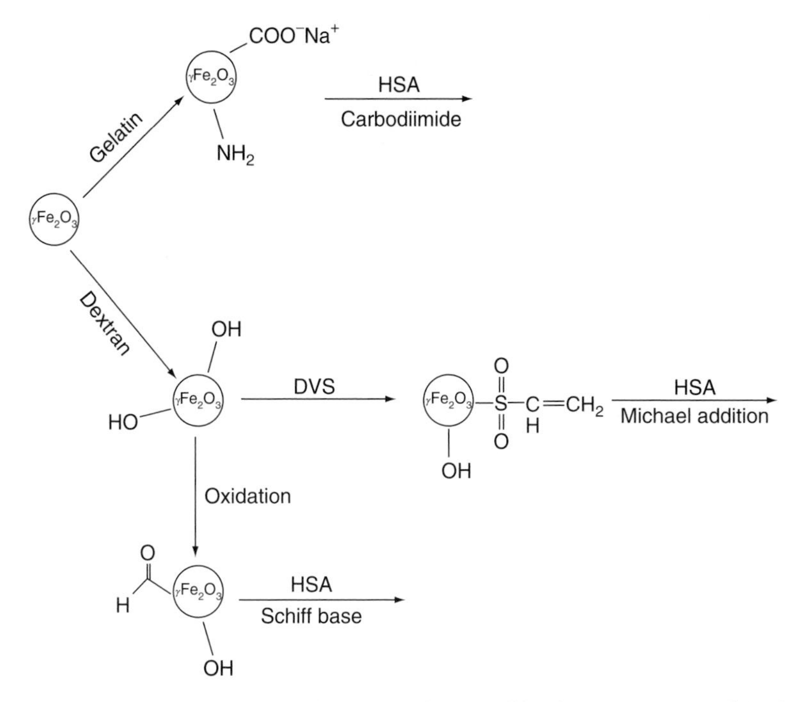

Fig. 6.8. A scheme demonstrating the surface modification steps accomplished for binding biomolecules (e.g., HSA) to the magnetic nanoparticles.

were obtained from the dextran coating by oxidizing a few of the geminal hydroxyl groups with sodium periodate, according to the literature [23–26, 43, 44]. Cross-linking and formation of activated double bonds were performed by interacting divinyl sulfone (DVS) or methacryloyl chloride (MAC) with part of the primary hydroxyl groups of the dextran coating. The various functional groups of the modified magnetic nanoparticles, for example aldehydes, residual double bonds, carboxylates, and primary amines, have been used for covalent binding, via different activation methods (e.g., carbodiimide, Michael addition, and Schiff base), of bioactive compounds such as proteins [e.g., human serum albumin (HSA)], drugs, and oligonucleotides to the surface of the magnetic nanoparticles [41–44].

6.2.2.2. Seeded emulsion polymerization

Maghemite nanoparticles of 15-nm average dry diameter have been coated with a shell of various polyacrylates, for example poly(divinyl benzene) [PDVB] and poly(2-methacryloyloxyethyl [2,3,5-triiodobenzoate]) (polyMAOETIB), by emulsion polymerization of the appropriate monomers in the presence of the maghemite nanoparticles dispersed in aqueous continuous phase. Briefly, polyMAOETIB/γ-Fe_2O_3 core-shell iodo nanoparticles were prepared by adding to 30 ml vial containing 20-ml maghemite nanoparticles dispersed in water (2 mg/ml) 30-μg sodium dodecyl sulfate (SDS), 8-mg potassium persulfate, and 0.4 g of the iodo monomer MAOETIB (Fig. 6.9) dissolved in 6-ml toluene. For the polymerization, the temperature was then raised to 73 °C for 3 hours. The formed core-shell polyMAOETIB/γ-Fe_2O_3 iodo nanoparticles were then washed by magnetic columns.

6.2.3. Biomedical applications

6.2.3.1. Immunogenicity of the bioactive maghemite nanoparticles

The maghemite nanoparticles prepared in this work contain encapsulated porcine gelatin which has been used during the nucleation step.

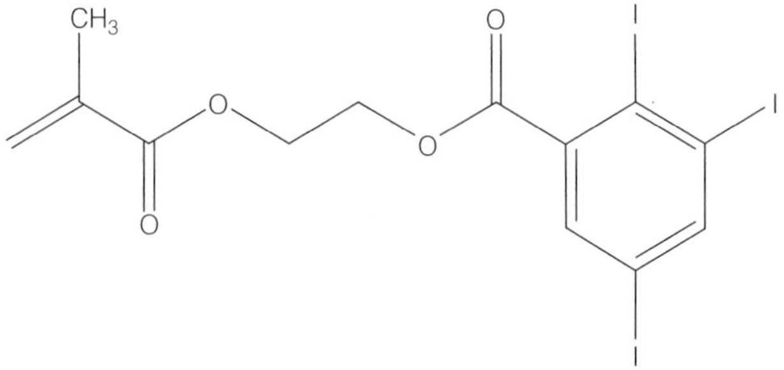

Fig. 6.9. Chemical structure of the monomer MAOETIB [2-methacryloyloxyethyl(2,3,5-triiodobenzoate)].

These nanoparticles were then coated with dextran, followed by a bioactive compound coating such as HSA. These bioactive maghemite nanoparticles are therefore conjugated to three potential antigens: porcine gelatin, dextran, and HSA. Since these conjugated nanoparticles are designated for various clinical applications, we examined the natural and acquired immunogenicity of these antigens in BALB-C mice model [61]. The mice were immunized with PBS dispersion of each of the following antigens: maghemite nanoparticles (containing gelatin), maghemite nanoparticles-dextran, and maghemite nanoparticles-dextran-HSA. The animals received three intraperitoneal (IP) injections of each of these potential antigens (0.1-ml PBS containing 0.5-mg nanoparticles for each mouse) every three weeks successively, and then bleed 10 days after the last immunization. Antibody titers of naïve (nonimmunized) and immunized mice were measured by the enzyme-linked immunosorbent assay (ELISA). This work demonstrated that plasma of naïve mice already contains basal levels of natural antibodies against gelatin, dextran, and HSA. The highest antibody titer was against HSA, followed by dextran and finally gelatin. IP injection of the coated and noncoated nanoparticles containing gelatin did not raise the anti-gelatin antibody titer above the basal level of naïve mice. In contrast, IP injection of the dextran-coated maghemite nanoparticles significantly increased the anti-dextran antibody titer above the basal level of naïve mice. Also, IP injection of the nanoparticles coated with dextran and additional coating of HSA resulted in considerable amplification of the anti-dextran antibody titer as well as in the increase of the anti-HSA antibody titer. The anti-dextran and anti-HSA antibody titers returned close to its basal levels within 7 weeks.

These findings have implications on the half-life span of the nanoparticles in blood and other tissues, since the antibodies are involved in opsonization processes, inflammation, and side effects of hypersensitivity.

6.2.4. MRI contrast agents

The potential of the maghemite nanoparticles coated with dextran as contrast agents for MRI was studied by measurements of the effects on proton relaxation in vitro, and by in vivo MRI of New Zealand white rabbits injected (IV) with the nanoparticles. In both in vivo and in vitro experiments a comparison to those of Feridex^{TM}, the commercial magnetic nanoparticles which is already in clinical use, has been performed.

6.2.4.1. Effects on proton relaxation

T_1 and T_2 relaxation times were measured with 2T Prestige (Elscint) whole body MRI system operating at a field strength of 1.9 T. T_1 was measured from 11 data points generated by an inversion-recovery pulse sequence. T_2 was measured from 8 data points generated by a multi-echo and spin-echo pulse sequence.

The R_1 relaxivity of the present maghemite nanoparticles dispersed in aqueous phase was determined to be 5.8 (mM sec)$^{-1}$ and the R_2 relaxivity was 367 (mM sec)$^{-1}$, compared to 6 and 220 (mM sec)$^{-1}$, respectively of the Feridex^{TM} dispersion.

6.2.4.2. Pharmacokinetics in healthy rabbits

For pharmacokinetics studies, six New Zealand white rabbits (1.5–2 kg) were anesthetized, and imaged using head coil in the 2T Prestige MRI. The effect of the maghemite nanoparticles was evaluated from the signal intensity of the liver before and after the injection of the contrast agent, and from the calculated T_2 maps of the same slices. Contrast agent (10 μmol/kg in 5% dextrose) was injected into one of the ear veins through a catheter so that the rabbit could be kept in the magnet during the long kinetic measurements.

The imaging sequences that were used were: (1) fast spin-echo sequence with one echo (TR/TE 4000/96 msec, axial slices of 6 mm); (2) FSE sequence with eight echoes (TR/TE 2000/24, 48, 72, 104, 124, 148, 172, and 204 msec).

After contrast injection, the original liver intensity drops to half of its initial value in about 5 min, and remains constant for at least 1 hour (Fig. 6.10). A week after the IV injection the contrast agent seemed to be cleared from the liver since the signal intensity of the liver returns to normal values. This kinetics was similar to that of the commercial agent—Feridex^{TM}. T_2 values of control liver of white rabbits in 2T field was 62 msec, 30 min after contrast agents injections T_2 values were 47 msec for the maghemite nanoparticles and 35 msec for Feridex^{TM}.

Maghemite nanoparticles were also injected into the striatum of a rat's brain. Axial T_2*-weighted gradient-echo MRI were then taken immediately (Fig. 6.11A), 6 days (B), and 27 days (C) after injection with a 0.2-mg/ml nanoparticles in aqueous continuous phase at 4 μl/min over 15 min. The presence of the maghemite nanoparticles is indicated by dark spots caused by loss of signal of MRI. Figure 6.11 demonstrates that the clearance kinetics of these nanoparticles from the rat's brain is significantly slower than that obtained for the rabbit's liver.

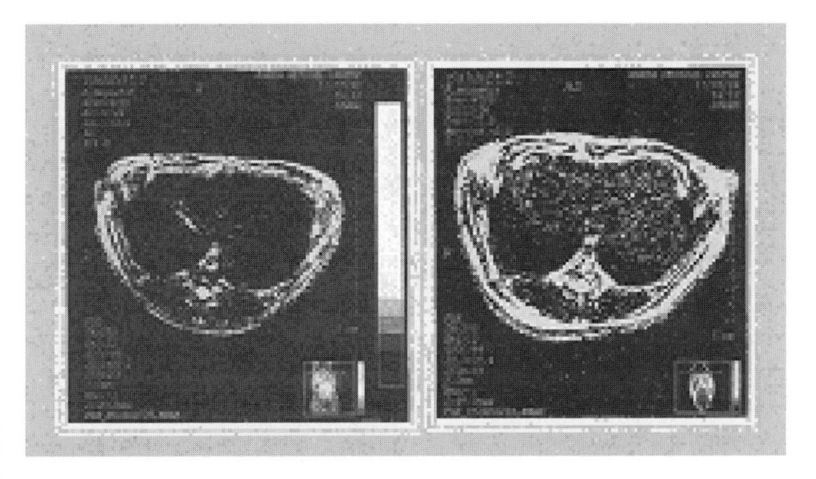

Fig. 6.10. MRI of rabbit liver before (left) and 1 hour after (right) IV injection of the maghemite nanoparticles.

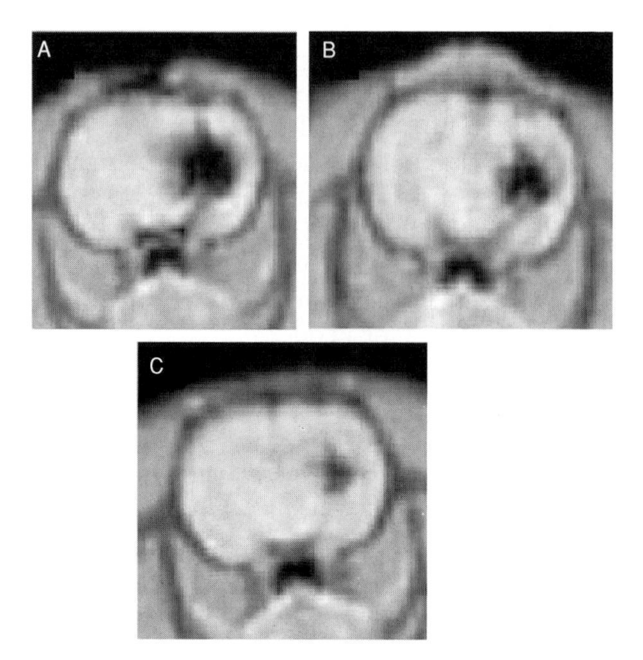

Fig. 6.11. Clearance of magnetic nanoparticles from a rat's brain as function of time.

6.2.5. *X-ray contrast agents*

Due to the presence of iodine atoms in the polyMAOETIB/γ-Fe_2O_3 core-shell nanoparticles, it is expected that they will posses radiopaque nature. The in vitro radiopacity of these particles was demonstrated by CT: an imaging technique based on X-ray absorption usually used in hospitals. Figure 6.12 illustrates the CT image visibility of an empty ependorf tube (A) and an ependorf tube containing dried pure polyMAOETIB powder (B), polyMAOETIB/γ-Fe_2O_3 core-shell nanoparticles (C), and γ-Fe_2O_3.nanoparticles (D). The CT image of the dried polyMAOETIB powder placed in ependorf (C) shows an excellent radiopaque nature compared to the image of the empty ependorf (A), which is almost transparent to X-ray

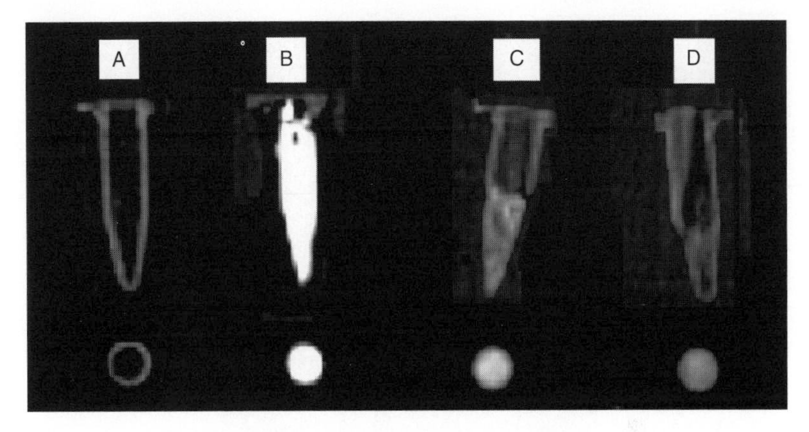

Fig. 6.12. CT images of empty ependorf tube and its section (A) and ependorf tube and related sections filled with dried polyMAOETIB powder (B), polyMAOETIB/γ-Fe_2O_3 core-shell nanoparticles (C), and γ-Fe_2O_3 nanoparticles (D).

irradiation. Figure 6.12 also shows that the radiopacity of the polyMAOETIB/γ-Fe_2O_3 core-shell nanoparticles (C) is significantly higher than that of the γ-Fe_2O_3 nanoparticles (D). The bottom of Fig. 6.12A, B, C, and D shows the CT images of sections of the expenders. These images provide clearer evidence that the X-ray visibility increases as the percentage of iodine of the materials is raised.

6.3. *Air-stable iron nanocrystalline particles*

PS template microspheres dispersed in aqueous solution have been used for entrapping $Fe(CO)_5$, by a single-step swelling process of methylene chloride emulsion droplets containing $Fe(CO)_5$ within these particles [30–32, 43, 44, 62]. Air-stable Fe/Fe_3C nanocrystalline particles have been prepared by thermal decomposition of the $Fe(CO)_5$ swollen template particles at 600 °C in an inert atmosphere. These nanocrystalline particles have a core-shell structure where a coating of Fe_3C and carbon protects the core body-centered cubic Fe from oxidation.

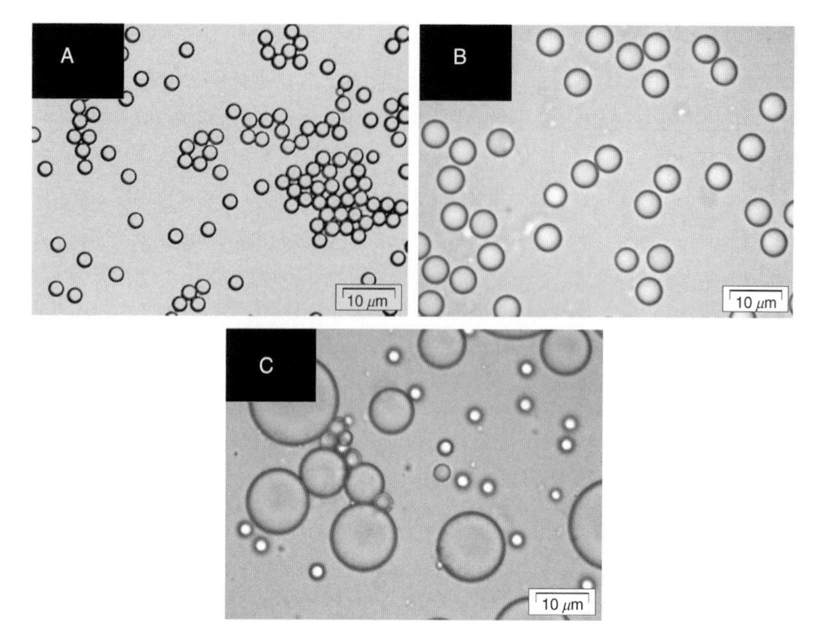

Fig. 6.13. Light microscopy pictures of PS template microspheres before (A) and after swelling with 2 ml of methylene chloride (B) or $Fe(CO)_5$ (C).

In a typical experiment, PS template microspheres of 2.4 ± 0.2 μm were swollen with a mixture of methylene chloride (1.3 ml) and $Fe(CO)_5$ (0.3 ml) up to 4.9 ± 0.3 μm, by adding to a 20-ml vial, 10 ml of SDS aqueous solution [1.5% (w/v)] and 1.6 ml of the swelling solvent [e.g., a mixture of methylene chloride and $Fe(CO)_5$]. Emulsion droplets of the swelling solvent were then formed by sonication of the former mixture at 4 °C for 30 sec. Three and a half milliliter of an aqueous suspension of the PS template microspheres (7% w/v) was then added to the stirred methylene chloride emulsion. After the swelling was completed, and the mixture did not contain any small emulsion droplets of the swelling solvent, as verified by optical microscopy, the diameter of the swollen microspheres was measured. PS swollen microspheres of various diameters were prepared by changing various parameters of the swelling process, for example volume and type [methylene chloride, $Fe(CO)_5$, and mixtures of different volume ratio between methylene chloride and $Fe(CO)_5$] of the swelling solvents.

Hemispherical particles were formed by evaporating the methylene chloride from the swollen template particles containing volume ratio of [methylene chloride]/[$Fe(CO)_5$] < 1. This was performed by purging nitrogen at room temperature for 3 hours through the shaken open vial containing the swollen particles aqueous mixture.

The swollen PS particles containing $Fe(CO)_5$, after removal of methylene chloride, were washed from excess reagents by several centrifugation cycles with water, and then water dried by nitrogen flow for several hours. Air-stable Fe nanoparticles were then formed by heating the water dried $Fe(CO)_5$ swollen PS particles in a quartz tube at 600 °C under flowing Ar gas for 3 hours.

Figure 6.13 shows light microscope pictures that allow one to compare the swelling ability of the template PS particles by methylene chloride and $Fe(CO)_5$. The PS microspheres before swelling have a size-and-size distribution of 2.4 ± 0.2 μm (Fig. 6.13A). As a consequence of their swelling with 2-ml methylene chloride their size distribution was retained, while their diameter increased from 2.4 ± 0.2 μm to 5.3 ± 0.3 μm, ca. 220% increase in the average diameter (Fig. 6.13B). On the other hand, a similar swelling process, substituting the 2-ml methylene chloride for 2-ml $Fe(CO)_5$ (Fig. 6.13C), resulted in nonuniform swelling of the template particles, that is the size-and-size distribution changed from 2.4 ± 0.2 μm to 7 ± 5.0 μm. Approximately 74% of the template particles were hardly swollen by $Fe(CO)_5$, while ca. 26% of these particles were swollen by $Fe(CO)_5$ to a larger extent than by methylene chloride. These results may indicate that methylene chloride is a good swelling solvent for PS particles, while $Fe(CO)_5$ is rather poor. Since the goal of these studies was to fill the swollen PS particles with $Fe(CO)_5$ while retaining their narrow size distribution, trials to use mixtures of methylene chloride and $Fe(CO)_5$ as swelling solvent, followed by evaporation of the methylene chloride from the swollen particles, have been performed.

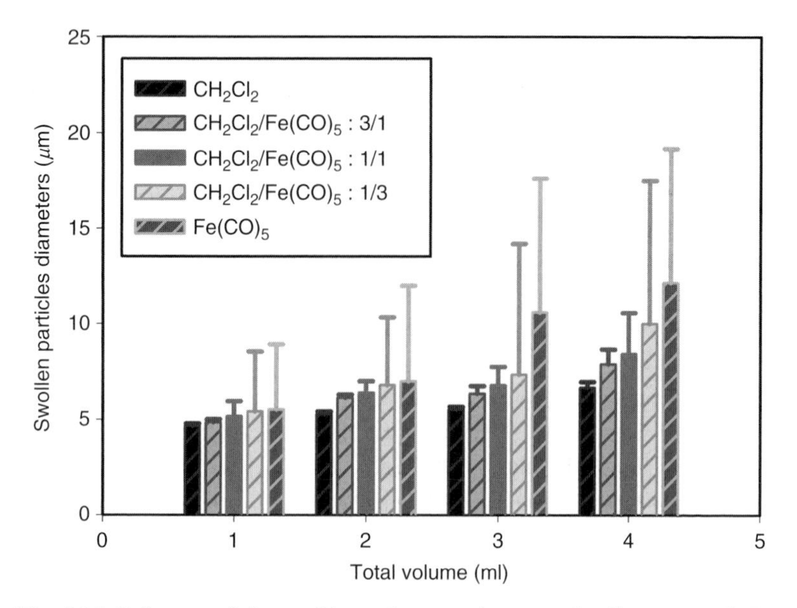

Fig. 6.14. Influence of the swelling solvents volume on the diameter and size distribution of the template PS particles. Five types of swelling solvents have been tested: methylene chloride and $Fe(CO)_5$ alone and three mixtures of these solvents: [methylene chloride]/[$Fe(CO)_5$] in ratios 3:1, 1:1, and 1:3 (v/v).

Figure 6.14 demonstrates the influence of different volumes of the swelling solvents on the diameter and size distribution of the template PS particles. For each volume, five types of swelling solvents have been tested: methylene chloride and $Fe(CO)_5$ alone, and three mixtures of these solvents: [methylene chloride]/ [$Fe(CO)_5$] = 3/1, 1/1, and 1/3 (v/v). Figure 6.14 illustrates that increasing the volume of all types of the swelling solvents resulted, as expected, in increased average diameter of the swollen particles. For example, in the absence of methylene chloride, and in the presence of 1-, 2-, 3-, and 4-ml methylene chloride, the diameter of the swollen particles increased from 2.4 ± 0.2 μm to 4.7 ± 0.2, 5.3 ± 0.2, 5.5 ± 0.2, and 6.7 ± 0.3 μm, respectively. A further increase in the volume of methylene chloride significantly damages the uniformity of the swollen particles. Addition of 7-ml methylene chloride resulted in the dissolution of the PS particles by methylene chloride dispersed in the aqueous phase. Figure 6.14 also shows that increasing the volume ratio [$Fe(CO)_5$]/[methylene chloride] resulted in a significant increase in the size distribution of the swollen particles. For example, in the absence and in the presence of 3 ml of the different swelling solvents: methylene chloride alone, [$Fe(CO)_5$]/[methylene chloride] = 1/3 and 3/1, and $Fe(CO)_5$ alone, the size distribution of the swollen particles increased from 2.4 ± 0.2 μm to 5.5 ± 0.2, 6.4 ± 0.4, 7.4 ± 6.8, and 10.6 ± 7 μm, respectively. Kinetics studies of the swelling of the PS template microspheres by 2 ml of the different swelling solvents indicated that under the experimental conditions the swelling process is completed within ca. 20 min. It should also be noted, as shown in Fig. 6.14, that the increase in the diameter (and volume) of the swollen particles was not linearly proportional to the volume of the added swelling solvent. For example, addition of 1.0- or 4.0-ml methylene chloride leads to an increase in the average diameter of the template particles of 195% and 278%, respectively. The first 1 ml of methylene chloride increased the diameter of the PS particles significantly more than the additional 3 ml. This nonlinear behavior is probably due to the packing arrangement of the PS chains within the template particles. The degree of entanglement of these chains determines the size (and volume) of the particles. The swelling solvents swell the template particles by penetrating within the PS chains of the particles, decreasing their degree of entanglement, and thereby increasing the counter length of the PS polymeric chains. As a consequence, the particles are less compact, and their size and volume increasing according to their swelling degree.

Methylene chloride was evaporated from the PS swollen particles containing both methylene chloride and $Fe(CO)_5$ by purging nitrogen through the shaken open vial containing the swollen particles aqueous mixture. Figure 6.15 shows light microscope pictures of the PS template microspheres (A) swollen with different volumes of $Fe(CO)_5$: 0.3 (B), 0.9 (C), and 1.2 (D) ml. The swelling process

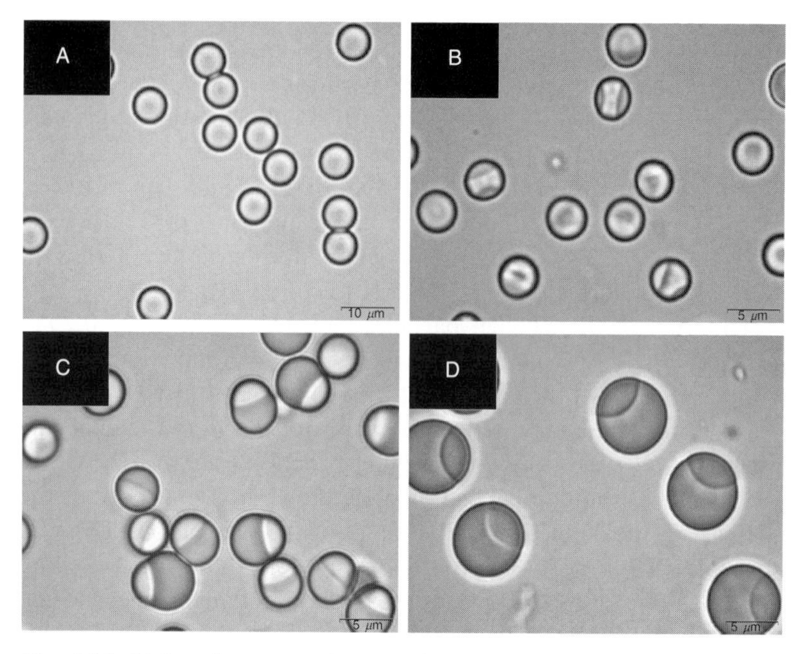

Fig. 6.15. Light microscope pictures of the PS template microspheres (A) swollen with different volumes of $Fe(CO)_5$: 0.3 ml (B), 0.9 ml (C), and 1.2 ml (D). The swelling process was accomplished with 1.6 ml of the swelling solvents (different mixtures of methylene chloride and $Fe(CO)_5$, followed by evaporation of methylene chloride from the swollen particles.

was accomplished with 1.6 ml of swelling solvents composed of different mixtures of methylene chloride and $Fe(CO)_5$, followed by evaporation of the methylene chloride from the swollen particles. Figure 6.15 demonstrates, as expected, an increase in the diameter of the PS swollen particles with increasing volume of the encapsulated $Fe(CO)_5$. Figure 6.15 also indicates that PS particles with biphase hemispherical morphology consist of PS and $Fe(CO)_5$ phases. This figure clearly demonstrates the relative increase in the $Fe(CO)_5$ phase with increasing volume of the encapsulated $Fe(CO)_5$. This biphase hemispherical shape is similar to that reported for PS/polybutyl methacrylate composite particles [30–32]. It should, however, be noted that biphase hemispherical particles were also observed by light microscopy when the swelling of the PS template particles was accomplished with mixtures of methylene chloride and $Fe(CO)_5$, where the volume ratio of [methylene chloride]/[$Fe(CO)_5$] < 1.0. On the other hand, when this volume ratio > 1.0, single-phase spherical particles were observed, and transition to biphase hemispherical particles was observed only after evaporation of the methylene chloride from the swollen particles. Methylene chloride is an excellent swelling solvent for PS particles, while $Fe(CO)_5$ is a poor one. Therefore, the swelling ability of mixtures of methylene chloride and $Fe(CO)_5$ decreases as the volume ratio between these two solvents decreases. We assume that during the swelling process methylene chloride carries the $Fe(CO)_5$ into the PS particles. However, on removal, or decreasing the relative concentration of methylene chloride, the $Fe(CO)_5$ phase separates from the PS phase.

Air-stable Fe nanoparticles were formed by heating the dried $Fe(CO)_5$ swollen PS particles in a quartz tube at 600 °C under Ar atmosphere. Typical microscopy pictures of the Fe/C composite particles are shown in Fig. 6.16. A light microscopy image of the Fe/C composite particles is shown in Fig. 6.16A, indicating some chains of particles, presumably oriented in a magnetic field. When prepared and fixed for electron microscopy, particles can be seen in their true size range, as shown in Fig. 6.16B and C. Low-resolution TEM image is presented in Fig. 6.16B. It can be assumed that iron nanoparticles formed from $Fe(CO)_5$ are included in a carbon formed from the PS matrix (bright regions is carbon, dark is iron). The high-resolution TEM image, depicted in Fig. 6.16C, provides further evidence for the identification of the product as Fe coated by carbon. The image illustrates the perfect arrangement of the atomic layers of bcc Fe, Fe_3C (1 1 1) plane, and carbon shell with thickness about 4 nm. The d-spacing of bcc Fe and Fe_3C (1 1 1) planes are very close: 0.287 and 0.302 nm, respectively. Figure 6.16C demonstrates d-spacing of 0.292–0.294 nm in the inner core of the particle and 0.298–0.3 nm in the outer part of

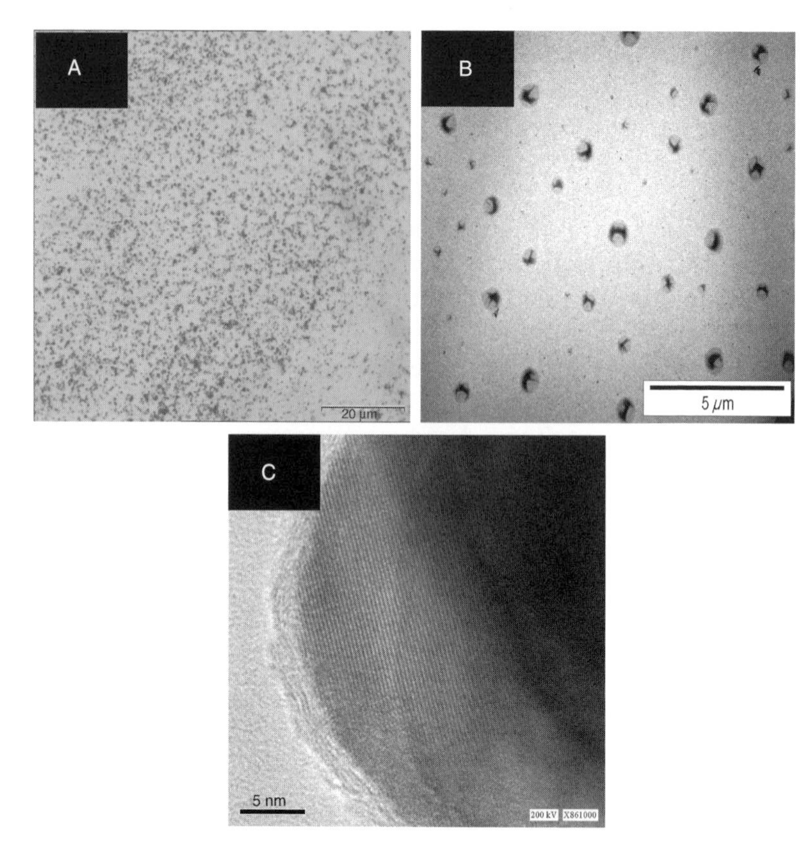

Fig. 6.16. Light microscope picture (scale bar 20 μm) (A), low-resolution TEM (B), and HRTEM (C) of the Fe/C nanoparticles. Fe/C nanoparticles were formed by swelling the template PS microspheres with swelling solvent emulsion droplets composed of 0.7 ml of methylene chloride and 0.9 ml of $Fe(CO)_5$.

the core (close to the carbon shell). Due to the close values of the d-spacing of bcc Fe and Fe_3C (1 1 1) plane, it is quite difficult to distinguish between these two phases. Figure 6.16 shows that the formed Fe/C composite particles possess broad size distribution ranging approximately from 10 to 600 nm. These results are quite disappointing since before burning off, the $Fe(CO)_5$ swollen particles had very narrow size distribution (~5%). We assume that the main reason for this significant change in size distribution and shape is the noncrosslinked character of the PS template particles, which probably melted during the heating process, before the pyrolysis.

The x-ray diffraction (XRD) for the Fe/C particles is shown in Fig. 6.17. The pattern was dominated by bcc Fe as shown by $2\theta =$ 44.8 and $2\theta = 65.2$. The smaller peaks between $2\theta = 43$ to $2\theta = 46$ match Fe_3C.

The magnetization curve presented in Fig. 6.18 shows that the Fe/C particle exhibits ferromagnetic behavior. Saturation magnetization (M_S) $=$ 75 emu g^{-1}, remnant magnetization (M_R) $=$ 13.5 emu g^{-1}, and coercivity (H_C) $=$ 250 Oe.

XPS is a common tool for studying the elemental surface composition of the Fe/C composite particles. The sampling depth of

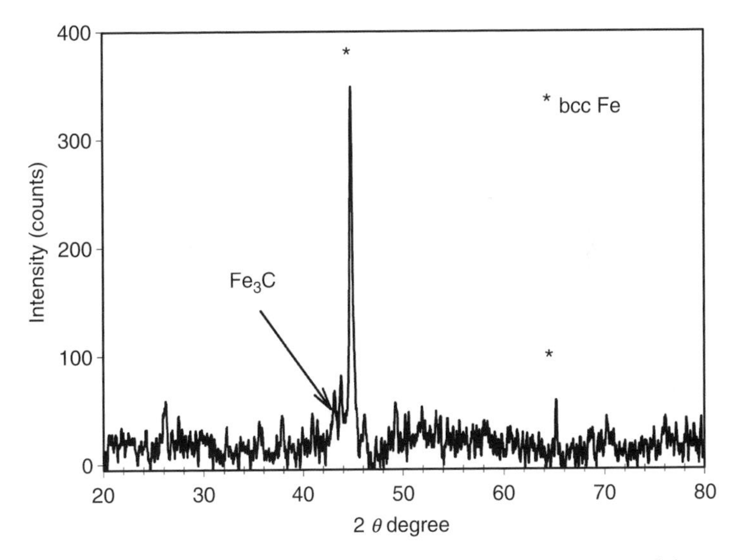

Fig. 6.17. XRD pattern of the Fe/C nanoparticles. Fe/C nanoparticles were formed according to the experimental procedures by swelling the template PS microspheres with swelling solvent emulsion droplets composed of 0.7 ml of methylene chloride and 0.9 ml of $Fe(CO)_5$.

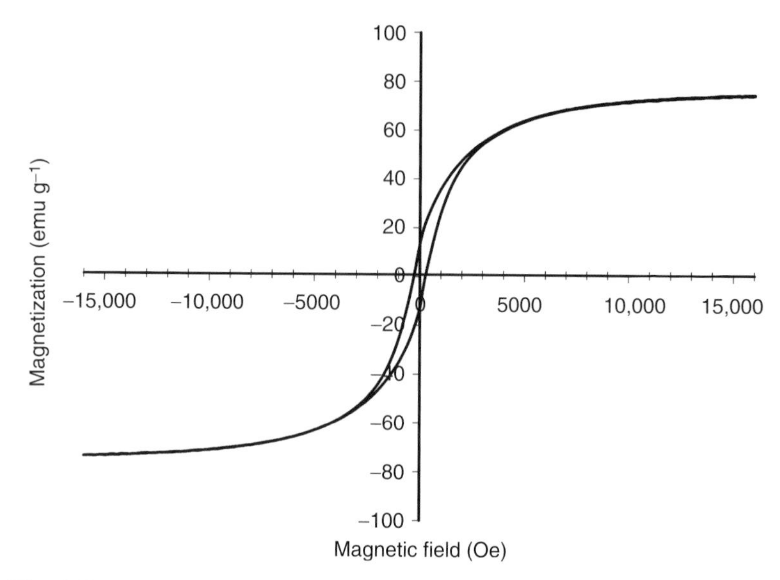

Fig. 6.18. Magnetization curve for the Fe/C nanoparticles. Fe/C nanoparticles were formed by swelling the template PS microspheres with swelling solvent emulsion droplets composed of 0.7 ml of methylene chloride and 0.9 ml of $Fe(CO)_5$.

XPS is limited by the effective mean free path of electrons escaping from the surface. XPS survey spectrum shows intense carbon (C_{1s}) peak at ca. 286, oxygen (O_{1s}) at ca. 532, and iron (Fe_{2p}) at ca. 710 eV. The integration of these peaks indicates the atomic percentage surface fraction of Fe/C particles that contains 89.9% C, 9.1% O, and 1% Fe.

Elemental analysis of the Fe/C particles confirms that the particles contain 48.8 atom % C, 0.8 atom % H, 3.1 atom % O, and 47.3 atom % Fe. The relative small amount of Fe (47.3%) in the particle explains the low-saturation magnetization value of these Fe/C particles.

TGA measurements have been performed in order to study the stability of annealed Fe/C particles (600 °C for 3 hours) in air atmosphere. Figure 6.19 shows that the annealed sample Fe/C has good stability against oxidation below 300 °C, due to the formation

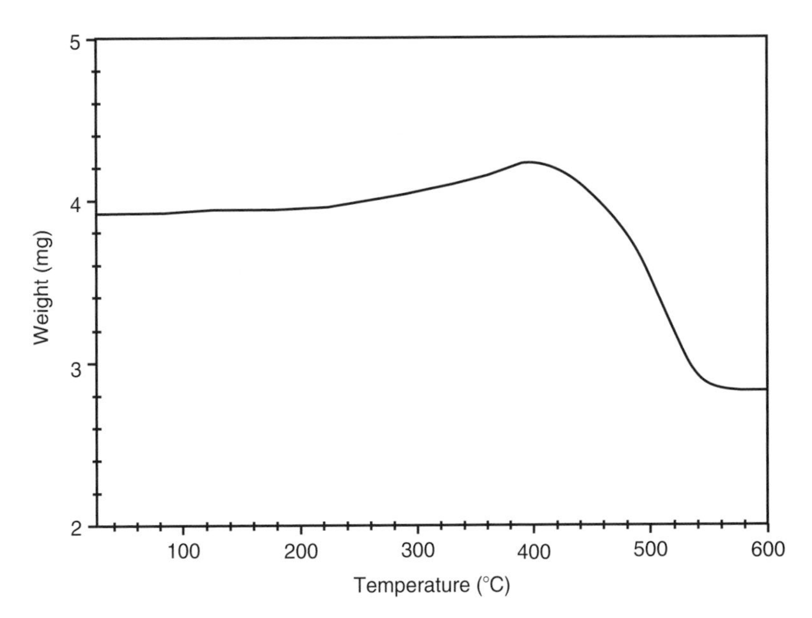

Fig. 6.19. TGA curve for the Fe/C nanoparticles under air atmosphere. Fe/C nanoparticles were formed by swelling the template PS microspheres with swelling solvent emulsion droplets composed of 0.7 ml of methylene chloride and 0.9 ml of $Fe(CO)_5$.

of a carbon protective shell on the surface of the alloy nanoparticles during the annealing process. The loss of weight at 400 °C is related to the burning of the carbon shell.

6.3.1. Conclusions

This investigation describes a new method for preparing air-stable Fe crystalline magnetic nanoparticles. The first step consists of encapsulation of the precursor $Fe(CO)_5$ within uniform PS template microspheres by a single-step swelling process. Hemispherical biphase microspheres were observed by light microscopy after evaporation of methylene chloride from the swollen template particles containing both methylene chloride and $Fe(CO)_5$. Annealing of the as-prepared $Fe(CO)_5$ swollen PS particles in argon at 600 °C leads to the growth of nonuniform air-stable iron nanocrystalline particles coated by a carbon and iron carbide protective layer. These air-stable nanocrystalline bcc-Fe particles exhibit ferromagnetic behavior with saturation magnetization of 75 emu g^{-1} and a large hysteresis with coercivity of 250 Oe. Studies concerning the precise composition of each phase of the hemispherical particles and the influence of various parameters (e.g., size and type of the template particles, type and concentration of the swelling solvent, annealing temperature) on the swollen template particles and the Fe/C composite nanoparticles are ongoing in our laboratory. Special care is taken to produce air-stable Fe nanoparticles of narrow size distribution.

6.4. Solid and hollow maghemite/polystyrene and silica/maghemite/polystyrene micron-sized composite particles of narrow size distribution

6.4.1. Synthesis and characterization of solid and hollow microspheres

Organic-inorganic magnetic composite micron-sized particles of narrow size distribution composed of cores of uniform micron-sized PS microspheres (1- to 6-μm diameter) and shells of maghemite nanoparticles of ca. 40-nm average diameter were formed by seeded polymerization of iron salts on the PS microspheres. Leaching of traces of the magnetic coating, or iron salts, from the γ-Fe_2O_3/PS magnetic composite particles into the continuous aqueous phase was prevented by a further coating of the magnetic hybrid composite microspheres with silica nanoparticles of ca. 40-nm average diameter. The silica coating on the magnetic composite particles was accomplished by seeded polymerization of $Si(OEt)_4$ on the γ-Fe_2O_3/PS composite microspheres, according to the Stober method [21]. Magnetic hollow silica microspheres were formed by burning off the organic PS core of the SiO_2/γ-Fe_2O_3/PS composite microspheres. Composite magnetic microspheres provided with functional groups, other than inorganic hydroxyls belonging to the silica, were prepared by reacting the silica-coated particles with ω-alcoholic reagents and/or ω-alkylsilane compounds such as $SiR_3(CH_2)_nX$; R = Cl, OCH_3, OC_2H_5; $n = 3$–17; X = CH_3, CN, CO_2Me, ΦCH_2Cl. Particles containing ω-hydroxyl or primary amine groups were prepared by interacting the silica surfaces with alkylalkoxysilane compounds such as $Si(OEt)_3(CH_2)_3CO_2CH_3$ or $Si(OEt)_3(CH_2)_3NH_2$; or with trichloroalkylsilane compounds such as $SiCl_3(CH_2)_3CO_2CH_3$ or $SiCl_3(CH_2)_3CN$, followed by diborane reduction of the terminal ester or nitrile surfaces, respectively [56, 57]. Polyaldehyde-derivatized magnetic microspheres were formed by reacting the ω-amine-derivatized particles with polyacrolein nanoparticles of ~60-nm diameter. Immunomagnetic microspheres were prepared by covalent binding of desired antibodies to the polyaldehyde magnetic particles. Preliminary studies demonstrated the potential use of the immunomagnetic microspheres for in vitro specific removal of anti-sperm antibodies (ASA) and sperm cells containing anti-sperm antibodies from semen of infertile males [63].

Briefly, 1 g of uniform PS microspheres of 2.3 ± 0.3-μm diameter (Figs. 6.20 and 6.21A) was added to a flask containing 20-ml distilled water. The formed mixture was sonicated for a few minutes and then mechanically stirred at ca. 200 rpm. The temperature was then preset to 60 °C. Nitrogen was bubbled through the suspension during the coating process to exclude air. 0.1 ml of iron chloride tetrahydrate aqueous solution (1.2 mmol in 10-ml H_2O) and 0.1 ml of sodium nitrite aqueous solution (0.02 mmol in 10-ml H_2O) were then successively introduced into the reaction flask. Thereafter, sodium hydroxide aqueous solution (0.5 mmol in 10-ml H_2O) was added until a pH of ca. 10 was reached. This procedure was repeated four times. During this coating process the microspheres became brown-black colored. The suspension was then cooled to room temperature. The formed magnetic composite microspheres (γ-Fe_2O_3/PS) were then washed extensively in water with a magnet,

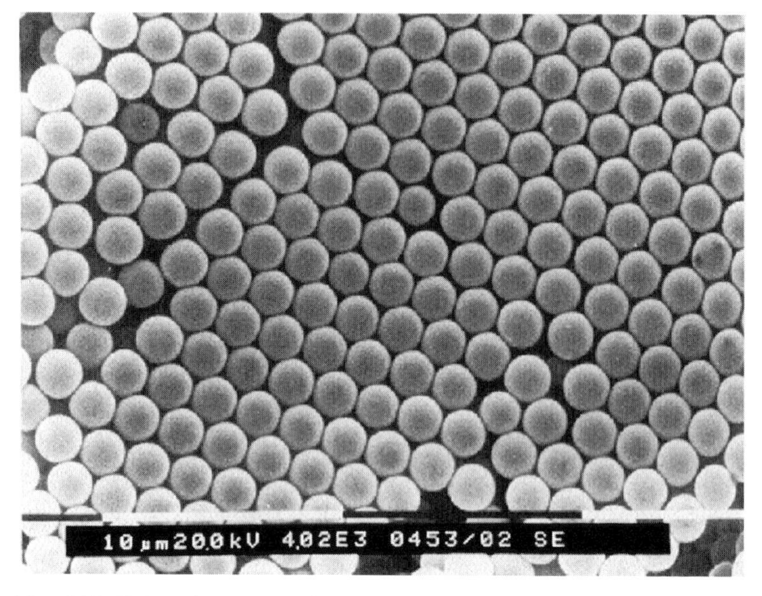

Fig. 6.20. SEM micrograph of PS microspheres of 2.3 ± 0.3-μm diameter.

and then dried in a vacuum oven. X-ray diffraction studies demonstrated that the coating on the PS microspheres was composed of crystallized γ-Fe_2O_3. Figure 6.21B visualizes the islands coating of the maghemite nanoparticles (ca. 40 nm) on the PS microspheres. Thermogravimetric and elemental analysis illustrated that the percentage of iron oxide coating on the microspheres was ~10% (w/w). Unfortunately, the iron oxide coating was not always stable, due to leaching of traces of the coating or iron ion salts from the particle surfaces into the working aqueous solution at high or low pH (pH $>$ 10 or $<$2). This problem, however, was overcome by a further coating of the γ-Fe_2O_3/PS composite microspheres with silica nanoparticles (Fig. 6.21C). Briefly, 1 g of the magnetic PS microspheres was added to a flask containing ethanol (93.6 ml) and distilled water (1.9 ml), and the mixture was sonicated to disperse the particles. Ammonium hydroxide (1.3 ml) and $Si(OEt)_4$ (3.2 ml)

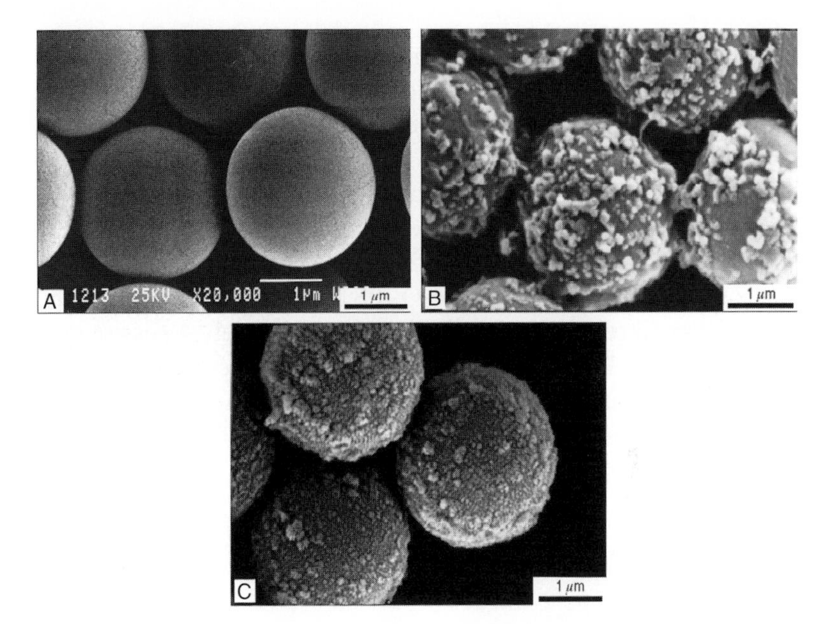

Fig. 6.21. SEM micrographs of PS microspheres of 2.3-μm average diameter, before (A), and after coating with maghemite nanoparticles (B), and then with silica nanoparticles (C).

were then added, and the resulting suspension was shaken at room temperature for 12 hours. The resulting $SiO_2/\gamma\text{-}Fe_2O_3$/PS composite particles were freed from free silica nanoparticles (ca. 40-nm diameter) by repeated centrifugation cycles. The silica-coated magnetic PS microspheres were then dried in a vacuum oven. A fully packed silica nanoparticles coating onto the magnetic PS particles was obtained after 12 hours coating under the experimental conditions (Fig. 6.21C). BET measurements demonstrated a surface area of 2.7 m^2/g for the PS microspheres and 15.8 m^2/g for the $SiO_2/\gamma\text{-}Fe_2O_3$/PS microspheres obtained after coating for 12 hours. The Brunauer-Emmet-Teller, or BET, method is based on gas adsorption measured against known adsorption of a standard. Thermogravimetric and elemental analysis showed that the coverage of the magnetic PS microspheres by the silica nanoparticles was $\sim$10% (w/w).

It should be noted that the magnetic susceptibility of the SiO_2/γ-Fe_2O_3/PS composite particles could be controlled by changing several parameters, for example iron ions concentration, $Si(OEt)_4$ concentration, and coating time.

Magnetic hollow silica uniform micron-sized microspheres were formed by burning off the PS core (ca. 600 °C) of the SiO_2/γ-Fe_2O_3/PS composite particles. Figure 6.22(A and B) shows a TEM photomicrograph of a solid SiO_2/γ-Fe_2O_3/PS composite microsphere and of a hollow silica magnetic microsphere, respectively. Cross-section picture of these hollow silica magnetic microspheres is illustrated in Fig. 6.22C, indicating that the thickness of the magnetic silica shell is ~100–150 nm.

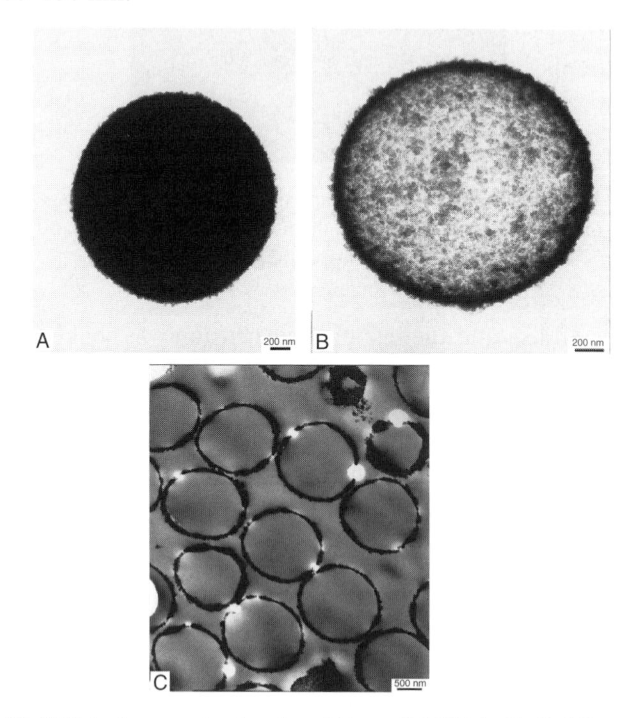

Fig. 6.22. TEM micrographs showing: (A) a solid SiO_2/γ-Fe_2O_3/PS composite microsphere and (B) a hollow silica magnetic microsphere obtained by burning off at approximately 600 °C the PS core of the former microsphere; (C) cross-section micrographs of these hollow silica magnetic microspheres.

SiO_2/γ-Fe_2O_3/PS composite particles containing terminal aldehyde groups were prepared by interacting these particles with $Si(OEt)_3(CH_2)_3NH_2$ and then with polyacrolein nanoparticles [45–50, 56, 57]. In a typical experiment, 8 ml of $Si(OEt)_3(CH_2)_3NH_2$ was added into a flask containing 1 g of the SiO_2/γ-Fe_2O_3/PS microspheres dispersed in 100-ml buffer acetate, 0.1 M at pH 5.5. The suspension was stirred at 60 °C for 18 hours. Thereafter, the amino-derivatized microspheres were washed intensively by several centrifugation cycles with buffer acetate and distilled water, respectively. Polyaldehyde microspheres were then formed by stirring the amino-derivatized microspheres for 5 hours at room temperature in an aqueous solution containing 1% of polyacrolein nanoparticles. The formed polyaldehyde-derivatized microspheres were then washed extensively with distilled water by repeated centrifugation cycles and then dried in a vacuum oven.

6.4.2. Synthesis of immunomagnetic microspheres for specific removal of ASA and sperm cells

6.4.2.1. Preparation of GαHIgG-conjugated microspheres

GαHIgG (goat anti-human immunoglobulins)-conjugated microspheres were prepared by covalent binding of GαHIgG to the polyaldehyde magnetic particles. This binding is based on the polyvalent Schiff base bonds formed by the primary amine groups of the antibodies and the aldehyde groups of the particles. In a typical experiment, 10 mg of the former polyaldehyde derivatized microspheres were added to PBS solution (1 ml) containing GαHIgG (anti-Fc, 0.1 mg). The formed suspension was then shaken at room temperature for 4 hours. Unbound protein was then removed by extensive washing of the conjugated particles in PBS with a magnet. Residual aldehyde groups on the microspheres were then blocked by shaking the GαHIgG-conjugated microspheres at room temperature for 30 min with ethanolamine aqueous solution at pH 7.0. Unbound ethanolamine was then removed by extensive washing of the conjugated particles in PBS with a magnet. The antibody-conjugated microspheres were then kept in PBS (1 ml) at 4 °C.

6.4.2.2. Specific removal of ASA and semen cells from the semen using GαHIgG-conjugated microspheres

Immunological infertility is estimated as the cause for infertility in 5–8% of infertile males [64]. Immunoinfertility can result from destruction of the spermatozoa by ASA, by inhibition of sperm motility and cervical mucus penetration, by inhibition of sperm–egg binding by ASA and/or prevention of embryo cleavage and early embryo development. Although, in the last 30 years, many therapies have tried to overcome this problem (immunosuppressive, sperm manipulation, and intrauterine insemination), the results are controversial. This article investigates a relatively new technique for solving the problem of immunological infertility by immunomagnetic polymeric microspheres.

Preliminary studies, performed according to the scheme described in Fig. 6.23 (left), demonstrated the potential use of the GαHIgG-conjugated magnetic microspheres for in vitro specific removal of ASA and sperm cells containing ASA from the semen of infertile males. In a typical experiment, GαHIgG-conjugated magnetic microspheres were added to a mixture containing 300-μl sperm cells and 300-μl seminal plasma containing ASA. Under these experimental conditions the ratio between the number of sperm cells and microspheres was 1:5. The formed mixture was then shaken at room temperature for 20 min. ASA and sperm cells containing ASA bound to the immunomicrospheres were then attracted to a magnet and the supernatant was removed. Figure 6.23 (right) demonstrates the binding of the immunomicrospheres to the tail of sperm cells and the agglutination of the immunomicrospheres due to their binding with ASA. ASA levels, measured by the sperm MAR test [65], decreased by this process from ca. 90% to ca. 28%.

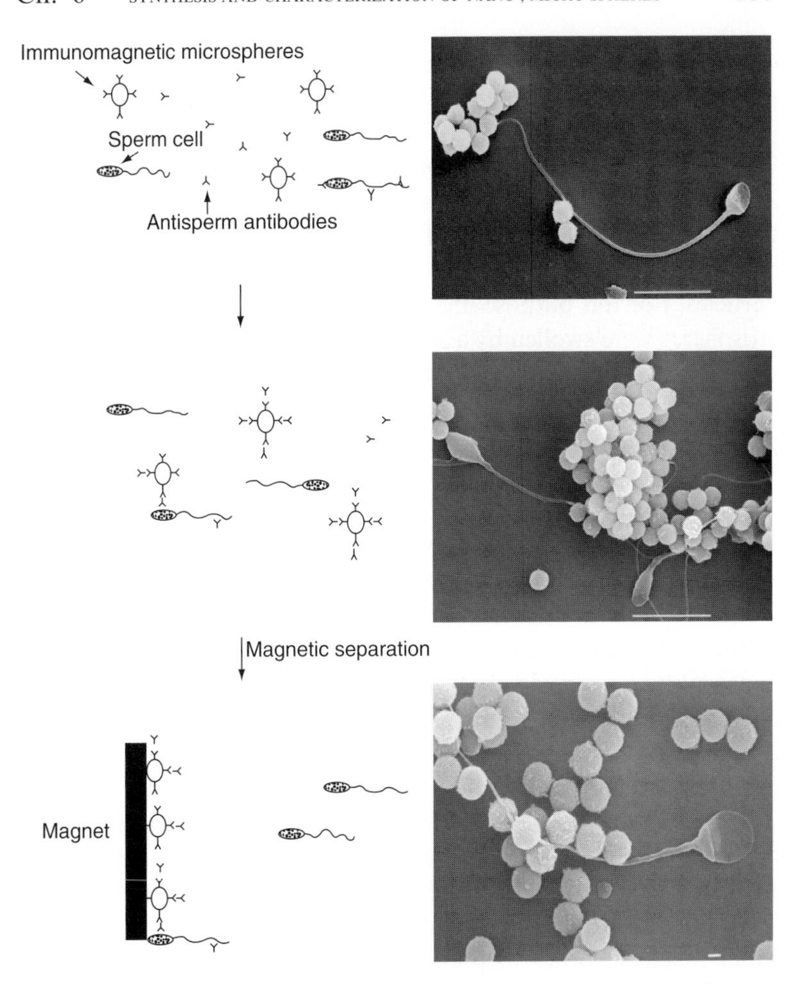

Fig. 6.23. SEM micrographs demonstrating the removal of ASA and sperm cells by GαHIgG conjugated magnetic microspheres.

Control experiments which were performed similarly, substituting the GαHIgG-conjugated microspheres with rabbit anti-sheep red blood cells (SRBC) conjugated microspheres, did not indicate any significant decrease in ASA level nor in sperm cells containing ASA.

6.5. Magnetic/nonmagnetic polystyrene/poly(methyl methacrylate) hemispherical composite micron-sized particles

The magnetic γ-Fe_2O_3/PS composite microspheres (ca. 2.4-μm average diameter) described previously (Fig. 6.21B) were used as template for the preparation of magnetic/nonmagnetic hemispherical polystyrene/poly(methyl methacrylate) (PS/PMMA) composite particles. For this purpose, the magnetic PS composite template microspheres were swollen by a single-step swelling process [30–32, 66] with methyl methacrylate (MMA) containing the initiator BP, followed by polymerization at 73 °C. Briefly, the dried magnetic γ-Fe_2O_3/PS template composite microspheres were dispersed in 30-ml aqueous solution containing 1.5% (w/v) SDS. In a separate vial (30 ml), 16.5 ml of 1.5% (w/v) SDS aqueous solution were added to 0.6-ml MMA containing 6 mg (1% w/v) BP. MMA emulsion (droplets size < 0.4 μm) was then formed by sonication of the former mixture at room temperature for 1 min. Thereafter, 3.5 ml of the suspension of the magnetic γ-Fe_2O_3/PS composite template microspheres were added to the stirred MMA emulsion. Kinetics studies of the swelling of the γ-Fe_2O_3/PS composite template microspheres by MMA indicated that under the experimental conditions the swelling rate was fast, and completed within ca. 5 min. The completion of the swelling process was verified by the disappearance of the small droplets of the emulsified MMA from the swollen particles dispersion, as observed by light microscopy. Before polymerization, sodium nitrite (0.1% w/v) was dissolved in the water media in order to prevent polymerization of MMA in the aqueous media. For polymerization of MMA within the swollen particles, the vial containing these particles was shaken at 73 °C for 20 hours. Excess MMA was then evaporated for 48 hours at room temperature, and the particles were then washed by extensive centrifugation cycles with water, and then air dried.

Figure 6.24 illustrates SEM micrographs of typical samples of the magnetic PS composite microspheres (A) and of the formed magnetic/nonmagnetic PS/PMMA hemispherical composite

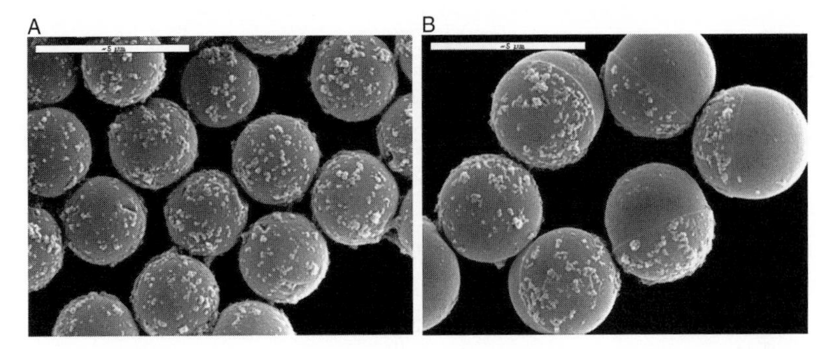

Fig. 6.24. SEM micrographs of magnetic PS composite microspheres (A) and magnetic/nonmagnetic PS/PMMA composite particles (B).

particles (B). The 3.8-μm diameter composite particles shown in (B) clearly appear to have hemispherical morphology with two distinct phases: magnetic PS and nonmagnetic PMMA. Thermogravimetric and elemental analyses illustrate that the formed particles are composed of 51.6% PMMA, 45.0% PS, and 3.4% iron oxide coating (w/w). The magnetization of the magnetic PS composite microspheres and the magnetic/nonmagnetic PS/PMMA composite particles as a function of external field strength is illustrated in Fig. 6.25A and B, respectively. The saturation magnetization of the magnetic/nonmagnetic particles at ca. 0.5 T is 1.4 emu g^{-1} (Fig. 6.25B), that is ca. 1/3 the magnetization of the template magnetic PS microspheres (Fig. 6.25A), as expected from the reduction of the percentage of magnetic coating of the PS/PMMA composite relative to the magnetic PS template particles.

We have demonstrated an effective route to microscale nonspherical magnetic/nonmagnetic composite particles of hemispherical morphology. We plan in the future to extend this approach also to PS template nanoparticles (diameter < 100 nm). The ability to change the percentage ratio of the magnetic/nonmagnetic parts as well as the bulk and surface compositions in these particles, via controlling the percentage of iron oxide coating, the percentage ratio PS/PMMA and the other polymerization parameters (e.g., initiator

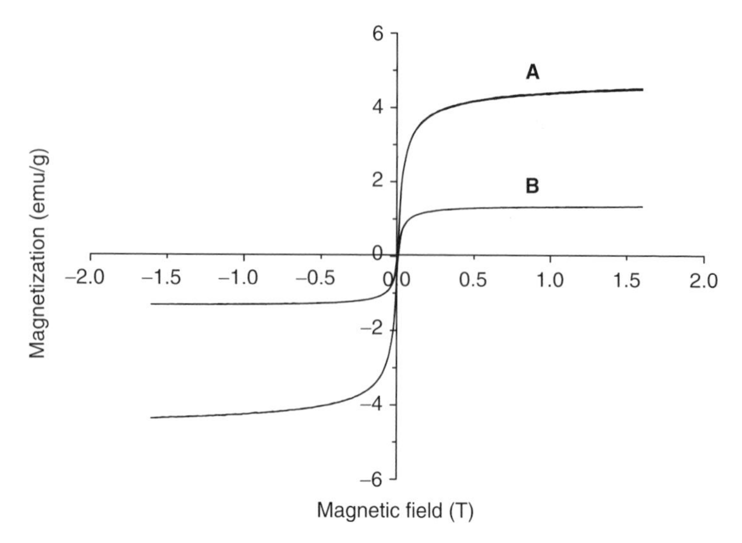

Fig. 6.25. Magnetization curves for magnetic PS composite microspheres (A) and magnetic/nonmagnetic PS/PMMA composite microspheres (B).

concentration) will provide additional ways to control the properties of these particles. Currently, more detailed studies of the synthesis, characterization and the use of these monodispersed magnetic/nonmagnetic composite particles are underway in our laboratory.

Summary

The present chapter describes the synthesis and characterization of a few unique magnetic nano- and micron-sized particles: (1) maghemite nanoparticles of ca. 15- to 100-nm diameter with narrow size distribution; (2) air-stable iron nanocrystalline particles; (3) solid and hollow maghemite/polystyrene and silica/maghemite/polystyrene micron-sized composite particles with narrow size distribution; (4) magnetic/nonmagnetic polystyrene/poly(methyl methacrylate) hemispherical composite micron-sized particles with narrow size distribution. Preliminary studies demonstrated a few potential biomedical uses of some of these magnetic particles. In the future we plan to focus our main efforts to the applications of these magnetic particles. Special efforts will be provided to the synthesis, surface modification, and biomedical uses of nano- and micron-sized Fe particles of narrow size distribution.

References

[1] Ugelstad, J., Söderberg, L., Berge, A. and Bergström, J. (1983). Monodisperse polymer particles – a step forward for chromatography. Nature *303*, 95–96.

[2] Ugelstad, J., Berge, A., Ellingsen, T., Schmid, R., Nilsen, T.-N., Mørk, P. C., Stenstad, P., Hornes, E. and Olsvik, Ø. (1992). Preparation and application of new monosized polymer particles. Prog. Poly. Sci. *17*, 87–161.

[3] Hergt, R., Andrä, W., d'Ambly, C. G., Hilger, I., Kaiser, W. A., Richter, U. and Schmidt, H.-G. (1998). Physical limits of hyperthermia using magnetite fine particles. IEEE Trans. Magn. *34*, 3745–54.

[4] Groman, E.V. and Josephson, L. (1988). Biologically degradable superparamagnetic particles for use as nuclear magnetic resonance imaging agents. United States Patent 4,770,183.

[5] Bockstaller, M. R., Mickiewicz, R. A. and Thomas, E. L. (2005). Block copolymer nanocomposites: Perspectives for tailored functional materials. Adv. Mater. *17*, 1331–49.

[6] Bockstaller, M. R., Lapetnikov, Y., Margel, S. and Thomas, E. L. (2003). Size-selective organization of enthalpic compatibilized nanocrystals in ternary block copolymer/particle mixtures. J. Am. Chem. Soc. *125*, 5276–77.

[7] Golosovsky, M., Neve-Oz, Y. and Davidov, D. (2003). Magnetic-field tunable photonic stop band in metallodielectric photonic crystals. Synth. Metals *139*, 705–09.

[8] Arshady, R. (1993). Microspheres for biomedical applications: Preparation of reactive and labelled microspheres. Biomaterials *14*, 5–15.

[9] Kreuter, J. (1991). Peroral administration of nanoparticles. Adv. Drug Deliv. Rev. *7*, 71–86.

[10] Hergt, R., Hiergeist, R., Hilger, I., Kaiser, W. A., Lapatnikov, Y., Margel, S. and Richter, U. (2004). Maghemite nanoparticles with very high AC-losses for application in RF-magnetic hyperthermia. J. Magn. Magn. Mater. *270*, 345–57.

[11] Hoyos, M., Moore, L. R., McCloskey, K. E., Margel, S., Zuberi, M., Chalmers, J. J. and Zborowski, M.. (2000). Study of magnetic particles pulse-injected into an annular SPLITT-like channel inside a quadrupole magnetic field. J. Chromatogr. A. *903*, 99–116.

[12] Sinyakov, M. S., Dror, M., Zhevelev, H. M., Margel, S. and Avtalion, R. R. (2002). Natural antibodies and their significance in active immunization and protection against a defined pathogen in fish. Vaccine *20*, 3668–74.

[13] Kreuter, J. (1988). Possibilities of using nanoparticles as carriers for drugs and vaccines. J. Microencapsul. *5*, 115–27.

[14] Abe, M. and Tamaura, Y. (1985). Ferrite plating in aqueous solution for preparing magnetic thin film, presented at *Fourth International Conference on Ferrites*, San Francisco, CA, 1985, 1984. In: Advances in Ceramics (Wang, F. F. Y., ed.), Vol. 15. American Ceramic Society, Inc., Columbus, OH, pp. 639–45.

[15] Abe, M., Tanno, Y. and Tamaura, Y. (1985). Direct formation of ferrite films in wet process. J. Appl. Phys. *57*, 3795–97.

[16] Abe, M., Itoh, T. and Tamaura, Y. (1992). Magnetic and biomagnetic films obtained by ferrite plating in aqueous solution. Thin Solid Films *216*, 155–61.

[17] Margel, S. and Rembaum, A. (1980). Synthesis and characterization of poly(glutaraldehyde). A potential reagent for protein immobilization and cell separation. Macromolecules *13*, 19–24.

[18] Almog, Y., Reich, S. and Levy, M. (1982). Monodisperse polymeric spheres in the micron size range by a single step process. British Polymer Journal *14*, 131–36.

[19] Rembaum, A. and Dreyer, W. J. (1980). Immunomicrospheres: Reagents for cell labeling and separation. Science *208*, 364–68.

[20] Badley, R. D., Ford, W. T., McEnroe, F. J. and Assink, R. A. (1990). Surface modification of colloidal silica. Langmuir *6*, 792–801.

[21] Stöber, W., Fink, A. and Bohn, E.. (1968). Controlled growth of monodisperse silica spheres in the micron size range. J. Colloid. Interface Sci. *26*, 62–69.

[22] Philipse, A. P., van Bruggen, M. P. B. and Pathmamanoharan, C. (1994). Magnetic silica dispersions: Preparation and stability of surface-modified silica particles with a magnetic core. Langmuir *10*, 92–99.

[23] Margel, S. and Bamnolker, H. (2000). Process for the preparation of microspheres and microspheres made thereby. Patent No. 6103379 (USA), August 15.

[24] Margel S. and Gura S. Nucleation and growth of magnetic metal oxide nanoparticles and its use. Patent No. 1088315 (EC).

[25] Bamnolker, H. and Margel, S. (1996). Dispersion polymerization of styrene in polar solvents: Effect of reaction parameters on microsphere surface composition and surface properties, size and size distribution, and molecular weight. J. Polym. Sci. Part A: Polym. Chem. *34*, 1857–71.

[26] Margel, S., Nov, E. and Fisher, I. (1991). Polychloromethylstyrene microspheres: Synthesis and characterization. J. Polym. Sci. Part A: Polym. Chem. *29*, 347–55.

[27] Boguslavsky, L., Baruch, S. and Margel, S. (2005). Synthesis and characterization of polyacrylonitrile nanoparticles by dispersion/emulsion polymerization process. J. Colloid. Interface Sci. *289*, 71–85.

[28] Bamnolker, H., Nitzan, B., Gura, S. and Margel, S. (1997). New solid and hollow, magnetic and non-magnetic, organic-inorganic monodispersed hybrid microspheres: Synthesis and characterization. J. Mater. Sci. Lett. *16*, 1412–15.

[29] Moore, L. R., Zborowski, M., Nakamura, M., McCloskey, K., Gura, S., Zuberi, M., Margel, S. and Chalmers, J. J. (2000). The use of magnetite-doped polymeric microspheres in calibrating cell tracking velocimetry. J. Biochem. Biophys. Methods *44*, 115–30.

[30] Akiva, U. and Margel, S. (2005). Surface-modified hemispherical polystyrene/polybutyl methacrylate composite particles. J. Colloid. Interface Sci. *288*, 61–70.

[31] Melamed, O. and Margel, S. (2001). Poly(N-vinyl α-phenylalanine) microspheres: Synthesis, characterization, and use for immobilization and microencapsulation. J. Colloid. Interface Sci. *241*, 357–65.

[32] Wizel, S., Margel, S. and Gedanken, A. (2000). The preparation of a polystyrene–iron composite by using ultrasound radiation. Polym. Int. *49*, 445–48.

[33] Green-Sadan, T., Kuttner, Y., Lublin-Tennenbaum, T., Kinor, N., Boguslavsky, Y., Margel, S. and Yadid, G. (2005). Glial cell line-derived neurotrophic factor-conjugated nanoparticles suppress acquisition of cocaine self-administration in rats. Exp. Neurol. *194*, 97–105.

[34] van Leemputten, E. and Horisberger, M. (1974). Immobilization of trypsin on partially oxidized cellulose. Biotechnol. Bioeng. *16*, 997–1003.

[35] Margel, S. (1984). Characterization and chemistry of polyaldehyde microspheres. J. Polym. Sci. Polym. Chem. *22*, 3521–3533.

[36] Margel, S. (1985). Polyacrolein microspheres. Methods Enzymol. *112*, 164–75.

[37] Wang, P.-C., Lee, C.-F., Young, T.-H., Lin, D.-T. and Chiu, W.-Y. (2005). Preparation and clinical application of immunomagnetic latex. J. Polym. Sci. Part A: Polym. Chem. *43*, 1342–1356.

[38] Wahlberg, J., Lundeberg, J., Hultman, T., Holmberg, M. and Uhlén, M. (1990). Rapid detection and sequencing of specific in vitro amplified DNA sequences using solid phase methods. Mol. Cell Probes *4*, 285–97.

[39] Seliger, H., Bader, R., Hinz, M., Rotte, B., Eisenbeiss, F., Gura, S., Nitzan, B. and Margel, S. (2000). Polymer-supported nucleic acid fragments: Tools for biotechnology and biomedical research. React. Funct. Polym. *43*, 325–39.

[40] Margel, S. (1989). Affinity separation with polyaldehyde microsphere beads. J. Chromatogr. *462*, 177–89.

[41] Kohn, J. and Wilchek, M. (1982). Mechanism of activation of sepharose and sephadex by cyanogen bromide. Enzyme. Microb. Technol. *4*, 161–63.

[42] Nilsson, K. and Mosbach, K. (1987). Tresyl chloride-activated supports for enzyme immobilization. Methods Enzymol. *135*, 65–78.

[43] Margel, S. and Sturchak, S. (1999). Bioactive conjugates of cellulose with amino compounds. United States Patent 5,855,987.

[44] Galperin, A., Margel, D. and Margel, S. (2006). Synthesis and characterization of uniform radiopaque polystyrene microspheres for X-ray imaging by a single-step swelling process. J. Biomed. Mater. Res. Part A. *79A*, 544–51.

[45] Margel, S., Sturchak, S., Ben-Bassat, E., Reznikov, A., Nitzan, B., Krasniker, R., Melamed, O., Sadeh, M., Gura, S., Mandel, E., Michael, E. and Burdygine, I. (1999). Functional microspheres: Synthesis and biological applications. In: Microspheres, Microcapsules & Liposomes: Medical & Biotechnology Applications (Arshady, R., ed.), Vol. 2. Citus Books, London, UK, (Chapter 1), pp. 11–42.

[46] Arshady, R., Margel, S., Pichot, C. and Delair, T. (1999). Functionalization of preformed microspheres. In: Microspheres, Microcapsules & Liposomes: Preparation & Chemical Applications (Arshady, R., ed.), Vol. 1. Citus Books, London, UK, (Chapter 6), pp. 165–95.

[47] Margel, S., Dolitzky, Y. and Sivan, O. (1992). Immobilized polymeric microspheres. Polyacrolein microspheres covalently bound in a monolayer structure onto glass surfaces. Colloids Surf. *62*, 215–30.

[48] Margel, S., Sivan, O. and Dolitzky, Y. (1991). Functionalized thin films. Synthesis, characterization, and potential use. Langmuir *7*, 2317–22.

[49] Margel, S., Cohen, E., Dolitzky, Y. and Sivan, O. (1992). Surface modification. I. Polyacrolein microspheres covalently bonded onto polyethylene. J. Polym. Sci. Part A: Polym. Chem. *30*, 1103–10.

[50] Moore, L. R., Milliron, S., Williams, P. S., Chalmers, J. J., Margel, S. and Zborowski, M. (2004). Control of magnetophoretic mobility by susceptibility-modified solutions as evaluated by cell tracking velocimetry and continuous magnetic sorting. Anal. Chem. *76*, 3899–3907.

[51] Bunker, B. C., Rieke, P. C., Tarasevich, B. J., Campbell, A. A., Fryxell, G. E., Graff, G. L., Song, L., Liu, J., Virden, J. W. and McVay, G. L. (1994). Ceramic thin-film formation on functionalized interfaces through biomimetic processing. Science *264*, 48–55.

[52] Ugelstad, J., Ellingsen, T., Berge, A. and Helgee, B. (1987). United States Patent No. 4,654,267.

[53] Margel, S., Vogler, E. A., Firment, L., Watt, T., Haynie, S. and Sogah, D. Y. (1993). Peptide, protein, and cellular interactions with self-assembled monolayer model surfaces. J. Biomed. Mater. Res. *27*, 1463–76.

[54] Carlsson, D. J. and Wiles, D. M. (1976). The photooxidative degeneration of polypropylene. Part 1. Photooxidation and photoinitiation processes. J. Macromol. Sci. Part C: Rev. Macromol. Chem. Phys. *14*, 65–106.

[55] Lacoste, J., Vaillant, D. and Carlsson, D. J. (1993). Gamma-, photo-, and thermally-initiated oxidation of isotactic polypropylene. J. Polym. Sci. Part A: Polym. Chem. *31*, 715–22.

[56] Brandriss, S. and Margel, S. (1993). Synthesis and characterization of self-assembled hydrophobic monolayer coatings on silica colloids. Langmuir *9*, 1232–40.

[57] Bronstein, L. M., Sidorov, S. N., Valetsky, P. M., Hartmann, J., Cölfen, H. and Antonietti, M. (1999). Induced micellization by interaction of poly(2-vinylpyridine)-*block*-poly(ethylene oxide) with metal compounds. Micelle characteristics and metal nanoparticle formation. Langmuir *15*, 6256–62.

[58] Partouche, E., Waysbort, D. and Margel, S. (2006). Surface modification of crosslinked poly(styrene-divinyl benzene) micrometer-sized particles of narrow size distribution by ozonolysis. J. Colloid. Interface Sci. *294*, 69–78.

[59] Marutani, E., Yamamoto, S., Ninjbadgar, T., Tsujii, Y., Fukuda, T. and Takano, M. (2004). Surface-initiated atom transfer radical polymerization of methyl methacrylate on magnetite nanoparticles. Polymer *45*, 2231–35.

[60] Vestal, C. R. and Zhang, Z. J. (2002). Atom transfer radical polymerization synthesis and magnetic characterization of $MnFe_2O_4$/polystyrene core/shell nanoparticles. J. Am. Chem. Soc. *124*, 14312–13.

[61] Sinyakov, M. S., Dror, M., Lublin-Tennenbaum, T., Salzberg, S., Margel, S. and Avtalion, R. R. (2006). Nano- and microparticles as adjuvants in vaccine design: Success and failure is related to host natural antibodies. Vaccine *24*, 6534–41.

[62] Shpaisman, N. and Margel, S. (2006). Synthesis and characterization of air-stable iron nanocrystalline particles based on a single-step swelling process of uniform polystyrene template microspheres. Chem. Mater. *18*, 396–402.

[63] Margel, S., Gura, S., Bamnolker, H., Nitzan, B., Tennenbaum, T., Bar-Toov, B., Hinz, M. and Seliger, H. (1997). Synthesis, characterization, and use of new solid and hollow, magnetic and non-magnetic, organic-inorganic monodispersed hybrid microspheres. In: Scientific and Clinical Applications of Magnetic Carriers (Häfeli, U., Schütt, W., Teller, J. and Zborowski, M., eds.). Plenum Press, New York, NY, (Chapter 3) pp. 37–52.

[64] Bronson, R., Cooper, G. and Rosenfeld, D. (1984). Sperm antibodies: Their role in infertility. Fertil. Steril. *42*, 171–83.

[65] Jager, S., Kremer, J. and van Slochteren-Draaisma, T. (1978). A simple method of screening for antisperm antibodies in the human male. Detection of spermatozoal surface IgG with the direct mixed antiglobulin reaction carried out on untreated fresh human semen. Int. J. Fertil. *23*, 12–21.

[66] Akiva, U. and Margel, S. (2005). New micrometer-sized hemispherical magnetic/non-magnetic monodispersed polystyrene/poly(methyl methacrylate) composite particles: synthesis and characterization. J. Mater. Sci. *40*, 4933–35.

Laboratory Techniques in Biochemistry and Molecular Biology, Volume 32
Magnetic Cell Separation
M. Zborowski and J. J. Chalmers (Editors)

CHAPTER 7

The biocompatibility and toxicity of magnetic particles

Urs O. Häfeli, Jennifer Aue, and Jaime Damani

Division of Pharmaceutics and Biopharmaceutics, Faculty of Pharmaceutical Sciences, The University of British Columbia, Vancouver, B.C. V6T 1Z3, Canada

7.1. Introduction

Since the beginning of the "nanotechnology revolution" in recent years, micro- and nanoparticles have increasingly formed the basis of many novel discoveries in biomedicine and therapeutics. Such particles have also found their niche commercially, for instance, when incorporated into wrinkle resistant and waterproof clothing [1]. Micro- and nanoparticles, which have been made magnetic, are also part of this nanotechnology revolution. The first successful application of magnetic particles were on the technological side, with ferrofluids, for example, being used to increase the power of high end loudspeakers and seal high vacuum pumps [2]. On the biomedical side, the first successful and Food and Drug Administration (FDA) approved applications include their diagnostic use as contrast agents for magnetic resonance imaging (MRI) [3] and their application in extracting tumor cells from bone marrow transplants before reinjection into a patient [4].

Magnetic particles are being used in many exciting new technologies. These include the first clinical trials of magnetic hyperthermia for tumor therapy [5] and also magnetic stem cell imaging, a procedure which is valuable in monitoring the success of gene therapy [6].

DOI: 10.1016/S0075-7535(06)32007-4

Many other applications related to drug delivery, drug targeting, and cancer therapy are currently being developed [7]. In addition, a large proportion of biological in vitro assays for the analysis of cellular components, the diagnosis of diseases, and the separation of cells are nowadays based on magnetic separation using specific magnetic particles [8]. The next three chapters in this book will review these applications in detail.

With the rise in the development of magnetic nano- and micro-particles, strict measures have to be in place to ensure their safe use. Part of these safety requirements must ensure that particles are bio-compatible and do not induce any toxicity when in contact with the body. Tests must thus be performed to obtain the toxicity profile of the particles.

When the particles are intended for internal use, it is important to determine any toxic potential of the particles and the materials used to make them. As with drugs, it is a requirement of the International Organization of Standardization (ISO) to ensure the biocompatibility and toxicity profile of micro- or nanospheres. For regulatory purposes, such particles are medical devices [9].

Unlike in drug toxicity, where the bioactivation of drugs, the activation of apoptosis-inducing protein kinases (e.g., JNK and p38 kinases), and the whole field of metabonomics is extremely important, magnetic drug targeting is more concerned with local effects, single or at best a few repeated applications. Systemic effects are thus generally much smaller, with the exception of magnetic contrast agents.

This chapter looks into the characteristics of nano- and micro-particles that may affect their toxicity profile. The chapter will also provide insight on the different available in vitro toxicity tests and assess the suitability of each as well as give examples of their evaluation both in vitro and in vivo. In vitro studies are superseding most in vivo studies, due to their ease of use, accessibility, cost effectiveness, and lack of ethical restrictions. The next section of the chapter will deal with the immunogenicity of biological targeting reagents and discuss issues around allergic reactions or complement activation. Since toxicity determinations are a crucial part of any drug approval process, we will also finally provide an overview of the legal aspects of the process according to the regulatory bodies in the United States, Europe, and Japan.

7.2. Definition of toxicity and biocompatibility

Toxicology is "the science of poisons" or more specifically encompasses the chemical and physical properties of poisons, their physiological or behavioral effects on living organisms, qualitative and quantitative methods for their analysis, and the development of procedures for the treatment of poisoning [10]. As Paracelsus, whom many call the "father of toxicology," already pointed out in the sixteenth century, a central concept of toxicology is that all toxicologic effects are dose-dependent. In his own words, "All substances are poisons; there is none which is not a poison. The right dose differentiates a poison from a remedy."

In general, there are three types of toxicity: chemical, biological, and physical. For magnetic particles, all of these can be important.

The first type of toxicity, chemical toxicity, includes the effects from both inorganic and organic substances. For magnetic particles, inorganic toxins include mainly the effects from the metals iron and cobalt, their salts, and especially their oxides. Organic toxins include effects from the particle matrix materials or from coatings that have been applied in order to change or improve solubility, allow for chemical derivatization, or permit the loading of drugs.

The second type of toxicity, biological toxicity, refers to effects on biological systems. It is more difficult to define, since the biological entity could be a complex system as present in every mammalian animal (or a person), a single cell organism such as a bacterium or cell, or a virus. Typical for biological toxicity is a "threshold dose" which might be due to the organism's ability to fight the toxin, such as is the case where the toxin is excreted, metabolized, or isolated, for example by encapsulation into a vesicle. A threshold dose might also be due to the fact that the effect is too small to see, or the observation time scale is too short to notice an effect. To analyze biological toxicity, it is very important to choose the correct system, since no effects may otherwise be seen. The release of toxic particle coatings inside a cell, which then interact with chromosomal DNA and transform a cell into a cancer cell, is such a biological effect. To prove that such effects exist, especially when they stem from mixtures of compounds, may however be very difficult and complex.

The third type of toxicity is physical toxicity and includes heat, vibration, mechanical effects, and nonionizing electromagnetic radiation such as infrared and visible light or ionizing radiation. Magnetic particles might generate physical toxic effects while flowing through an artery (abrasion) or when being subjected to an alternating magnetic field (heat production).

Toxicity can be measured by the effects on the target organism, organ, or tissue. Since different targets typically have different levels of response to the same dose of a toxin, a population-level measure of toxicity, which relates the probability of an outcome for a given target in a population, is often used. A frequently cited measure is the LD_{50}, "LD" standing for "lethal dose," which is a concentration measure for a toxin at which 50% of the members of an exposed population die from exposure. When such data do not exist, estimates are made by comparison to known similar toxins, or to similar exposures in similar organisms. Then "safety factors" must be built in to protect against the uncertainties of such comparisons.

Toxicity of a substance can be affected by many different factors, such as the pathway of administration (application of the toxin to the skin, ingestion, inhalation, or injection), the time of exposure (a brief, acute encounter versus a long-term chronic exposure), the number of exposures (a single dose versus multiple doses over time), the physical form of the toxin (solid, liquid, or gas), the organ system involved (cardio, nephro-, hemo-, nervous, or hematopoietic system), and even the genetic makeup and robustness of the target cells or organisms. For more details, please consult a good toxicology textbook such as Mulder and Dencker [11].

7.2.1. Biomaterials classifications

A general definition of biomaterials is any natural or synthetic material that interfaces with living tissue and/or biological fluids. For in vivo magnetic particle therapy, we have to assure both the responsible regulatory body and the patient that the biomaterials to be used in the therapy are safe, or nontoxic. This does not mean that there will be no reaction to the biomaterials, since most synthetic or natural materials placed within the human body produce a tissue reaction toward the implant. But it does mean that the host response to the implant must be appropriate.

Materials producing an appropriate host response can generally be classified as biotolerant, bioinert, bioactive, and bioresorbable. Biotolerant refers to any material to which the body reacts by encapsulating it. A fibrous capsule forms around bioinert implants, stops any acute inflammatory response, and integrates the implant into the body. Typical biotolerant materials are polymethylmethacrylate (PMMA), silicone, and glass. Silica- and carbon-coated magnetic particles might also be biotolerant materials.

The term bioinert refers to materials that once placed in an organism have minimal interaction with its surrounding tissue. Typical bioinert implant materials are stainless steel, titanium, ultra-high molecular weight polyethylene, and aluminum oxide.

The term bioactive defines a material, which upon being placed within the human body interacts with the surrounding bone or soft tissue. This occurs through a time-dependent kinetic modification of the material surface, triggered by implantation within the living tissue. The best example for bioactive interactions are the ion-exchange reaction between a bone implant (e.g., bioglass® or synthetic hydroxyapatite) and the surrounding body fluids. In this process, a biologically active carbonate apatite layer which is chemically and crystallographically equivalent to the mineral phase in bone is formed on the implant. At the present, we are not aware of the existence of bioactive magnetic particles.

The last term bioresorbable refers to a material that upon placement within the human body starts to dissolve, is resorbed by specific cells and then slowly replaced by advancing tissue (such as bone). Common examples of bioresorbable materials include tricalcium phosphate [$Ca_3(PO_4)_2$], hydroxyapatite [$Ca_{10}(PO_4)_6(OH)_2$] (used in tissue engineering for bone cement), and polylactide-co-glycolide copolymers. The latter material has been used as the matrix material for the preparation of magnetic microspheres.

Biomaterials are biocompatible if the living tissue reacts to the nonliving materials with an appropriate and specific host response. Importantly, the response of an organism to a material is thus not automatically a toxic reaction. Biomaterials inducing a host response can be, and often are, still biocompatible. Before biomaterials such as magnetic particles can be released for medical or biological use, it is, however, important to decipher any significant effects on an organism, identify why they occur, determine if they are toxic, and perhaps learn how to modify them into a benign reaction. Only biocompatible biomaterials, which correspond to the ISO 10993 standards for toxicity testing can be used in vivo.

7.3. Particle characteristics that influence toxicity

Many investigations have shown that many parameters are responsible for the toxicities seen in magnetic particles. These parameters include material composition, processes including degradation, oxidation, and leakage of enclosed components, size, and morphology (shape, surface area, and surface microstructure). These parameters will be discussed here with explicit examples.

7.3.1. Particle size

Many studies have shown that nano- and a micro-sized particles made from the identical material can have different toxicities. The reason for this observation is that material properties such as solubility, transparency, color, wavelength of absorption or emission, conductivity, melting point, and catalytic behavior can change when size is altered [12]. From a more biological standpoint, smaller, nano-sized particles are often more (re)active and thus might also be more toxic than larger micro-sized particles. One study by Gilmour et al. [13] looked at the pulmonary and systemic effects of inhaling ultrafine and fine carbon black particles in adult male Wistar rats. The ultrafine carbon particles had a diameter of 14 nm, whereas the fine ones had a diameter of 260 nm [13]. The rats inhaled particle concentrations of 1 mg/m^3 for 7 hours before being sacrificed at intervals of up to 48 hours post exposure at which time the bronchoalveolar lavage fluid (BALF) was analyzed. The resulting inflammatory profile showed that more neutrophils were produced after exposure to ultrafine carbon black particles as compared to fine ones. The leukocyte number did not vary in the blood following exposure to fine carbon particles, whereas there was a significant increase of leukocytes in the blood following ultrafine exposure at zero and 48 hours post exposure.

Overall, nanoparticles seem more toxic than microparticles due to three main size factors [14]. First, nanoparticles have greater surface areas relative to mass and therefore greater ability to interact with cell membranes and deliver toxic substances. Second, smaller particles often have increased clearance times and are retained longer in the body. Third, nanoparticles can be deposited deeper in tissue compared to microparticles due to their size. Post exposure clearance is thus also slower. Such a distribution of particles after inhalation can be clearly seen in Fig. 7.1 [15]. Small nanoparticles may not be removed by the ciliated cells, but instead stay in contact with the lung epithelium for years, are taken up by the cells or alveolar macrophages, or are transported into the interstitial space. Transport into the blood stream has also been reported, although this route is still controversial. One article described less than 1% of translocation of 15–20 nm iridium nanoparticles [16], while another reported that about 3–5% of the inhaled carbonaceous nanoparticles were transported into the blood [17].

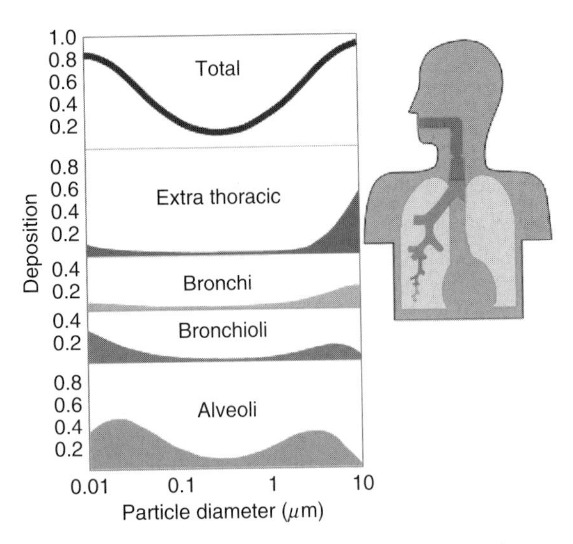

Fig. 7.1. Regional deposition of particles in the human respiratory tract after inhalation through the mouth (according to an ICRP Report [107] and adapted by Kreyling, Semmler-Behnke, and Möller [15], with permission).

Biological adverse effects seem to be driven by large and active surface areas, or in other words, as a consequence of the particles getting smaller and smaller [18]. As soon as nanoparticles turn into agglomerates, which often happens upon contact with the many proteins present in vivo or when surface properties are uncontrolled, then all (nano-)size dependent effects are lost.

One important system that also reacts differently to different particle sizes is the reticuloendothelial system (RES). The main RES cells include macrophages, monocytes, and organ-bound cells such as Kupffer cells in the liver. These cells are responsible for clearing cellular debris, aged cells, pathogens, and particles from the human circulation and tissues. The RES generally phagocytoses particles. The factors that induce this behavior are not well known. However, there is evidence from in vitro studies that nanoparticles stimulate the macrophages via reactive oxygen species and calcium signaling followed by the production of proinflammatory cytokines such as tumor necrosis factor alpha [19].

7.3.2. Particle surface properties

Surface morphology, charge, lipophilicity, and other coating properties have a large influence on the dispersibility, conductivity, catalytic behavior, and optical properties of small particles [12] and can considerably influence particle toxicity. Yin et al. [20] studied the effect of oleic acid coating on the cytotoxicity of nickel ferrite nanoparticles. Oleic acid is commonly used as surfactant in the synthesis of monodisperse nanoparticles. Its carboxyl functional group had also been used to immobilize proteins, oligonucleotides, and anticancer drugs on particle surfaces. Particles without a coating or with one or two layers of oleic acid particles were compared (Fig. 7.2). The uncoated and bilayer coated particles had a hydrophilic charged surface due to functional -OH or -COOH groups, respectively, which were exposed to the cells, while the one oleic layer coated particles were lipophilic with $-CH_3$ groups on the surface. Uncoated particles showed no significant decrease in cell viability with increasing particle mass concentration, whereas all coated particles showed a statistically significant decrease in viability. The hydrophobic particles were more cytotoxic than the hydrophilic ones because, as the study concluded, hydrophobic groups were more "accessible" to cells and therefore would cause more interaction between cells and invasive particles. In another study, oleic acid-coated magnetic nanoparticles were PEGylated using poly (poly(ethylene glycol) monomethacrylate) (P(PEGMA)) [21]. The toxicity similar to the one observed by Yin et al. disappeared once the nanoparticles were PEGylated.

Yin et al. also tested the cytotoxicity of the surfactant in the absence of particles to assess if the properties of the coating were altered by the particles. No toxicity was found in the experimental

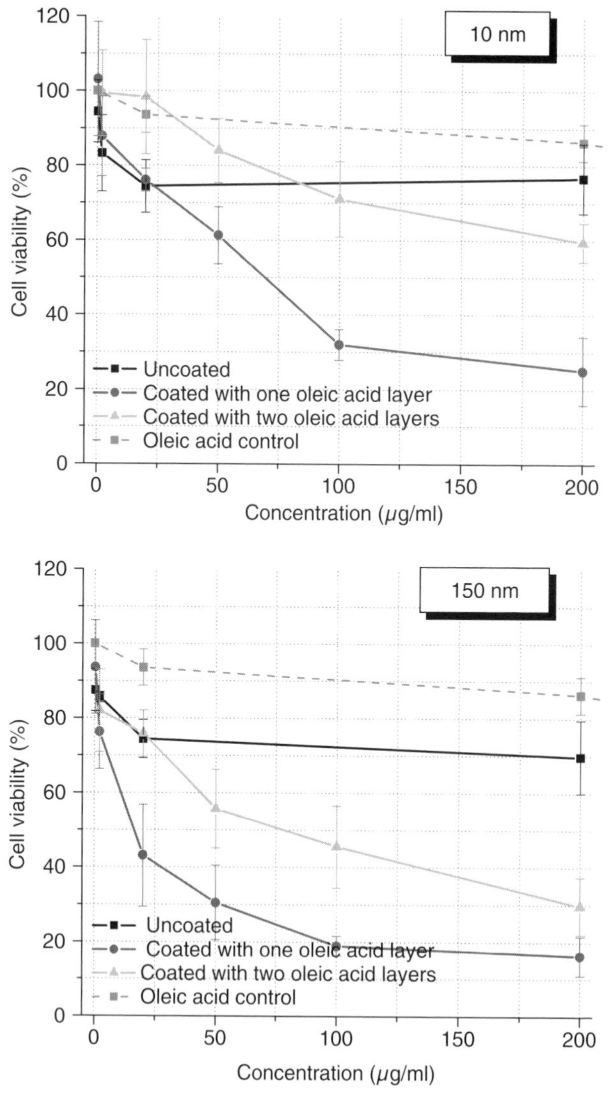

Fig. 7.2. Cell viabilities of magnetic nanoparticles before and after coating with oleic acid, redrawn from Yin, Too, and Chow [108]. The cell viability of pure oleic acid is also given for comparison.

range (Fig. 7.2). Oleic acid coated particles, however, were more cytotoxic than oleic acid alone. The authors attributed the toxic effects to conformational effects. They hypothesized that the surfactant adsorbed onto the nanoparticle surface resulted in the hydrophilic groups becoming spatially aligned and thus exerting a stronger toxic stimulus on the cells. Not only the absolute concentration of the surfactant is thus important, but also its conformation or alignment.

Other commonly used surfactants have also been tested for their toxicity [22]. Using 9 nm magnetite nanoparticles coated with starch and different surfactants, Park et al. showed that citric acid was nontoxic up to 8% of blood volume, while oleic acid was nontoxic up to 4%. Decanoic and nonanoic acid were already toxic at 0.5%. More hydrophilic coatings are generally less toxic than lipophilic ones. For example, PEGylated nanoparticles only slightly and nonsignificantly (86%) reduced fibroblast adhesion compared to control cells (100%), while uncoated magnetite nanoparticles led to a significant loss of adhesion (36%) to glass cover slips [23].

In a recent study the influence of the charge on the surface of superparamagnetic iron oxide nanoparticles (SPIONs) on the uptake in and toxicity to brain cells was investigated [24]. Amino-polyvinylalcohol (Amino-PVA) coated positively charged particles were taken up by isolated brain-derived endothelial and microglial cells at a much higher level than the other SPIONs, but in vivo only invaded the first cell layer and were well tolerated by mice in concentrations up to 3.5 mM. Neutral PVA-coated particles (30 nm), dextran-coated Sinerem® (30 nm) and Endorem® (80–150 nm) magnetic particles, and negatively charged carboxy- and thiol-PVA-coated particles (30 nm) were not cytotoxic nor did they induce the production of the inflammatory mediator Nitric Oxide (NO) used as a reporter for cell activation.

The adsorption of blood proteins to magnetic particles initiates the immune system response (see later in this chapter) and has been found to be related to the surface charge, where neutral or slightly negative surface charge decreases adsorption [25]. Charge is often expressed as zeta potential, which is the potential at the shear plane between the immobilized layer (resisting shear) and bulk solution. At constant pH 5, the zeta potential of coated nanoparticles was proportional to the basicity of the applied polymers (Fig. 7.3). The zeta potential of the particles and the basicity of the corresponding polymers decreased in the typical order from polyethylene imine (secondary amine) to chitosan (primary amine) further to the acid amide (polyvinylpyrrolidone) and finally to the primary alcohols starch and dextran [25]. Long PEG chains which are formed by a hydrated layer on the particle surface can also move the shear plane in the direction of the solution, and as a result the zeta potential decreases. Important is also the pH, the number of charged groups, or the existence of other ionizable groups like the DTPA group in Fig. 7.3. In addition to direct effects from the charged nanoparticle surface, insufficient zeta potentials might lead to the agglomeration of particles that exert toxic effects.

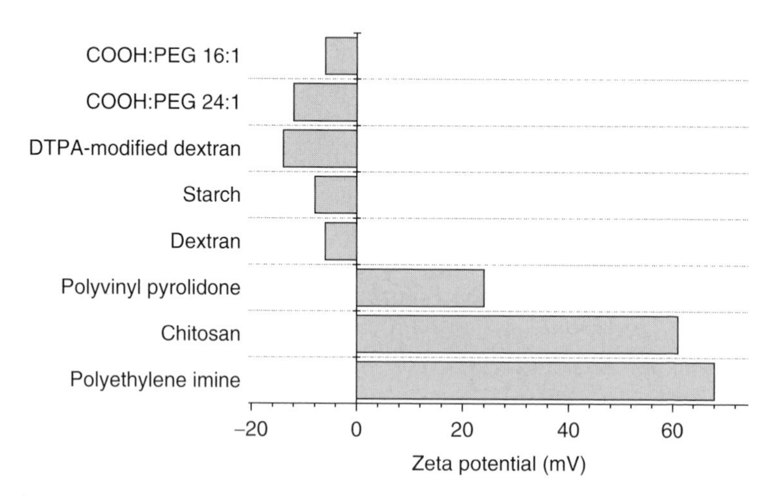

Fig. 7.3. Zeta potential of magnetic nanoparticles at constant pH (pH 8 for the first three nanoparticle coatings, and pH 5 for the next coatings). Modified from Grüttner et al. [25].

When comparing the surface properties of PLA and PLA-magnetite microspheres to those of magnetite, the magnetite particles show a distinct isoelectric point (or pH of zero zeta potential) in the vicinity of pH 7 (Fig. 7.4) [26]. In contrast, PLA microspheres always bear a net negative surface charge, and an isoelectric point below pH 2. If an efficient coverage of magnetite by PLA is achieved, it is reasonable to predict that composite particles will have an isoelectric point lower than pH 7 [26]. The electrokinetic properties of composite particles were found to be pH 5.2 and thus point to an incomplete coverage of the magnetite particle or an inhomogeneous polymer shell. Another prediction, that pH should have a smaller effect on zeta potential in the composite microspheres than in magnetite was confirmed in a later paper [27].

The investigation of particle surface properties is difficult and depends on many different parameters, especially pH, salt concentration,

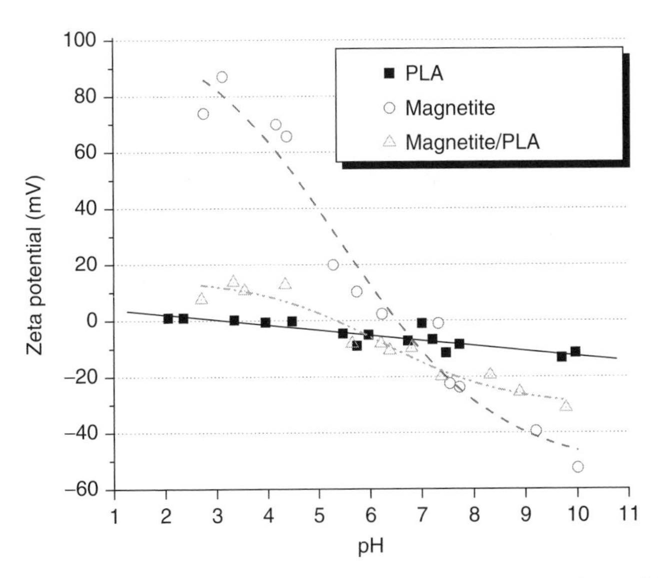

Fig. 7.4. Zeta potential of magnetite, PLA, and magnetic PLA particles as function of pH in presence of 10 mM KNO_3. Adapted from Gomez-Lopera, Plaza, and Delgado [109].

and the presence of other molecules. Besides zeta potential measurements, other useful methods include secondary ion mass spectroscopy (SIMS), X-ray photoelectron spectroscopy (ESCA), thermogravimetry, scanning electron microscopy (SEM), atomic force microscopy (AFM), and scanning tunneling microscopy (STM).

7.3.3. Promotion of oxidative processes

In aqueous solution and in contact with cells, some magnetic particles produce free radicals. This is possibly due to the metals present or the particle coating. The free radicals can then initiate oxidative processes which may go on to damage cellular components in the body or interfere with intra- and extracellular processes.

For iron, the main ferromagnetic component in magnetic particles, free radical forming processes are likely connected to Fenton chemistry, as described in a recent review of iron and oxidative stress [28]. Luckily, the human body is used to dealing with iron, since each healthy adult contains about 3–5 g of it, mainly in hemoglobin bound form. For a recent review, see Papanikolaou and Pantopoulos [29]. Normal iron serum levels in blood are 80–180 μg/dL and action levels are above 500 μg/dL, while overall toxic levels are reached above a threshold of 10 mg/kg body mass acutely added iron [30]. None of the therapeutic interventions with magnetic particles, be they cancer treatment, drug delivery, imaging, or detoxification, are expected to come even close to these iron concentrations. Toxicities that might be expected from magnetic particle therapies, however, are local effects from prolonged oxidative stress and inflammation.

Magnetic ultrafine cobalt and nickel particles were also shown to produce large concentrations of free radicals that were able to break down supercoiled plasmid strands [31] (Fig. 7.5). Drawing on a previous study, the authors linked the free radical concentration and the initiation of an inflammatory response by neutrophils postexposure to all tested particles. The neutrophil influx was

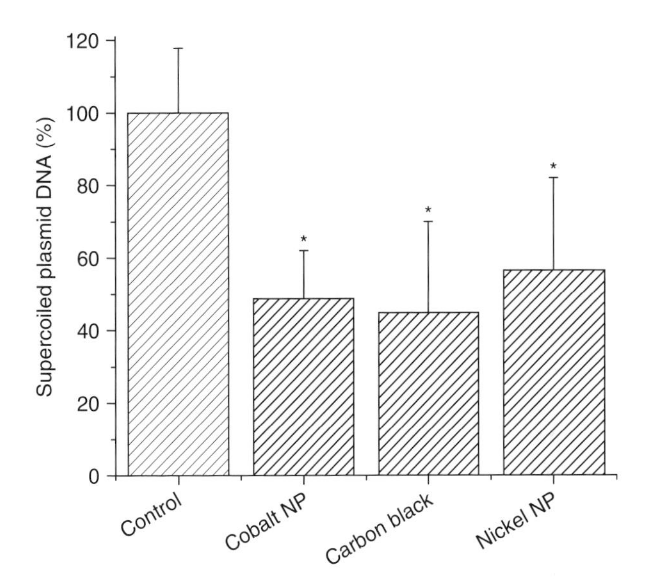

Fig. 7.5. Depletion of supercoiled DNA by ultrafine particles. The results with an asterix are statistically significant ($p < 0.05$) compared to the control. Redrawn from Dick et al. [110].

greatest with ultrafine nickel, then with cobalt and then carbon black, correlating with free radical release.

A method to investigate particle surface reactivities, such as radical or reactive oxygen species production, is electron paramagnetic resonance (EPR) combined with spin-trap measurements [32]. Other methods for investigation of lipid peroxidation and general oxidative stress after magnetic particle application are the determination of thiobarbituric acid-reactive substances (TBARS) and catalase activity [33]. Using these techniques, Freitas et al. found after intraperitoneal injection of waterbased magnetic nanoparticles a 4-fold TBARS increase in homogenates of liver and spleen, an effect which still persisted after 28 days. Catalase activity had declined initially after the injection of small amounts of nanoparticles, but had recovered after 28 days. These results might reflect an ongoing fibrotic process [33].

7.3.4. Leachables

In some cases the biocompatibility of particles can be affected by the diffusion or release of biologically active substances from magnetic particles to the surrounding tissue and the host. These substances are known as leachables and may lead to local or systemic toxicity. Primary leachables are the soluble salts Fe^{2+}, Fe^{3+}, Ni^{2+} and Co^{2+} from the magnetic particle components. Their toxicity will be discussed below. Secondary leachables come from the matrix material, additives used to prepare the magnetic particles, impurities, breakdown products, and residual solvents.

The opposite of leaching, swelling, is also known to occur if the surrounding fluid or components from the tissue is absorbed by the particle or biomaterial. Swelling more likely affects the function of the material, rather than cause toxicity to the host, due to an increase in volume resulting in a fully dense material [34].

7.4. Testing for particle toxicity and biocompatibility

Modern toxicology investigations result in a level of understanding much more advanced than that of Paracelsus [10]. Specifically, toxic effects on organs and cells are now being revealed at the molecular level, and DNA and various biomolecules that maintain cellular functions are being probed for unusual behavior. New entities such as chemicals, drugs, medical devices, and magnetic particles, therefore, now undergo a preliminary screening process to find acute reactions and other problems that will exclude them from further consideration (Table 7.1). Most of these preliminary tests, as well as the broad and involved determination of a substance or material's "toxicological profile," followed by the evaluation of the "relevant

TABLE 7.1
Overall approach to the toxicity and biocompatibility testing of new chemicals, drugs, and biomedical materials (e.g., magnetic particles) and devices

1. Preliminary Assessment	Perform initial screening Determine acute reactions
2. Toxicological Profile	Perform battery of in vitro tests to examine toxicity Broad tests, not limited to the use of the material Could lead to new uses
3. Relevance to Use	Repeat those tests which are relevant to use(s) determined in point 2 and perform some extended ones Tests should suggest what side effects are to be expected
4. In Use Assessment	Perform tests in relevant animal models chosen to mimic the clinical application of the device
5. Clinical Testing	Perform clinical phase I trial

to use properties," are performed in cell assays. The following section will review and discuss many of these in vitro assays particularly as they apply to magnetic particle toxicity investigations.

7.4.1. Toxicity testing in vitro, in vivo and in silico

Materials or particles intended for internal human use usually require testing both in vitro and in vivo. In vitro testing, which is accomplished outside of the body using cultured human cells, allows for careful study of the interactions between human cells and the biomaterial. In vitro tests provide information and help predict how the material may perform in vivo, or in the body [35]. In vitro cell culture methods testing for biocompatibility or toxicity are standard methods nowadays, even though they are known to be more sensitive to toxicity than body tissue. They also are important as an alternative to animal testing. Cell culture assays can often help to eliminate possible causes of toxicity, to define the actual mechanism by which cellular toxicity occurs and to show which cellular functions are affected, whether enhanced or harmed. In vitro methods may also be more economical than in vivo testing and more accessible to carry out multiple tests, thus at least reducing the number of animal tests needed later. Each method has not only advantages, but also disadvantages. For in vitro cell culture tests, disadvantages include the lengthy assays (often 3 to 5 days), the potential for contamination, and the difficulty of keeping the conditions constant (some cell lines only grow in a constant fashion for a certain number of passages).

In vivo animal tests have the advantages that overall biological systems are probed, while in vitro tests allow determining the mechanism of action. In vivo testing is routinely carried out during the preclinical trials of a drug or product, before it is released for human trials. This enables us to see if the product is active in vivo, how it affects the rest of the body, and if there are side effects from break down products. Disadvantages of animal testing include its cost as well as the fact that no two animals are identical and results therefore may vary. Sometimes, the materials must be tested in more than one species. Since ethical guidelines have become an important argument against the use of animals due to wastage and cruelty, it is important to keep the 3 R's in mind–replacement, reduction, and refinement [36].

As standards have now been produced for in vitro testing, its use has become important for initial toxicity tests and for the elimination of toxic factors before in vivo testing is done. The number of animal tests can thus be considerably reduced and overtime possibly abolished.

Another form of toxicity testing which is becoming increasingly popular is in silico testing. It refers to "computer simulations that mimic biological processes." [37] Although this method is not routinely integrated in toxicology assessments, it is becoming a useful technique of looking at the toxicity of drugs even before their synthesis during drug discovery. As nano- and microparticles have very different properties to bulk materials, it is not possible to use in silico testing alone to predict toxic properties and in vitro experiments are still required for assessing the particle toxicity.

7.4.2. In vitro biocompatibility and toxicity testing procedures

7.4.2.1. Cell culture assays for biocompatibility and toxicity testing

The biocompatibility of magnetic particles is usually initially determined using in vitro cytotoxicity tests. These tests look either at proliferation or morbidity; cell viability or death. Cytotoxicity can be reversible or irreversible, the latter of which leads to apoptosis (programmed cell death). Toxic effects compromise a number of cell functions, of which altered membrane porosity is the easiest to detect. Weakened membrane function can be detected by watching altered exclusion/uptake behavior of dyes or by measuring the changed release of substances.

Many biochemical cytotoxicity assays that exploit altered membrane behavior or the biochemistry of apoptosis have become available on the market. They differ in a number of ways including cost, ease of use, number of procedural steps to perform, time to obtain results, and suitability for the specific cell type or the toxic product being tested. Some of the most popular assays will be discussed here in more detail.

7.4.2.2. Assays measuring metabolic activity

Cytotoxic materials often impact the metabolic activity of cells, which is mainly expressed as mitochondrial function. The most prominent assay to determine mitochondrial damage includes a group of assays that uses different tetrazolium salts such as MTT [(3-(4,5 dimethyl thiazoloyl-2)-2,5-diphenyl tetrazolium bromide] [38], MTS [3-(4,5-dimethylthiazol-2-yl)-5-(3-carboxymethoxyphenyl)-2-(4-sulfophenyl)-2H-tetrazolium], and XTT [sodium(2,3-bis (2-methoxy-4-nitro-5-sulphophenyl)-2H-tetrazolium-5-carboxanilide] [35]. The tetrazolium salts are initially yellow and able to penetrate cell membranes. They therefore also easily penetrate the mitochondrial membranes. Mitochondrial activity then metabolizes the yellow dye to blue insoluble formazan (Fig. 7.6) which is trapped in the healthy cells and appears as huge crystals (Fig. 7.7B and E). As

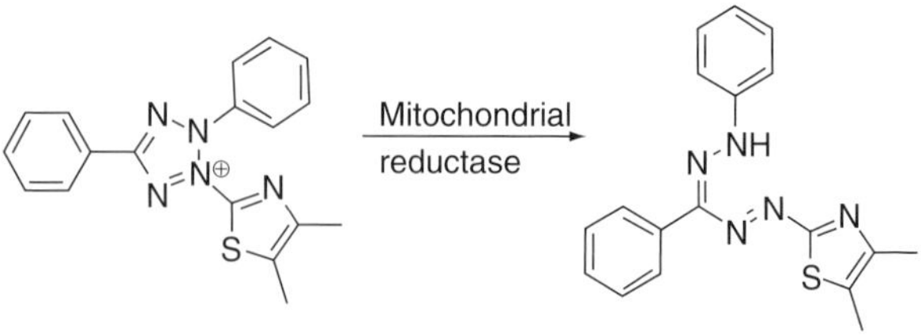

Mitochondrial reductase

Fig. 7.6. Diagram showing the reduction of the yellow dye MTT to blue formazan crystals by the cleavage of the tetrazolium ring [38].

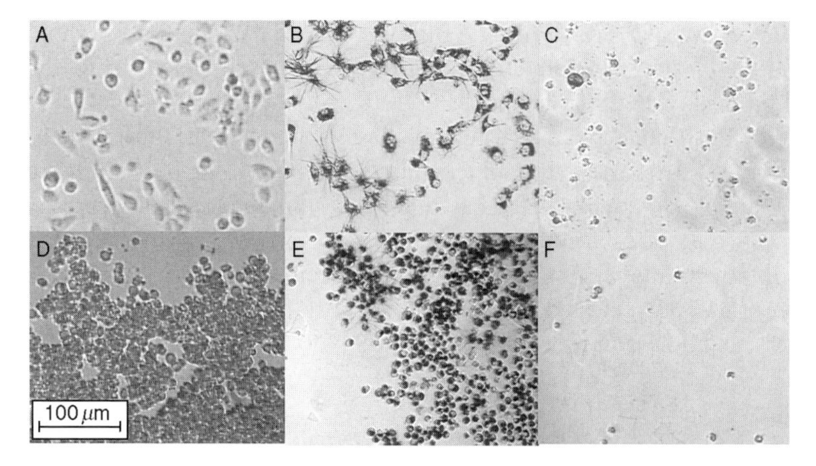

Fig. 7.7. (A–C) Adherent human prostate cancer cells (DU-145) and (D–F) murine suspension lymphoma cells (EL-4) growing in a 96-well plate immediately before (A and D) and 3 hours after (B and E) the addition of MTT. The formation of crystals within the healthy cells can be clearly seen. Cells in panels C and F were treated with 10 mg/ml $FeCl_3$ 3 days earlier and show the highly cytotoxic effects of such iron concentrations [39].

dead cells lack metabolic activity, the amount of formazan produced is directly proportional to the number of viable cells. After 3 hours, the crystals can be solubilized and the color quantified. Figure 7.7 shows adherent human prostate cancer cells DU-145 and murine lymphoma suspension cells EL-4 before and after the addition of MTT [39]. Incubation of the cells with $FeCl_3$ solutions showed no cell viability loss up to 0.005 mg/ml. Iron concentrations of 1 mg/ml showed an EL-4 cell viability of $80 \pm 20\%$, while DU-145 cells were more susceptible to iron with $45 \pm 25\%$ cell viability. Concentrations higher than 5 mg/ml iron resulted in no surviving cells in both cell lines (Fig. 7.7C and F).

The rationale behind performing an MTT assay for rapidly evaluating cell viability, cell survival, and cell growth is its ease and reproducibility. Some drawbacks include that nondividing cells may still be viable and thus show mitochondrial activity. Furthermore, the conversion of MTT to formazan is not only driven by mitochondrial dehydrogenases but can also be influenced by other enzymes, metabolic factors, and glucose consumption [35, 40]. For these reasons, the MTT assay should be regarded as a marker of metabolic activity and not only of mitochondrial action.

Any extreme results in the MTT assay (<40% cell viability) point at serious toxicity and should be investigated with other methods such as the agar overlay technique [41], histopathological observations, morphological studies such as the direct contact method where cells grow directly on nanoparticles or on films made from the particles [42], cytogenetic analysis, micronuclei assay (e.g., Lacava et al. [43], Liu et al. [44], and Sadeghiani et al. [45]), evaluation of genotoxicity using the Ames assay, and the chromosome aberration assay [46]. Another alternative method is the determination of the LD_{50}, although it is being replaced evermore by the tests described in this chapter. The LD_{50} of dextran-coated magnetite nanoparticles after intravenous or intraperitoneal injection has been determined as 5.0 g/kg in C57Bl/6 mice, 0.6 to 1.5 g/kg in rabbits, and 0.5 to 0.9 g/kg in dogs, while ferro-carbon nanoparticles have an LD_{50} of 0.7 g/kg in BALB/c mice [47].

The MTS and XTT assays are newer versions of the MTT assay and do not require the final solubilization step which many studies have shown to produce errors in results [48]. The solubilization step required for the MTT assay destroys the cells under investigation and therefore only allows for a single time-point measurement. MTS and XTT are water soluble formazans and therefore allow continuous monitoring of the cells without having to destroy them. However, unlike the MTT assay they both require the addition of an electron acceptor such as phenazine methosulphate (PMS) for reduction and formazan formation. In a comparison between MTS and XTT assays, the XTT/PMS reagent mixture was found to be unstable, resulting in PMS depletion over time and reduction of formazan formation [48]. The resulting response errors and poor precision were not encountered with the use of MTS/PMS mixtures. For the analysis of higher concentrations of strongly colored (dark) magnetic nanoparticle suspensions, however, it is sometimes necessary to add washing steps to the cell viability measurements. The MTT assay is in this case the preferred method.

A special redox assay is the Alamar blue™ assay in which the soluble blue dye resazurin is converted, by cells, to the pink fluorescent dye resorufin. The advantages of this assay include the water solubility of the dye, easy handling, and high sensitivity [49]. Incubation times are normally slightly longer than for MTT assays, however, its low toxicity makes this assay useful not only for final, but also for serial kinetic measurements. The excitation wavelength is 544 nm and the emission wavelength is 590 nm.

Another assay measuring metabolic activity is the adenosine triphosphate (ATP) assay. ATP is a biomolecule and has a crucial role in cell metabolism (energy source) as well as neurotransmitter. Cell damage or toxicity often causes a depression of the ATP metabolism. The ATP assay can then assess cell toxicity, as well as cell proliferation and cell number [50]. The metabolic activity is determined by bioluminescence measurements in which the enzyme luciferase catalyses the formation of light from ATP and luciferin. The light intensity at 560 nm correlates linearly with the ATP concentration [51]:

$$\begin{aligned} \mathrm{ATP} + \text{d-Luciferin} + \mathrm{O_2} \rightarrow \mathrm{Oxyluciferin} + \mathrm{AMP} \\ + \mathrm{Pyrophosphate} + \mathrm{CO_2} + \mathrm{Light} \end{aligned} \tag{7.1}$$

Many studies have looked at the usefulness of the ATP assay in comparison to other cytotoxicity assays. These studies have shown the ATP assay to be more sensitive than MTT, MTS, Neutral Red (NR), and the lactate dehydrogenase (LDH) assay [52, 53]. Compared to the LDH assay (see Section 7.4.2.3), the ATP assay is more sensitive. This, in turn, also has some disadvantages. The luminescence readout can be affected by sample quenching, which can be a significant problem when magnetic particles are present. In addition, luminescence intensity is time dependent. Despite these disadvantages, the ATP assay is one of the quickest and most efficient methods, requiring only 10 min of incubation time before readout.

A different type of assay measuring metabolic activity is the cytochrome C assay which detects extracellular release of superoxide radical anions [54]. Extracellular radicals reduce 100 μM ferricytochrome (cytochrome^{3+}) to cytochrome^{2+} that can be measured at 550 nm. Although Simko et al. [54] showed, with this technique, radical formation only in the presence of ultrafine carbon black particles of 12–14 nm diameter, it may also be useful for magnetic particle investigations.

7.4.2.3. Assays involving membrane integrity

The analysis of cell membrane integrity is another good marker for toxicity, since damaged cell membranes lose their structure and become more permeable, allowing the contents of the cell to leak out [53]. Lactate dehydrogenases are enzymes present in the cytosol of the cell and can only be measured extracellularly when membrane damage occurs. This is known as the LDH assay. The assay has evolved from its original method, which was simple and cheap but extremely time consuming, to a coupled assay whereby LDH catalyses the conversion of 2-*p*-(iodophenyl)-3-(*p*-nitrophenyl)-5-phenyltetrazolium chloride (INT), a tetrazolium salt, to a red formazan that can be quantified by optical absorbance, much like the concept of MTT:

$$\text{Lactate} + \text{NAD}^+ \rightarrow \text{Pyruvate} + \text{NADH} \qquad (7.2)$$

$$\text{NADH} + \text{INT} \rightarrow \text{NAD}^+ + \text{Formazan(red)} \qquad (7.3)$$

The LDH assay is best used when a toxic agent damages the cell membrane but is less sensitive and sometimes misleading when the toxic agent affects the cell intracellularly [53]. The LDH assay is an indicator of irreversible cell death due to membrane damage and is reliable, but is the least sensitive when compared to the tetrazolium assays [55]. The assay is not ideal for the detection of a threshold concentration where the toxicity starts to show but will reliably determine the concentration at which cells can no longer survive. Incubation times of the assay vary between less than an hour and up to more than 4 hours [56]. Weissleder et al. [57] used the LDH assay to determine the viability of lymphocytes after magnetic labeling for trafficking studies.

Another group of cytotoxicity assays involves the use of dyes that are able to cross the cell membranes and stain the cell. Depending on the nature of the stain, it can either be taken up or excluded by viable cells. If normally excluded, it will cross the membrane of dead or dying cells. Popular dye assays include the NR (3-amino-m-dimethylamino-2-methylphenazine hydrochloride) uptake assay or the trypan blue exclusion assay [35].

The Neutral Red assay can be used to quantify the number of viable cells [53]. This vital dye is taken up via active transport into the cell, accumulates in the lysosomes and can then be spectrophotometrically quantified after releasing the dye with acidified ethanol. Incubation times range from 2–4 hours. The NR assay is sensitive to a wide range of materials, is reliable compared to the MTT and calcein fluorescence assay, and is not influenced by occasional microbial contamination which would usually lead to inaccurately higher cell viability measurements. A disadvantage of the NR assay involves the fixation step that is required before measurements can be read. Cells that do not completely adhere or are loosely attached have a questionable viability status, but are still fixed and included in the final reading, possibly leading to inaccurate cell viabilities [35].

As a complement to the NR assay, trypan blue, erythrosine B, propidium iodide, ethidium homodimer, and 7-amino-actinomysin (7-AAD) assays are exclusion assays which only stain dead cells after crossing their damaged cell membranes [35]. The trypan blue exclusion assay can be used to measure cell viability and is the most popular method for cell counting. Limitations of this technique, where dead cells are stained blue, include the method being slow, problems with incomplete resuspension of adhering cells, and inability to stain cells undergoing autolysis. In addition, the counting needs to be done within 5 minutes of adding trypan blue in order to avoid getting too high a yield of stained cells. In general, the trypan blue assay should only be used to check overall cell viability or confirm the results of other toxicity assays [58]. Another exclusion assay, the propidium iodide dye assay, stains the nucleic acids of cells. It may be used simultaneously with the NR assay, since each assay works via different mechanisms and stains different cells. The resulting two-colored cell populations can easily be counted microscopically or by fluorescence activated cell sorting (FACS). In general, fluorescence-based assays are popular methods of assessing cytotoxicity. They are thought to be more sensitive than colorimetric assays and can be analyzed in plate reader format, by fluorescence microscopy and confocal microscopy, and also in FACS. Probably the best known among them is the calcein AM assay in which the soluble nonfluorescent dye enters live cells with intact membranes, and then undergoes an enzymatic conversion to an intensely green fluorescing product inside the cell. The cleaved product is the lipophilic acetoxymethyl (AM) blocking group, producing charged fluorescent calcein [59]. The calcein AM assay is often combined with other fluorescence-based assays that label dead cells instead of viable cells, such as the Live/Dead® assay by Invitrogen. It incorporates the calcein AM assay with an ethidium homodimer assay [60] (Fig. 7.8). Ethidium homodimer is a hydrophilic dye that is able to enter cells with damaged membranes and bind to nucleic acids, producing bright red fluorescence. The use of the Live/Dead® assay has successfully been used to investigate the impact of pancreatic islet labeling with Resovist® superparamagnetic particles and Dynabeads® in vitro [61]. The Live/Dead® assay confirmed that islet cells cultivated for 48 hours with Resovist or Dynabeads did

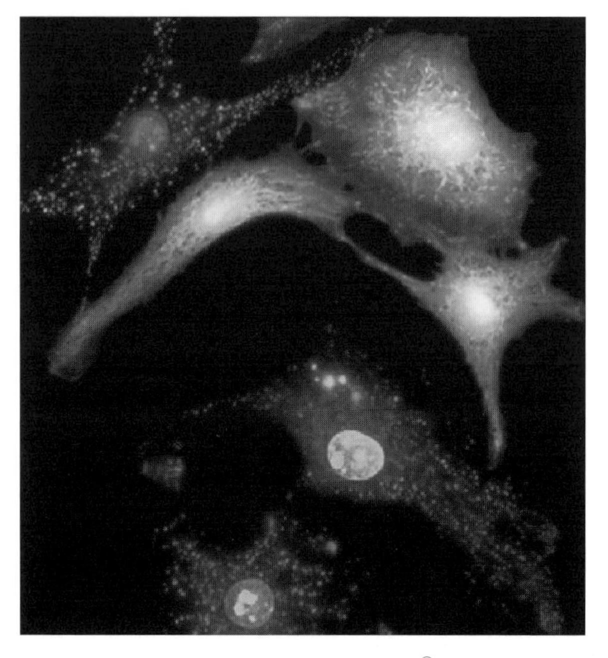

Fig. 7.8. Epithelial cells stained with the Live/Dead® assay. Live cells fluoresce bright green, whereas dead cells with compromised membranes show red-fluorescing nuclei. (See Color Insert.)

not differ from control cells (90.8 ± 1.9% for Resovist, 89.4 ± 2.0% for Dynabeads, and 91.1 ± 2.5% for control cells). Such labeled islet cells were then used to image the success of pancreatic cell transplantation in diabetes patients using MRI.

7.4.2.4. Assays to detect apoptosis

Different methods are available to detect apoptosis. One method includes the detection of the cellular caspases, which are activated once the cell decides to undergo apoptosis. The increase of caspase activity is detected by a highly sensitive luminescent assay for caspases 8, 9, or 3/7 (Caspase-Glo® assays; Promega) or by a fluorescent assay for caspases 3/7 (Apo-ONE® assay; Promega). Apoptosis can also be detected by a Terminal transferase dUTP nick end labeling (TUNEL) assay, which identifies the presence of oligonucleosomal DNA fragments later in the cell death process. Readout is possible by fluorometric or colorimetric end points (e.g., DeadEnd™ Promega).

Another common way to determine apoptosis is FACS using FITC-labeled annexin V which binds to phosphatidylserine at the surface of dying cells. Schwalbe et al. (unpublished results) prepared gold-coated magnetic nanoparticles and then tested their long-term toxicity after adding them to K562 chronic myeloid leukemia cells. Over 8 days, the gold-coated nanoparticles did not slow cell growth compared to control cells, both of which multiplied by about 100×. Gold particles coated with the chemotherapeutic tyrosine kinase inhibitor drug imatinib, however, were significantly cytotoxic (Fig. 7.9A). FACS results, in combination with 7-AAD exclusion, confirmed that cell death was by apoptosis (Fig. 7.9B–D).

The single cell gel electrophoresis assay known as the Comet assay is a simple, rapid, and sensitive technique for analyzing and quantifying DNA damage in individual mammalian cells. Initially, there was much discussion about whether the assay could really distinguish apoptosis from necrosis, a topic which has been resolved in recent years with an improved protocol. The Comet assay can be performed by coating a glass slide with agarose and adding the test cells in a separate agarose layer on top [62]. Using this approach, successful magnetic hyperthermia treatment was confirmed after the direct injection of magnetic nanoparticles into an MX-1 tumor and application of an alternating magnetic field (frequency 400 kHz and amplitude 6.5 kA/m) [63]. Tumor cells treated with magnetic nanoparticles showed prominent DNA damage visible as distinct comet trails (see Fig. 7.10).

7.4.2.5. Assays that follow the intracellular fate of magnetic particles
An old and essential science and useful tool of biology is histopathology, the study of thin slices of diseased tissue. Histopathology, which can be described as microscopic anatomy, is a good first

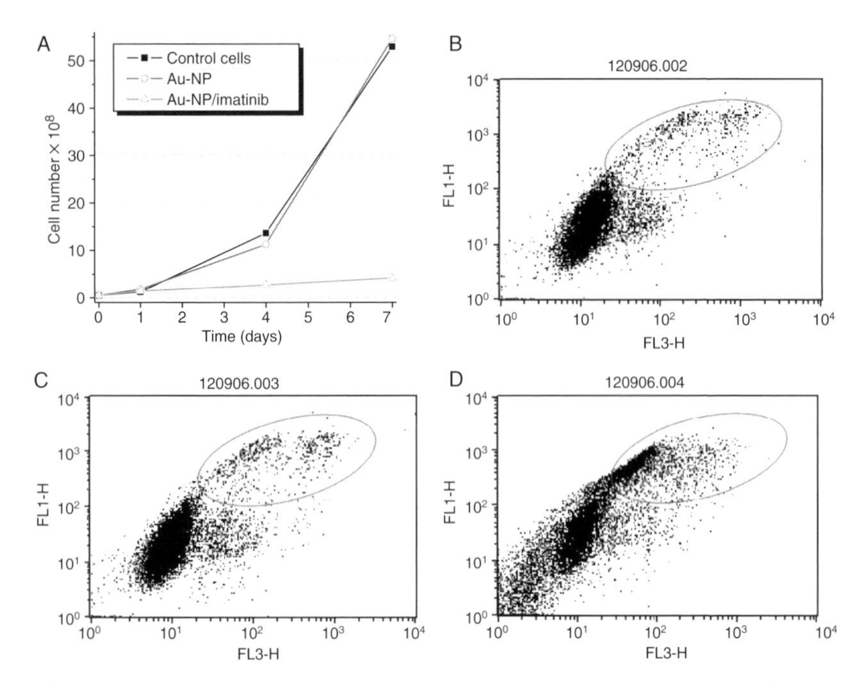

Fig. 7.9. (A) Cell growth curves of K562 leukemia tumor cells (control), with added gold-coated magnetic nanoparticles (NP) and with Au-NP additionally containing the chemotherapeutic drug imatinib. (B–D) FACS analysis after 1 day of incubation confirmed that Au-NP did not influence cell growth (5.8% gated apoptosis vs 5.9% in control cells), while drug-coated Au-NP induced significant apoptosis (29.3%) unpublished results by Schwalbe et al., Friedrich-Schiller-University, Jena, Germany, with permission.

choice in the evaluation of toxic reactions of cells to biomaterials. Early on, it was discovered that magnetic nanoparticles are avidly taken up by fibroblasts, macrophages, and tumor and many other cells. In an extensive study, the uptake kinetics of different magnetic nanoparticles coated with dextran or aminosilane were observed [64]. Depending on the cell type and the coating of the magnetic nanoparticles, very high iron concentrations, of up to 500 pg per tumor cell, were reached. Particle uptake begins immediately after their addition to the cells by endocytosis. The particles encapsulated in phagosomes can be observed by transmission electron

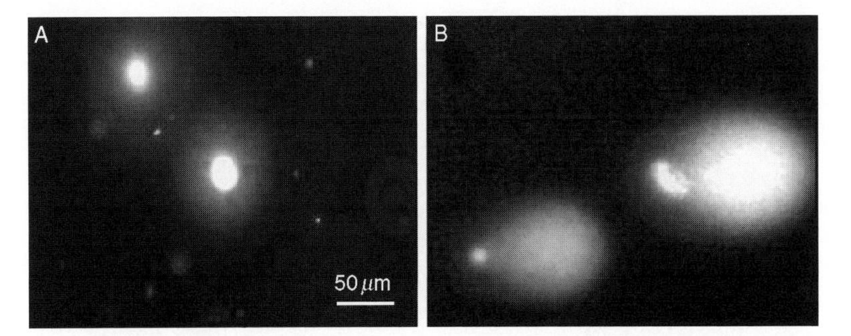

Fig. 7.10. Fluorescence images of nuclei from MX-1 tumor cells after exposure to an alternating magnetic field (magnetic thermoablation). (A) Nuclei of control cells without application of magnetic nanoparticles. (B) Cell nuclei with application of magnetic nanoparticles showing DNA damage (comet tails) after reaching 72°C during the treatment [111], reprinted with permission.

microscopy (TEM) (Fig. 7.11). Despite the enormous amounts of magnetic nanoparticles inside the phago- and lysosomes, no toxicity was noticed. The uptake of magnetic nanoparticles has also been followed by confocal microscopy (Fig. 7.12). Within minutes, magnetic nanoparticles show up inside the cell plasma. In healthy cells, they are never found inside the nucleus.

Internalization into cells seems to be a saturatable process. Using microspheres with up to 50% magnetite encapsulated into the biodegradable polymers poly(lactic acid) (PLA) and poly (lactide-co-glycolide) (PLGA), some cytotoxicity to granulocytes was found at higher concentrations [65]. Interestingly, once an intracellular particle concentration of 0.5% had been reached, no additional increase in toxicity was observed, even after further increasing the particle concentration. Müller et al. hypothesized that cell toxicity was only conferred after internalization into the cells. At 0.5%, the cells are filled with particles (maximum uptake), and are then slowly degraded in the lysosomes. The shorter the polyesters are, the faster the degradation and the lower the cell viability. This saturation hypothesis seems feasible, since no extracellular effects were found from additionally added nanoparticles.

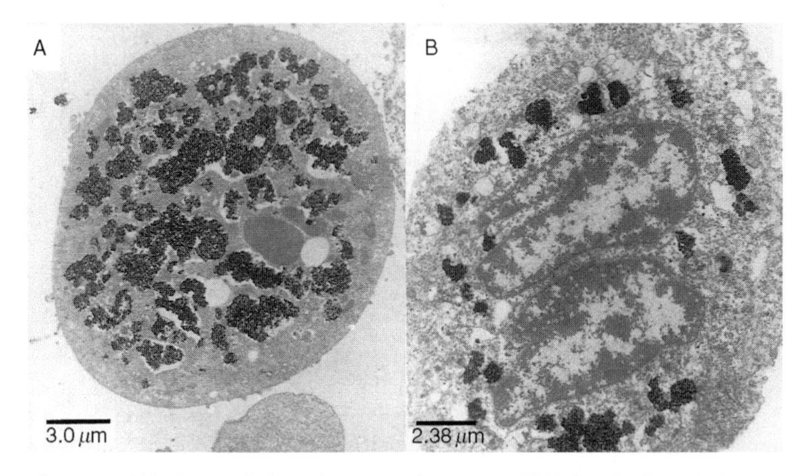

Fig. 7.11. (A) Transmission electron microscopy (TEM) of adherent human BT20 mammary tumor cells grown for 7 days in medium containing 0.6 mg/ml aminosilane-coated magnetic nanoparticles of 13 nm diameter. (B) During cell division, the particles contained in the phagosomes are equally distributed to both daughter cells, and no cytotoxic effects are visible. With permission from Jordan et al. [64].

Furthermore, Müller et al. [65] confirmed particle internalization into the granulocytes by labeling the particles with luminol, a chemiluminescent dye, which nicely correlated with intracellular iron uptake.

7.4.3. *In vivo biocompatibility and toxicity testing procedures*

Before treating patients or performing a clinical trial, magnetic particles must be evaluated in animal studies. The first magnetic particles to have undergone toxicity tests and that have received FDA approval are superparamagnetic iron oxides (SPIO's) for use as MRI contrast agents. Other magnetic micro- and nanospheres were approved for clinical trials of cancer agents, but are not fully approved for patient use yet.

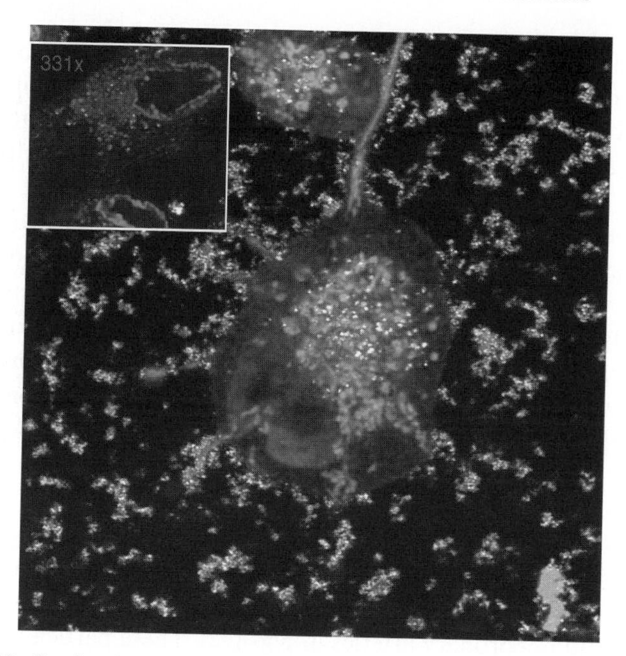

Fig. 7.12. Confocal microscopy of prostate cancer C4–2 cells after incubation for 8 hours with cobalt nanoparticles. Large amounts of the particles are taken up into the cell plasma, some are also seen adhering to the outside. The insert shows the same cells after incubation with magnetic nanoparticles coated with a cytotoxic copolymer. The cells are dying, as seen by the nuclei showing the typical picture of condensing chromatin while undergoing apoptosis. (See Color Insert.)

7.4.3.1. In vivo evaluations of particles to be used as MRI contrast agent

The early toxicity animal tests for the approval of MRI contrast agents were done with dextran-coated magnetic nanoparticles. The particles were entirely taken up by the reticuloendothelial cells, with no apparent acute, subacute, nor mutagenic toxicity for particle amounts of up to 250 mg of iron per kilogram of rat or 3 mmol Fe per kilogram of dog or rat [57, 66–68]. These properties made the magnetic nanoparticles ideal liver cancer imaging agents. No evidence of hepatic mitochondrial or microsomal lipid peroxidation or organelle dysfunction was found. The magnetic particles seemed to be partially degraded, started to show up in red blood cells within 6 weeks, and then slowly cleared from the liver within about 3 months, as shown by radiotracer studies with ^{59}Fe [57, 67, 69]. Feridex (Endorem^{TM}) was then tested in humans and described as safe and efficacious [70]. The most frequent adverse reaction was back pain. It occurred in nine patients (4%) and necessitated interruption of the infusion of ferumoxides in five of these. Although lumbar pain has been associated with administration of a variety of colloids and emulsions, the physiologic causes are unknown, since no significant changes in chemistry values, vital signs, and electrocardiographic findings were found. Morphological studies confirmed the low toxicity of dextran-coated magnetic nanoparticles. Even 200× higher doses than used for MR imaging only altered the histology of spleen and liver in a minor way [71].

Very small magnetic nanoparticles with citrate coating can also be used for the imaging of thoracic and abdominal vasculature in rats and of coronary arteries in pigs [72]. These particles showed a good tolerance and safety profile with an $LD_{50} > 17.9$ mmol Fe per kg. Another application, lymph node imaging, works well with ferumoxtran-10, an ultrasmall magnetic nanoparticle [73]. Toxicity was observed only at very high exposure levels and was linked to a massive iron overload after repeated injections. The contrast agent was not mutagenic but teratogenic in rats and rabbits. In view of its proposed use as a single-dose diagnostic agent for human MR imaging of lymph nodes, its seems satisfactory. Another application of partially dextran coated SPIO's (ferumoxide), completely dextran coated USPIO's (ferumoxtran-10), or USPIO's coated completely with a semisynthetic carbohydrate (ferumoxytol) is brain and intracerebral tumor imaging [74]. Histology studies showed no pathological brain cell or myelin changes, making these particles candidates for brain imaging contrast agents.

Recently, different MRI contrast agents (SPIO's) have been used for magnetic cell labeling ex vivo. The magnetically labeled cells were reinjected into patients for cellular imaging of cell therapy following transplantation, transfusion, or gene therapy [75, 76]. The nontoxicity of this approach was determined beforehand by measuring the iron uptake of the cells (up to 6 pg/cell), cell viability by MTT assay, apoptosis by annexin V assay, and reactive oxygen species (ROS) using the fluorescent probe CM-H_2DCFDA [77]. Magnetic cellular labeling with the magnetic nanoparticle complex had no short or long-term toxic effects on tumor or stem cells.

For stem cell therapy, the extracted stem cells must be purified from tumor cells before reinjection. A very efficient method to do this is magnetic separation (see other chapters in this book). The nontoxicity of the magnetic particles used for this separation approach was evaluated by at least one group [78]. They compared the well established Dynabeads to 16 different small magnetic nanoparticles. In their test, porous silica beads of 250 nm size fared best in terms of yield, purity, and viability of isolated epithelial tumor cells.

7.4.3.2. In vivo evaluations of particles to be used as cancer agents
Most in vivo toxicity tests of magnetic particles as carriers of chemotherapeutic drugs, hyperthermia, or radiation were performed in rodents, and only one in swine.

One of the first well defined magnetic microspheres designed for intravascular chemotherapy application were magnetite-filled albumin microspheres of 1–2 μm in size [79]. In preparation of clinical trials, the investigators injected intravenously 0.04–400 mg/kg of the drug-free carrier containing 20% Fe_3O_4 into BDF1 mice. At the highest dose (400 mg/kg), 2 acute deaths occurred in a group of 10 animals. On histopathological examination, these deaths were due to pulmonary embolization. This appeared to result from clumping of the microspheres, which occurs when they are suspended at extremely high concentrations. No other major toxic signs or findings were observed in the entire 90-day study.

Another type of microspheres, biodegradable magnetic PLA microspheres, was tested in a rat model in preparation of using them clinically to treat intraspinal glioblastoma by delivering local radiation [80]. After the injection of 0.5 mg of microspheres through an intrathecal catheter into 12 rats, all survived for 15 months with no apparent side effects or weight loss. Histological examinations at the end of the experiment showed no necrosis or chronic inflammation, making this type of magnetic particles biocompatible.

To evaluate the toxicity of smaller magnetite nanoparticles of 100 nm in diameter as an adequate carrier fluid (ferrofluid) for cancer therapy, this ferrofluid was injected intravenously into mice [81]. Injecting increasing volumes of a 1.5 weight% ferrofluid, it was found that injection of 0.1, 1, and 5% of the estimated blood volumes of the animals resulted in no obvious toxicities and all animals survived for 12 weeks, the length of the experiment. Although the particle size of these ferrofluids was higher than that of the MR contrast agents, no particle accumulation in the lungs or any sign of respiratory problems was seen. The use of very high ferrofluid volumes of 10 and 20% of the blood volume, however, led to a mortality of 20 and 25% of the animals, respectively. Such ultrahigh ferrofluid amounts resulted in acute iron overload, lethargy, and discoloration of the entire animal for about a week, symptoms that have been described earlier [82]. Using the lower ferrofluid concentrations, Lübbe et al. adsorbed the anticancer drug epirubicin to the magnetic nanoparticles and tested them clinically in locally advanced tumors [83].

Another type of magnetic particles to undergo toxicity testing was made from 10–50 nm magnetite and polyalkylcyanoacrylate able to adsorb the chemotherapeutic drug dactinomycin [84]. When tested in mice, no toxicity of the magnetite suspension was found, and the toxicity of magnetic and nonmagnetic polyalkylcyanoacrylate microspheres was virtually identical with an LD_{50} of 245 and 242 mg/kg, respectively.

The only reported toxicity test in swine was performed with carbon-coated iron particles of 1 to 5 μm in diameter (MTC) [85]. These particles were infused into the hepatic artery of pigs using 45 ml of a 10% mannitol/0.5% carboxymethylcellulose solution while a magnet was placed above a distinct target area. All three animals that received 225 mg of iron particles survived the 29-day-long experiment and showed normal weight gain of 2.8 ± 0.9 kg. Histologic examination of the liver showed a few areas of hepatic lobule necrosis which may have been due to the incomplete extravasation of MTC particles out of the hepatic artery and subsequent blockage of the hepatic artery branches, concurrent with occlusion of the portal vein due to portal fibrosis and inflammation. Necrosis as well as bile pigment, peribiliary fibrosis, neutrophilic inflammation of bile ducts, and bile duct rupture was observed only in groups receiving 75 mg of MTC particles or greater.

Concluding from all these in vivo toxicity tests, the relatively small amounts of magnetite contained in the micro- and nanoparticles necessary for magnetic targeting, cancer and hyperthermia therapy are benign, biocompatible, and seem appropriate for clinical use. The coating and matrix materials used to prepare the magnetic particles, however, are often more of a concern, determine largely the overall toxicity of the particles and must be examined in detail.

7.5. Immunogenicity of biological targeting reagents

The immune system of mammals is composed of a complex constellation of cells, organs, and tissues, arranged in an elaborate and dynamic communications network and equipped to optimize the response against invasion by pathogenic organisms. The immune system is, in its simplest form, a cascade of detection and adaptation, culminating in a system that is remarkably effective at spotting foreign substances and reacting to them. The more different a substance is from our own tissue components, the more likely it will elicit an immune response. This principle causes problems when extrapolating animal test results to the human situation as humans may respond to foreign materials in a different way than animals and vice versa. Animal testing may not detect all substances which can cause sensitivity reactions in humans.

7.5.1. Immune response system elements

In general there are two broad categories of lymphocytes (white blood cells) involved in the human body's immune system (Fig. 7.13). The first category consists of the large granular lymphocytes, better known as natural killer (NK) cells. These cells bridge the adaptive immune system with the innate immune system as they do not require activation in order to kill cells. The second category contains small lymphocytes and is subdivided into B- and T cells.

B cells are produced in the bone marrow of most mammals and play a major role in humoral immunity as the principle function of B-cells is to make antibodies against soluble antigens. These antibodies are located on the surface of the B cells and react with antigens. Each B cell is programmed to make one specific antibody.

T lymphocytes constitute the "cellular" arm of acquired/specific immunity by controlling the acquired immune response and serving as crucial effector cells through antigen specific cytotoxic activity and the production of soluble mediators called lymphokines (the cytokines produced by lymphocytes). Like B cells, T cells express a clonal antigen-specific receptor which is similar to an immunoglobulin.

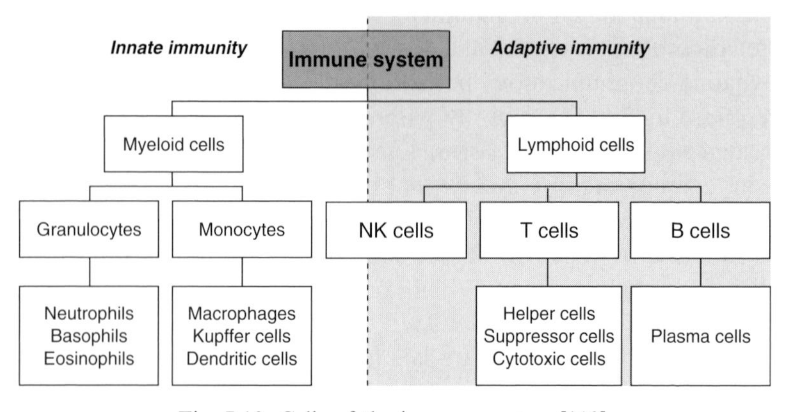

Fig. 7.13. Cells of the immune system [112].

Identifying B- and T cells is not possible by morphology, but can be done through identification of surface markers. For the B-cells, these are IgM and IgD immunoglobulins, while all T cells contain the cluster differentiation markers (CD markers) CD3, and most also CD2. T-helper cells also express CD4, while cytotoxic T cells express CD3 and CD8 but not CD4. The general health of the immune response is determined by drawing a blood sample and performing four tests: a count of the total number of white cells and lymphocytes, and a determination of the fraction of cells expressing immunoglobulins and CD3 [86].

7.5.2. *Types of hypersensitivity reactions*

The consequence of the adaptive immune system recognizing a biomaterial as foreign is a hypersensitivity reaction. Such reactions to magnetic particles are of concern since they may seriously harm a patient. A substance stimulating an immune response is called an antigen and is usually a protein or polysaccharide, but can be any small molecule or metal coupled to a carrier-protein. The overall mechanisms and manifestations of hypersensitivity reactions were classified at the end of the 1960s into four major types [87] (Table 7.2).

Type 1 hypersensitivity is an allergic reaction provoked by reexposure to a specific antigen. Exposure may be by ingestion, inhalation, injection, or direct contact. The reaction is mediated by IgE antibodies, which are usually attached to the surface of specialized cells in the tissue, and produced by the immediate release of histamine, tryptase, arachidonate, and derivatives by basophiles and mast cells. The reaction may be either local or systemic. Symptoms vary from mild irritation to sudden death from anaphylactic shock.

In type 2 hypersensitivity, the reaction is similar in end result to that of type 1 but the mechanism is different as the antibodies produced by the immune response bind to antigens on the patient's own cell surfaces. IgG and IgM antibodies bind to these antigens to

Table 7.2
Summary of hypersensitivity reactions [87]

Characteristics	Type I (anaphylactic)	Type II (cytotoxic)	Type III (immune complex)	Type IV (delayed type)
Antibody	IgE	IgG, IgM	IgG, IgM	None
Antigen	exogenous	cell surface	soluble	tissues & organs
Response time	15–30 minutes	minutes-hours	3–8 hours	48–72 hours
Appearance	weal and flare	lysis and necrosis	erythema and edema, necrosis	erythema and induration
Histology	basophiles and eosinophil	antibody and complement	complement and neutrophils	monocytes and lymphocytes
Transferred with	antibody	antibody	antibody	T cells
Examples	allergic asthma, hay fever	myasthenia gravis, autoimmune hemolytic anemia	serum sickness, arthritis	contact dermatitis to nickel, poison ivy

form complexes that activate the classical pathway of complement activation for eliminating cells presenting foreign antigens. That is, mediators of acute inflammation are generated at the site and membrane attack complexes cause cell lysis and death. Type 2 hypersensitivity rarely occurs with biomaterials, but is of concern with drugs that bind to platelets.

In type 3 hypersensitivity, which is rarely seen, soluble immune complexes (aggregations of antigens and IgG and IgM antibodies) form in the blood and are deposited in various tissues (typically the skin, kidney, and joints) where they may trigger an immune response according to the classical pathway of complement activation. For this sensitivity reaction to occur, the antigen and the antibody must both be present at the same time, in the correct proportion. This is not common with biomaterials and it is most likely to occur in slow release or slow degradation materials with constant antigenic stimulation of the system over the hours to days it takes for the type 3 hypersensitivity reaction to develop.

The delayed type 4 hypersensitivity is a cell-mediated response. CD8 cytotoxic T cells and CD4 T-helper cells recognize antigens in a complex with either type 1 or 2 major histocompatibility (MHC) complex. The antigen-presenting cells in this case are macrophages and they release interleukin 1, which stimulates the proliferation of further CD4 cells. These cells release interleukin 2 and interferon gamma, which together regulate the immune reaction. Activated CD8 cells destroy target cells on contact while activated macrophages produce hydrolytic enzymes and, on presentation with certain intracellular pathogens, transform into multinucleated giant cells. Type 4 hypersensitivity is important in metals-related immune reactions [88]. The most recognized of these reactions is contact dermatitis to nickel, cobalt, and chromium, prevalent in about 10% of females and 1–2% of males [89], and first described more than 50 years ago [90]. Exposure to these metal salts in daily life is common, for example, when handling cookware and wearing jewelry.

Due to the well-known allergic reactions to nickel [91] and less well-known ones to cobalt [92], these two out of the three ferromagnetic elements are not generally considered being useful as magnetic particles for in vivo use. Although Ni^{2+} and Co^{2+} seem to be able to directly activate C3 complement [93], the mechanism of toxicity is not clear. Some authors have reported that toxicity may be transferred through cobalt and nickel induced radicals, as measured directly with EPR [28, 94]. When large amounts of nickel or cobalt are released, such as from implants, the result might be necrosis, while small amounts produce apoptosis [95]. Other authors report that Ni^{2+} and Co^{2+} lead to a dose-dependent expression of proinflammatory proteins, specifically MCP-1, which is chemotactic for monocytes and nucleophil granulocytes (Fig. 7.14) [96].

Concerning the third ferromagnetic metal used for the preparation of magnetic particles, iron, no direct immunotoxic effects have been reported. There is one study, however, which reports that high

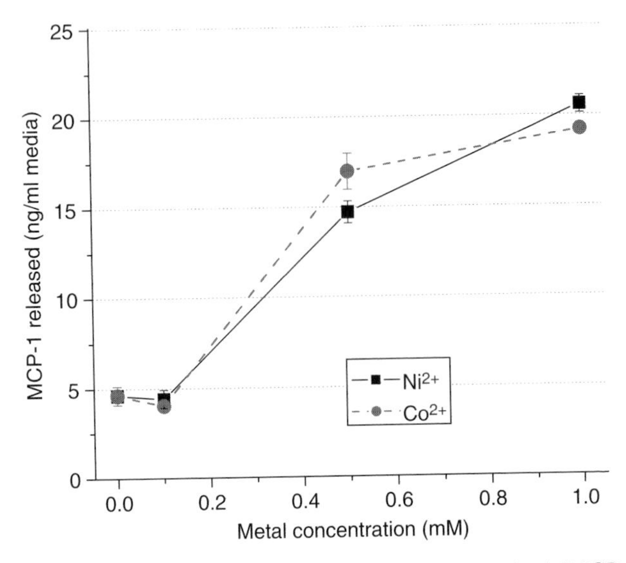

Fig. 7.14. Release of the monocyte chemoattractant protein-1 (MCP-1) after incubation of human endothelial cells with different concentrations of nickel and cobalt salts [113].

iron concentrations (0.1 mM) of ferrous (Fe(II)) and ferric (Fe(III)) salt solutions, and also iron-containing coal (40 to 120 μg/ml) suppressed the immune response significantly in Mishell-Dutton culture, an in vitro system for the production of IgM [97]. The authors hypothesized that lipid peroxidation catalyzed by iron might be involved. Recently, the inhalation of ferrous salt solutions together with ambient particulate matter (undefined properties) was shown to weaken the immune response of rats against pneumonia [98]. More research is clearly needed to pinpoint the mechanisms involved.

Performing assays that test the influence of magnetic particles on the immune system are relatively new and not common. In a recent paper, however, magnetic particles to be used for the delivery of thrombolytic factors to blood clots were tested for medical biocompatibility and underwent a SC5b-9 complement assay [99]. In this assay, the terminal complement complex (TCC) is generated

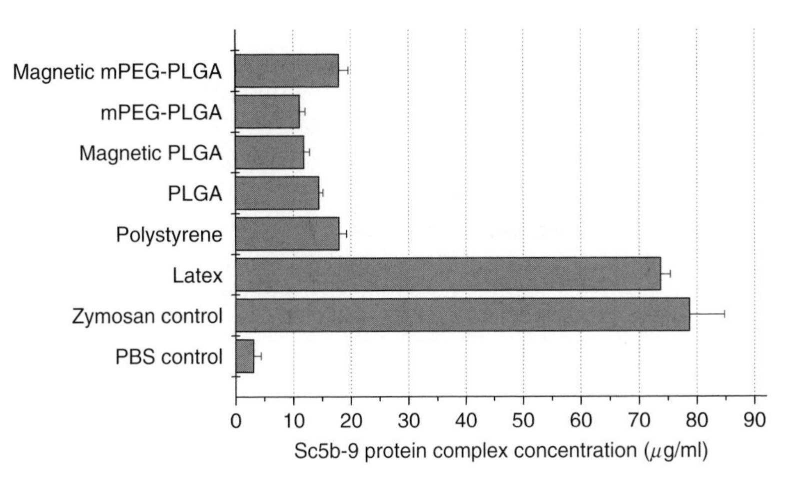

Fig. 7.15. Activation of the complement system by different magnetic and nonmagnetic particles measured by the SC5b-9 complement assay. Phosphate buffered saline (PBS) was the negative and Zymosan, an insoluble cell wall polysaccharide of yeast, the positive control. Redrawn from Torno et al. [99].

by the assembly of C5 through C9 as a consequence of activation of the complement system by either the classical or alternative pathway [100] and then quantified in a human SC5b-9 enzyme immunoassay (Quidel, San Diego, CA). Complement activation was highest in (nonmagnetic) latex particles, while polystyrene particles were biologically tolerant (Fig. 7.15). PLA-microspheres containing mPEG groups on the surface produced the lowest levels, while making the particles magnetic increased the levels slightly. A study confirmed the importance of surface charge density on complement activation [101]. The removal of the negative charge on the surface of pegylated particles totally prevented complement activation.

7.6. Legal standards for toxicity testing

To ensure that the evaluation of the biocompatibility of medical materials or devices is carried out efficiently and systematically, guidelines have evolved over the years. In 1987 a document was released by the governments of the UK, Canada, and the United States called Tripartite Biocompatibility Guidance [102]. This document provided an approach to the toxicity testing of medical devices and was intended for use by manufacturers and government health authorities. Although the document was never finalized, most manufacturers followed the document as if it were an official FDA regulation. Some time later, the ISO developed similar guidelines, as part of a series of regulations, known as the Biological Evaluation of Medical Devices ISO 10993. Manufacturers were expected to use the ISO guidelines for submission in Europe and the Tripartite guidelines for submission in America. The FDA soon adopted the ISO 10993 guidelines, with some modifications, for submission in America (Table 7.3).

Subpart 5 of the ISO guidelines deals with in vitro cytotoxicity assays and other tests needed to assess the adverse biological effects of materials used in medical devices. The methods discussed include the elution method which involves culturing cells in a nutrient medium and then replacing some of the medium with the test extracts, under standard conditions (typically 24 h at 37°C, using pharmacopeia based media volumes per biomaterial surface area). The cells can then be analyzed for signs of toxicity, reduction in cell number, or altered appearance. The cell viability or proliferation can be quantified with biochemical assays as described earlier in this chapter. A different method described in ISO 10993 is to apply test materials directly to a semisolid nutrient agar layer which can minimize the physical stress that may occur to cells when test materials are added. The test materials can migrate through the agar or medium to the cells underneath. The cells are then analyzed for any effect under and around the sample, often seen as discoloration due to loss of stain from the agar gel. An additional, less popular ISO 10993, method involves the inhibition of cell growth after adding a saline extract to cell suspensions. The cell mass is measured after incubation with the test extracts and toxicity calculated by comparing with healthy controls.

The ISO 10993 guidelines apply to all medical devices and are thus not very specific to magnetic or nonmagnetic nano- and microparticles, or other nanomaterials. As toxicity of such nanomaterials has been discussed a lot in recent years [103], the reaction of the governments of the United States, Japan, and the European Union (EU) to these potential threats will be discussed in the following and is intended to provide a basis for movement towards appropriate and useful toxicity and biocompatibility guidelines for magnetic particle materials.

7.6.1. Efforts to develop risk-based safety evaluations for nanomaterials in the United States

The National Nanotechnology Initiative (NNI) is a federal R&D program that was established to coordinate the multiagency efforts in Nanoscale Science, Engineering, and Technology (NSET). The goals of the NNI are to maintain a world-class research and development program aimed at realizing the full potential of nanotechnology; to facilitate transfer of new technologies into products for economic growth, jobs, and other public benefit; to develop educational resources, a skilled workforce, and the supporting infrastructure and tools to advance nanotechnology and to support responsible development of nanotechnology. The NSET subcommittee is the coordination body for the US government's efforts to evaluate the human health and environmental impacts of nanotechnology. Since 2001, various federal agencies have sponsored extramural and intramural research to evaluate the potential applications of nanotechnology and the associated human health and environmental implications. These agencies include the US Environmental Protection Agency (EPA) which is taking a holistic approach to studying nanotechnology, targeting research toward the identification of the beneficial applications of nanotechnology, seeking additional exposure and fate/transport data, developing appropriate

TABLE 7.3
Suggested biocompatibility and toxicity tests per FDA and ISO 10993 guidelines

Device categories			Biological effect											
Body contact		Contact duration A = Limited (≤24 Hours) B = Prolonged (24 Hours-30 Days) C = Permanent (>30 days)	Cytotoxicity	Sensitization	Irritation/intracutaneous reactivity	Acute systemic toxicity	Subchronic toxicity	Genotoxicity	Implantation	Hemocompatibility	Chronic toxicity	Carcinogenicity	Reproductive/developmental	Biodegradation
Surface devices	Skin	A	X	X	X									
		B	X	X	X									
		C	X	X	X									
	Mucosal membrane	A	X	X	X									
		B	X	X	X	O	O		O					
		C	X	X	X	O	X	X	O		O			
	Breached or compromised surfaces	A	X	X	X	O								
		B	X	X	X	O	O		O					
		C	X	X	X	O	X	X	O		O			

Externally communicating devices	Blood path, indirect	A	X	X	X	X				X				
		B	X	X	X	X	O			X				
		C	X	X	O	X	X	X	O	X	X	X		
	Tissue/bone/dentin communicating[1]	A	X	X	X	O								
		B	X	X	O	O	O	X	X					
		C	X	X	O	O	O	X	X		O	X		
	Circulating blood	A	X	X	X	X		O[2]		X				
		B	X	X	X	X	O	X	O	X				
		C	X	X	X	X	X	X	O	X	X	X		
Implant devices	Tissue/bone	A	X	X	X	O								
		B	X	X	O	O	O	X	X					
		C	X	X	O	O	O	X	X		X	X		
	Blood	A	X	X	X	X			X	X				
		B	X	X	X	X	O	X	X	X				
		C	X	X	X	X	X	X	X	X	X	X	X	X

X = tests per ISO 10993-1.
O = additional tests, which may be applicable in the U.S.
Note[1] = tissue includes tissue fluids and subcutaneous spaces.
Note[2] = for all devices used in extracorporeal circuits.
Source: Table Based Upon FDA Blue Book Memorandum G95-1.

risk assessment/management strategies, pursuing novel pollution prevention and environmentally benign techniques for the technology, and assisting in the development of novel treatment and remediation technologies using nanotechnology. The critical research needs for the development of comprehensive risk assessments for nanomaterials are given in EPA's draft Nanotechnology White Paper and include:

- *Characterization*: Research needs in this area include compiling data on the unique chemical and physical characteristics of nanomaterials, particularly the impact of size, morphology, charge, and surface coatings on reactivity, toxicity, and mobility.
- *Transformation and Interaction*: Data concerning the transformation of specific compounds—individually and in complexes with other compounds—and the interactions that may occur between a compound and its surroundings are critical to developing environmental protection policies.
- *Environmental Fate and Transport*: It is essential to determine the fate of nanomaterials in the environment and the availability of these materials to living organisms. Accordingly, more research is needed on the transport and potential transformation of nanomaterials in soil, subsurface, surface waters, water treatment systems, and the atmosphere.
- *Exposure Assessment*: Needs include determining the adequacy and accuracy of current exposure assessment techniques for nanomaterials, determining those nanomaterial properties that have relevance for hazard (e.g., size, surface characteristic, charge, and morphology), determining potential exposure scenarios for sensitive populations (such as children, the elderly, and people with health conditions such as asthma), and understanding the impacts of varying the physical and chemical properties of nanomaterials on exposure outcomes.
- *Ecological Effects*: The use of nanomaterials in the environment may result in novel by-products or metabolites that may pose significant risks. Furthermore, research is done on determining the

distribution of nanomaterials in ecosystems; determining adsorption, distribution, metabolism, and excretion (ADME) parameters of various nanomaterials for ecological receptors; and examining the interaction of nanomaterials with model ecosystems.

- *Human Health Effects Assessment*: As a result of their size, nanoparticles may pass into cells directly through cell membranes or via cellular transport mechanisms and may penetrate the skin and distribute throughout the body.
- *Life Cycle Analysis*: Life cycle analysis is an approach to evaluating the environmental consequences of a material through all of the stages along its life cycle, including production, use, recycling, and disposal.

7.6.2. Efforts to develop risk-based safety evaluations for nanomaterials in Europe

Products based on Nanosciences and Nanotechnologies (N&N) are already in use (e.g., sunscreen), and analysts expect markets to grow by billions of euros during this decade. The risk assessment of these nanoengineered materials has become the focus of increasing international attention. Some dedicated EU–funded research within the previous (FP5) and present (FP6) Framework Program in the field of nanotechnology is underway to assess these potential risks and determine the most appropriate basis for developing a science-based regulatory program for nanomaterials. Because nanomaterials lack geographical boundaries, it would be advantageous to systematically pool knowledge at the international level. Listed here are the EU projects that are ongoing and have some relevance for magnetic particles:

- *Nanosafe*: This project assesses the risks associated with the production, handling, and use of nanoparticles in industrial processes that produce commercial and consumer products.
- *Nanoderm*: The objectives of this project are to apply and develop methods for evaluating the effectiveness of skin as a barrier to

nanoparticles and to assess the biological activity of nanoparticles in skin.

- *Nanopathology*: The goal for this project is to identify innovative diagnostic methods for micro- and nanoscale particles, to investigate the pathological mechanisms of possible particle-included diseases, and to determine the pathological significance of the nanoparticles.

7.6.3. Efforts to develop risk-based safety evaluations for nanomaterials in Japan

In Japan, research on the toxicity of engineered nanomaterials was initiated in the early 1990s, but only took off in 2005 with the establishment of the Advisory Council for Formulating Nanotechnology policy, operated by the Ministry of Economics, International Trade, and Industry (METI). The basis of the council's nanotechnology priorities were: (1) the national goal of promoting nanotechnology, (2) the feasible application of nanotechnology, (3) industrial nanotechnology policy, and (4) the social impact of nanotechnology. It was reported that nanotechnology was not yet at the stage in which definite toxicity inherent to nanomaterials could be determined but that it is crucial that continuous efforts be devoted to evaluating the safety of nanomaterials. The first project funded by METI in 2005 embraced "research on the development and standardization of in vitro toxicity testing methods for nanomaterials," while the following 5 years are to see a comprehensive research project on the safety and risk assessment/management of nanomaterials (see Fig. 7.16).

The United States, Europe, and Japan have devoted substantial resources to promoting nanotechnology for economic, commercial, and societal benefit by allocating significant resources to identifying and developing promising applications of the technology. Unfortunately, it does not appear that sufficient data currently exist to accommodate a life cycle evaluation of nanomaterials because

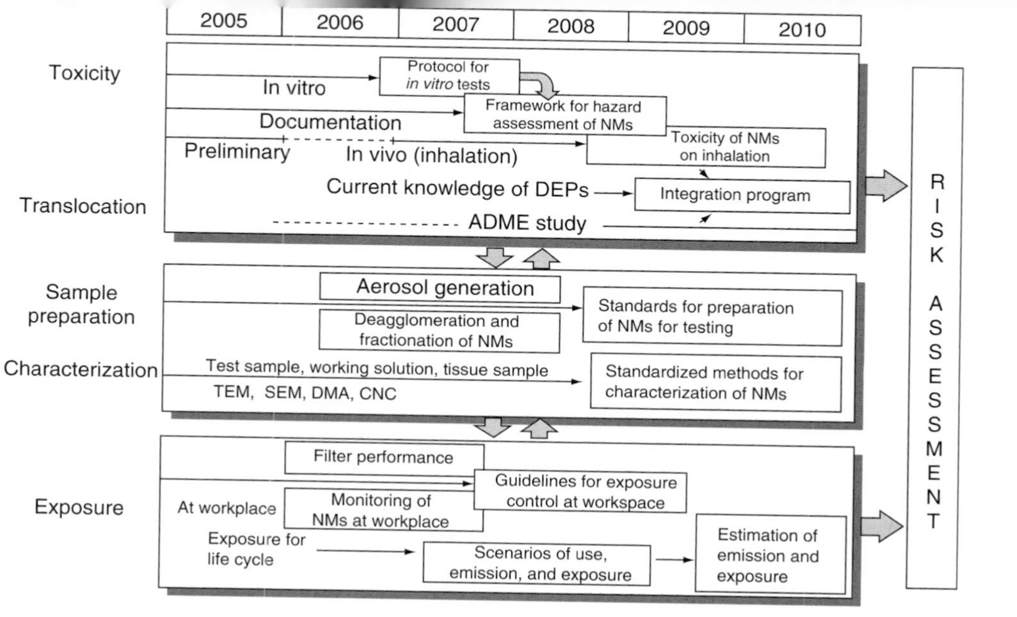

CNC = Condensation nucleus counter
DMA = Differential mobility analyzer
SEM = Scanning electron microscope
TEM = Transmission electron microscope
DEPs = Diesel exhaust particles

Fig. 7.16. Framework for hazard assessment of nanomaterials and toxicity tests, as planned by the Japanese government [104].

most current research focuses on inhalation exposure. While this route is certainly important, particularly given what is known about the systemic hazards associated with exposure to fine-sized particles, it is also important to develop comprehensive programs to address dermal and oral exposure to nanomaterials.

The bottom-line is that international cooperation for standardization of assessment methods and harmonization of risk evaluation techniques will be critical for the development of scientifically rational standards for public health decision making. Given the similarities of the goals associated with the human health and environmental research underway, the development of formal collaborations and consortiums could facilitate the generation of data for risk assessments in a more efficient manner and minimize duplication of effort. Government agencies in the United States, Europe, and Japan should encourage these collaborations in a way that is transparent and that allow input from a broad spectrum of stakeholders [104].

7.7. Conclusions

Nano- and microparticles intended for biomedical applications must be biocompatible and nontoxic. Keeping in mind the intended application, an extensive biocompatibility assessment of magnetic particles may involve testing for local tissue response, systemic toxic effects, blood compatibility, and immunogenic and carcinogenic reactions. It is also important to test the toxicity of any degradation products or leachables from magnetic particles.

This review has discussed the properties of particles that may contribute to their toxicity as well as some methods of assessing this toxicity in vitro. The importance of in vitro toxicity testing has increased in recent times, mainly due to its desirable qualities over in vivo testing. Specifically, in vitro tests are easier to manipulate, more cost effective, faster to carry out, reproducible, have fewer ethical restrictions and controversy associated with them, and are easier to interpret.

When choosing a cytotoxicity assay it is important to bear a number of issues in mind. In particular, these issues include what the intended use of the magnetic particles are, which cells they may come into contact with, what the route of administration is, if there are any of the particle components known to be toxic, what form of toxicity we are looking for, and if all the equipment to carry out the desired tests are available.

Fluorescence-based assays are especially beneficial for the evaluation of toxicity, once the type of cell culture and the assay parameters have been adjusted to the particles at hand. The fluorochromes will stain the viable cell one color and the dead cell another so that toxicity can be observed visually using a fluorescence or confocal microscope [105]. The fluorescence produced may also be read using a fluorescence spectrophotometer and cell viability can be quantified by measuring the absorbance. Fluorescent cell assays are valuable because many different biological and analytical principles can be used and because they can be kept close to the in vivo situation. Disadvantages are the cost involved for the equipment and cell culture supplies, larger standard deviations than those from physical methods, and the time involved.

Assays measuring metabolic activity are generally cheap, fast, and easy to use. They also have a high degree of accuracy as the functional integrity of cells is evaluated. In addition, they are simple to repeat and reproduce. Many studies have shown that the ATP assay is a quick and simple test for cell viability giving results more quickly than other assays within this group [50]. The ATP assay has proven to be more sensitive than the MTS, MTT and LDH assays as any form of cell injury leads to a rapid depletion of ATP in the cytoplasm. Therefore measuring intracellular ATP levels is a key in determining reduced cell viability [52].

In conclusion, the cytotoxicity of particles must be quantified in order to define their nontoxic threshold. Although in vitro tests are only one, and generally the first one, of the required stages in biocompatibility testing, they are valuable in providing an initial toxicity profile, before the materials become further developed or require more costly testing procedures. In vitro tests can help to determine mechanisms of cell toxicity and general suitability of the particles. In addition to cytotoxicity tests, it is important to always search the relevant literature for already performed tests. An Environmental Health and Safety (EHS) database at Rice University covers nanotechnology toxicity articles and facilitates such toxicity searches considerably for the magnetic particle community [106].

References

[1] Chen, L., Zhao, X., Chen, M. and Li, H. (2004). Wrinkle-resistant, self-cleaning and good feeling wool fabrics. Patent No. 1624224 (China). (Dec. 13).

[2] Berkovski, B. and Bashtovoy, V. (1996). Magnetic fluids and applications handbook. Begell House, Inc., New York, NY.

[3] Wang, Y. X., Hussain, S. M. and Krestin, G. P. (2001). Superparamagnetic iron oxide contrast agents: Physicochemical characteristics and applications in MR imaging. Eur. Radiol. *11*, 2319–31.

[4] DeRosa, L., Montuoro, A., Pandolfi, A., Lanti, T., Pescador, L., Morara, R. and DeLaurenzi, A. (1991). Immunomagnetic purging procedure for autologous bone marrow transplantation in lymphoid malignancies. Haematologica *76(Suppl. 1)*, 37–40.

[5] Gneveckow, U., Jordan, A., Scholz, R., Eckelt, L., Maier-Hauff, K., Johannsen, M. and Wust, P. (2005). Magnetic force nanotherapy: with nanoparticles against cancer. Experiences from three clinical trials. Biomed. Tech. *50*, 92–3.

[6] Tallheden, T., Nannmark, U., Lorentzon, M., Rakotonirainy, O., Soussi, B., Waagstein, F., Jeppsson, A., Sjögren-Jansson, E., Lindahl, A. and Omerovic, E. (2006). In vivo MR imaging of magnetically labeled human embryonic stem cells. Life Sci. *79*, 999–1006.

[7] Häfeli, U. O. (2006). Magnetic nano- and microparticles for targeted drug delivery. In: Smart Nanoparticles in Nanomedicine – the MML Series (Arshady, R. and Kono, K., eds.), Vol. 8. Kentus Books, London, UK, pp 77–126.

[8] Andrä, W., Häfeli, U. O., Hergt, R. and Misri, R. (2006). Application of magnetic particles in medicine and biology. In: The Handbook of Magnetism and Advanced Magnetic Materials (Kronmüller, H. and Parkin, S., eds.), Vol. 4 – Novel Materials. John Wiley & Sons Ltd, London, UK.

[9] Wallan, R. F. and Arscott, E. F. (1998). A practical guide to ISO 10993–5: Cytotoxicity. Med. Dev. Diagnost. Industr. *April*, 96–98.

[10] Langman, L. J. and Kapur, B. M. (2006). Toxicology: Then and now. Clin. Biochem. *39*, 498–510.

[11] Mulder, G. J. and Dencker, L. (2006). Pharmaceutical toxicology. Pharmaceutical Press, London, UK.

[12] Borm, P. J., Robbins, D., Haubold, S., Kuhlbusch, T., Fissan, H., Donaldson, K., Schins, R., Stone, V., Kreyling, W., Lademann, J., Krutmann, J., Warheit, D. et al. (2006). The potential risks of nanomaterials: A review carried out for ECETOC. Part. Fibre Toxicol. *3*, 11.

[13] Gilmour, P. S., Ziesenis, A., Morrison, E. R., Vickers, M. A., Drost, E. M., Ford, I., Karg, E., Mossa, C., Schroeppel, A., Ferron, G. A., Heyder, J. and Greaves, M. (2004). Pulmonary and systemic effects of short-term inhalation exposure to ultrafine carbon black particles. Toxicol. Appl. Pharmacol. *195*, 35–44.

[14] Zhang, Z., Kleinstreuer, C., Donohue, J. F. and Kim, C. S. (2005). Comparison of micro- and nanosized particle deposition in a hunal upper airway model. J. Aerosol. Sci. *36*, 211–33.

[15] Kreyling, W. G., Semmler-Behnke, M. and Möller, W. (2006). Health implications of nanoparticles. J. Nanopart. Res. *8*, 543–62.

[16] Kreyling, W. G., Semmler-Behnke, M., Erbe, F., Mayer, P., Takenaka, S., Schulz, H., Oberdorster, G. and Ziesenis, A. (2002). Translocation of ultrafine insoluble iridium particles from lung epithelium to extrapulmonary organs is size dependent but very low. J. Toxicol. Environ. Health A *65*, 1513–30.

[17] Nemmar, A., Vanbilloen, H., Hoylaerts, M. F., Hoet, P. H., Verbruggen, A. and Nemery, B. (2001). Passage of intratracheally instilled ultrafine particles from the lung into the systemic circulation in hamster. Am. J. Respir. Crit. Care. Med. *164*, 1665–8.

[18] Borm, P. J. and Kreyling, W. (2004). Toxicological hazards of inhaled nanoparticles—potential implications for drug delivery. J. Nanosci. Nanotechnol. *4*, 521–31.

[19] Brown, D. M., Donaldson, K., Borm, P. J., Schins, R. P., Dehnhardt, M., Gilmour, P., Jimenez, L. A. and Stone, V. (2004). Calcium and ROS-mediated activation of transcription factors and TNF-alpha cytokine gene expression in macrophages exposed to ultrafine particles. Am. J. Physiol. Lung. Cell. Mol. Physiol. *286*, L344–53.

[20] Yin, H., Too, H. P. and Chow, G. M. (2005). The effects of particle size and surface coating on the cytotoxicity of nickel ferrite. Biomaterials *26*, 5818–26.

[21] Hu, F., Neoh, K. G., Cen, L. and Kang, E. T. (2006). Cellular response to magnetic nanoparticles "PEGylated" via surface-initiated atom transfer radical polymerization. Biomacromolecules *7*, 809–16.

[22] Park, S. I., Lim, J. H., Kim, J. H., Yun, H. I., Roh, J. S., Kim, C. G. and Kim, C. O. (2004). Effects of surfactant on properties of magnetic fluids for biomedical application. Phys. Stat. Sol. (b) *241*, 1662–4.

[23] Gupta, A. K. and Curtis, A. S. (2004). Surface modified superparamagnetic nanoparticles for drug delivery: Interaction studies with human fibroblasts in culture. J. Mater. Sci. Mater. Med. *15*, 493–6.

[24] Cengelli, F., Maysinger, D., Tschudi-Monnet, F., Montet, X., Corot, C., Petri-Fink, A., Hofmann, H. and Juillerat-Jeanneret, L. (2006). Interaction of functionalized superparamagnetic iron oxide nanoparticles with brain structures. J. Pharmacol. Exp. Ther. *318*, 108–16.

[25] Grüttner, C., Teller, J., Schütt, W., Westphal, F., Schümichen, C. and Paulke, B. R. (1997). Preparation and characterization of magnetic nanospheres for in vivo application. In: Scientific and Clinical Applications of Magnetic Carriers (Häfeli, U., Schütt, W., Teller, J. and Zborowski, M., eds.). Plenum Press, New York, pp. 53–67.

[26] Gomez-Lopera, S. A., Plaza, R. C. and Delgado, A. V. (2001). Synthesis and characterization of spherical magnetite/biodegradable polymer composite particles. J. Colloid. Interface. Sci. *240*, 40–7.

[27] Gomez-Lopera, S. A., Arias, J. L., Gallardo, V. and Delgado, A. V. (2006). Colloidal stability of magnetite/poly(lactic acid) core/shell nanoparticles. Langmuir *22*, 2816–21.

[28] Valko, M., Morris, H. and Cronin, M. T. D. (2005). Metals, toxicity and oxidative stress. Curr. Med. Chem. *12*, 1161–208.

[29] Papanikolaou, G. and Pantopoulos, K. (2005). Iron metabolism and toxicity. Toxicol. Appl. Pharmacology *202*, 199–211.

[30] Perrone, J. Iron. (2002). In: Goldfrank's Toxicologic Emergencies. Part D. The Clinical Basis of Medical Toxicology. Section, I. Case Studies in Toxicologic Emergencies (Goldfrank, L. R., Flomenbaum, N. E., Lewin, N. A., Howland, M. A., Hoffman, R. S. and Nelson, L. S., eds.). 7th ed. McGraw-Hill Medical, New York, NY, pp. 548–62 (Chapter 36).

[31] Dick, C. A., Brown, D. M., Donaldson, K. and Stone, V. (2003). The role of free radicals in the toxic and inflammatory effects of four different ultrafine particle types. Inhal. Toxicol. *15*, 39–52.

[32] Knaapen, A. M., Shi, T., Borm, P. J. and Schins, R. P. (2002). Soluble metals as well as the insoluble particle fraction are involved in cellular DNA damage induced by particulate matter. Mol. Cell. Biochem. *234–5*, 317–26.

[33] Freitas, M. L. L., Silva, L. P., Freitas, J. L., Azevedo, R. B., Lacava, Z. G. M., Homem de Bittencourt, P. I., Curi, R., Buske, N. and Morais, P. C. (2003).

Investigation of lipid peroxidation and catalase activity in magnetic fluid treated mice. J. Appl. Phys. *93*, 6709–11.

[34] Black, J. (2006). Biological performance of materials: Fundamentals of biocompatibility. 4th ed. CRC Press, New York, NY.

[35] Ciapetti, G., Granchi, D., Arciola, C. R., Cenni, E., Savarino, L., Stea, S., Montanaro, L. and Pizzoferrato, A. (2000). In vitro testing of cytotoxicity of materials. In: Biomaterials and Bioengineering Handbook Marcel Dekker Inc., NewYork, NY, pp. 179–98.

[36] Russell, W. M. S. and Burch, R. L. (1959). The principles of humane experimental technique (Vol. [Reissued 1992, Universities Federation for Animal Welfare Herts, England.]). Methuen & Co, London, UK.

[37] Yu, H. and Adedoyin, A. (2003). ADME-Tox in drug discovery: Integration of experimental and computational technologies. Drug. Discov. Today *8*, 852–61.

[38] Pieters, R., Huismans, D. R., Leyva, A. and Veerman, A. J. P. (1989). Comparison of the rapid automated MTT-assay with a dye exclusion assay for chemosensitivity testing in childhood leukaemia. Br. J. Cancer *59*, 217–20.

[39] Häfeli, U. O. and Pauer, G. J. (1999). In vitro and in vivo toxicity of magnetic microspheres. J. Magn. Magn. Mater. *194*, 76–82.

[40] Marshall, N. J., Goodwin, C. J. and Holt, S. J. (1995). A critical assessment of the use of microculture tetrazolium assays to measure cell growth and function. Growth. Regul. *5*, 69–84.

[41] Devineni, D., Klein-Szanto, A. and Gallo, J. M. (1995). Targeting anticancer drugs to the brain, III: Tissue distribution of methotrexate following administration as a solution and as a magnetic microsphere conjugate in rats bearing brain tumors. J. Neuro. Oncol. *24*, 143–52.

[42] Zange, R. and Kissel, T. (1997). Comparative in vitro biocompatibility testing of polycyanoacrylates and PLGA using different mouse fibroblast (L929) biocompatibility test models. Eur. J. Pharm. Biopharm. *44*, 149–57.

[43] Lacava, Z. G. M., De Azevedo, R. B., Lacava, L. M., Martins, E. V., Garcia, V. A. P., Rebola, C. A., Lemos, A. P. C., Sousa, M. H., Tourinho, F. A., Morais, P. C. and Da Silva, M. F. (1999). Toxic effects of ionic magnetic fluids in mice. J. Magn. Magn. Mater. *194*, 90–95.

[44] Liu, L., Tang, M., He, Z., Ma, M. and Gu, N. (2004). Studies on toxicity and mutagenicity of nanoparticles of Fe3O4 and Fe3O4 coated with glutamic acid. Huanjing Yu Zhiye Yixue *21*, 14–17.

[45] Sadeghiani, N., Barbosa, L. S., Silva, L. P., Azevedo, R. B., Morais, P. C. and Lacava, Z. G. M. (2005). Genotoxicity and inflammatory investigation in mice treated with magnetite nanoparticles surface coated with polyaspartic acid. J. Magn. Magn. Mater. *289*, 466–8.

[46] Kim, J. S., Yoon, T. J., Yu, K. N., Kim, B. G., Park, S. J., Kim, H. W., Lee, K. H., Park, S. B., Lee, J. K. and Cho, M. H. (2006). Toxicity and tissue distribution of magnetic nanoparticles in mice. Toxicol. Sci. *89*, 338–47.
[47] Kuznetsov, O. A., Brusentsov, N. A., Kuznetsov, A. A., Yurchenko, N. Y., Osipov, N. E. and Bayburtskiy, F. S. (1999). Correlation of the coagulation rates and toxicity of biocompatible ferromagnetic microparticles. J. Magn. Magn. Mater. *94*, 83–89.
[48] Goodwin, C. J., Holt, S. J., Downes, S. and Marshall, N. J. (1995). Microculture tetrazolium assays: A comparison between two new tetrazolium salts, XTT and MTS. J. Immunol. Methods *179*, 95–103.
[49] Fields, R. D. and Lancaster, M. V. (1993). Dual-attribute continuous monitoring of cell proliferation/cytotoxicity. Am. Biotechnol. Lab. *11*, 48–50.
[50] Cree, I. A. and Andreotti, P. E. (1997). Measurement of cytotoxicity by ATP-based luminescence assay in primary cell cultures and cell lines. Toxicol. In Vitro *11*, 553–6.
[51] Weyermann, J., Lochmann, D. and Zimmer, A. (2005). A practical note on the use of cytotoxicity assays. Int. J. Pharm. *288*, 369–76.
[52] Eirheim, H. U., Bundgaard, C. and Nielsen, H. M. (2004). Evaluation of different toxicity assays applied to proliferating cells and to stratified epithelium with glycocholate. Toxicol. In Vitro *18*, 649–57.
[53] Weyermann, J., Lochmann, D. and Zimmer, A. (2005). A practical note on the use of cytotoxicity assays. Int. J. Pharm. *288*, 369–76.
[54] Simko, M., Hartwig, C., Lantow, M., Lupke, M., Mattsson, M. O., Rahman, Q. and Rollwitz, J. (2006). Hsp70 expression and free radical release after exposure to non-thermal radio-frequency electromagnetic fields and ultrafine particles in human Mono Mac 6 cells. Toxicol. Lett. *161*, 73–82.
[55] Fotakis, G. and Timbrell, J. A. (2006). In vitro cytotoxicity assays: Comparison of LDH, neutral red, MTT and protein assay in hepatoma cell lines following exposure to cadmium chloride. Toxicol. Lett. *160*, 171–7.
[56] Olbrich, C., Bakowsky, U., Lehr, C. M., Muller, R. H. and Kneuer, C. (2001). Cationic solid-lipid nanoparticles can efficiently bind and transfect plasmid DNA. J. Control. Release *77*, 345–55.
[57] Weissleder, R., Stark, D. D., Engelstad, B. L., Bacon, B. R., Compton, C. C., White, D. L., Jacobs, P. and Lewis, J. (1989). Superparamagnetic Iron Oxide: Pharmacokinetics and toxicity. Am. J. Roentgenol. *152*, 167–73.

[58] Lappalainen, K., Jääskeläinen, I., Syrjänen, K., Urtti, A. and Syrjänen, S. (1994). Comparison of cell proliferation and toxicity assays using two cationic liposomes. Pharm. Res. *11*, 1127–31.

[59] Yang, A., Cardona, D. L. and Barile, F. A. (2002). In vitro cytotoxicity testing with fluorescence-based assays in cultured human lung and dermal cells. Cell. Biol. Toxicol. *18*, 97–108.

[60] Haugland, R. P., MacCoubrey, I. C. and Moore, P. L. (May 24, 1994). Dual-fluorescence cell viability assay using ethidium homodimer and calcein AM, Patent No. 5314805 (USA).

[61] Berkova, Z., Kriz, J., Girman, P., Zacharovova, K., Koblas, T., Dovolilova, E. and Saudek, F. (2005). Vitality of pancreatic islets labeled for magnetic resonance imaging with iron particles. Transplant. Proc. *37*, 3496–98.

[62] Singh, N. P., McCoy, M. T., Tice, R. R. and Schneider, E. L. (1988). A simple technique for quantitation of low levels of DNA damage in individual cells. Exp. Cell. Res. *175*, 184–91.

[63] Hilger, I., Rapp, A., Greulich, K. O. and Kaiser, W. A. (2005). Assessment of DNA damage in target tumor cells after thermoablation in mice. Radiology *237*, 500–6.

[64] Jordan, A., Scholz, R., Wust, P., Schirra, H., Schiestel, T., Schmidt, H. and Felix, R. (1999). Endocytosis of dextran and silan-coated magnetite nanoparticles and the effect of intracellular hyperthermia on human mammary carcinoma cells in vitro. J. Magn. Magn. Mater. *194*, 185–96.

[65] Müller, R. H., Maassen, S., Weyhers, H., Specht, F. and Lucks, J. S. (1996). Cytotoxicity of magnetite-loaded polylactide, polylactide/glycolide particles and solid lipid nanoparticles. Int. J. Pharm. *138*, 85–94.

[66] Bacon, B. R., Stark, D. D., Park, C. H., Saini, S., Groman, E. V., Hahn, P. F., Compton, C. C. and Ferrucci, J. T. (1987). Ferrite particles: A new magnetic resonance imaging contrast agent. Lack of acute or chronic hepatotoxicity after intravenous administration. J. Lab. Clin. Med. *110*, 164–71.

[67] Fahlvik, A. K., Holtz, E., Schroder, U. and Klaveness, J. (1990). Magnetic starch microspheres, biodistribution and biotransformation. A new organ-specific contrast agent for magnetic resonance imaging. Invest. Radiol. *25*, 793–7.

[68] Kawamura, Y., Endo, K., Watanabe, Y., Saga, T., Nakai, T., Hikita, H., Kagawa, K. and Konishi, J. (1990). Use of magnetite particles as a contrast agent for MR imaging of the liver. Radiology *174*, 357–60.

[69] Lawaczeck, R., Bauer, H., Frenzel, T., Hasegawa, M., Ito, Y., Kito, K., Miwa, N., Tsutsui, H., Vogler, H. and Weinmann, H.-J. (1997). Magnetic iron oxide particles coated with carboxydextran for parenteral

administration and liver contrasting. Pre-clinical profile of SH U555A. Acta. Radiol. *38*, 584–97.

[70] Ros, P. R., Freeny, P. C., Harms, S. E., Seltzer, S. E., Davis, P. L., Chan, T. W., Stillman, A. E., Muroff, L. R., Runge, V. M. and Nissenbaum, M. A. (1995). Hepatic MR imaging with ferumoxides: A multicenter clinical trial of the safety and efficacy in the detection of focal hepatic lesions. Radiology *196*, 481–8.

[71] Okon, E. E., Pouliquen, D., Pereverzev, A. E., Kudryavtsev, B. N. and Jallet, P. (2000). Toxicity of magnetite-dextran particles: Morphological study. Tsitologiya *42*, 358–66.

[72] Wagner, S., Schnorr, J., Pilgrimm, H., Hamm, B. and Taupitz, M. (2002). Monomer-coated very small superparamagnetic iron oxide particles as contrast medium for magnetic resonance imaging: Preclinical in vivo characterization. Invest. Radiol. *37*, 167–77.

[73] Bourrinet, P., Bengele, H. H., Bonnemain, B., Dencausse, A., Idee, J. M., Jacobs, P. M. and Lewis, J. M. (2006). Preclinical safety and pharmacokinetic profile of ferumoxtran-10, an ultrasmall superparamagnetic iron oxide magnetic resonance contrast agent. Invest. Radiol. *41*, 313–24.

[74] Muldoon, L. L., Sandor, M., Pinkston, K. E. and Neuwelt, E. A. (2005). Imaging, distribution, and toxicity of superparamagnetic iron oxide magnetic resonance nanoparticles in the rat brain and intracerebral tumor. Neurosurgery *57*, 785–96 discussion 785–96.

[75] de Vries, I. J., Lesterhuis, W. J., Barentsz, J. O., Verdijk, P., van Krieken, J. H., Boerman, O. C., Oyen, W. J., Bonenkamp, J. J., Boezeman, J. B., Adema, G. J., Bulte, J. W. Scheenen, T. W. et al. (2005). Magnetic resonance tracking of dendritic cells in melanoma patients for monitoring of cellular therapy. Nat. Biotechnol. *23*, 1407–13.

[76] Zhang, Z., van den Bos, E. J., Wielopolski, P. A., de Jong-Popijus, M., Duncker, D. J. and Krestin, G. P. (2004). High-resolution magnetic resonance imaging of iron-labeled myoblasts using a standard 1.5-T clinical scanner. MAGMA *17*, 201–9.

[77] Arbab, A. S., Bashaw, L. A., Miller, B. R., Jordan, E. K., Lewis, B. K., Kalish, H. and Frank, J. A. (2003). Characterization of biophysical and metabolic properties of cells labeled with superparamagnetic iron oxide nanoparticles and transfection agent for cellular MR imaging. Radiology *229*, 838–46.

[78] Sieben, S., Bergemann, C., Lübbe, A. S., Brockmann, B. and Rescheleit, D. (2001). Comparison of different particles and methods for magnetic isolation of circulating tumor cells. J. Magn. Magn. Mater. *225*, 175–9.

[79] Widder, K. J., Senyei, A. E. and Ranney, D. F. (1979). Magnetically responsive microspheres and other carriers for the biophysical targeting of antitumor agents. Adv. Pharmacol. Chemother. *16*, 213–71.

[80] Häfeli, U. O., Pauer, G. J., Roberts, W. K., Humm, J. L. and Macklis, R. M. (1997). Magnetically targeted microspheres for intracavitary and intraspinal Y-90 radiotherapy. In: Scientific and Clinical Applications of Magnetic Carriers (Häfeli, U., Schütt, W., Teller, J. and Zborowski, M., eds.). 1st. Ed. Plenum Press, New York, NY, pp. 501–16.

[81] Lübbe, A. S., Bergemann, C., Huhnt, W., Fricke, T., Riess, H., Brock, J. W. and Huhn, D. (1996). Preclinical experiences with magnetic drug targeting: tolerance and efficacy. Cancer Res. *56*, 4694–701.

[82] Van Hecke, P., Marchal, G., Decrop, E. and Baert, A. L. (1989). Experimental study of the pharmacokinetics and dose response of ferrite particles used as contrast agent in MRI of the normal liver of the rabbit. Invest. Radiol. *24*, 397–9.

[83] Lübbe, A. S., Bergemann, C., Riess, H., Schriever, F., Reichardt, P., Possinger, K., Matthias, M., Dörken, B., Herrmann, F., Gürtler, R., Hohenberger, P. Haas, N. et al. (1996). Clinical experiences with magnetic drug targeting: a phase I study with 4′- epidoxorubicin in 14 patients with advanced solid tumors. Cancer Res. *56*, 4686–93.

[84] Ibrahim, A., Couvreur, P., Roland, M. and Speiser, P. (1983). New magnetic drug carrier. J. Pharm. Pharmacol. *35*, 59–61.

[85] Goodwin, S. C., Bittner, C. A., Peterson, C. L. and Wong, G. (2001). Single-dose toxicity study of hepatic intra-arterial infusion of doxorubicin coupled to a novel magnetically targeted drug carrier. Toxicol. Sci. *60*, 177–83.

[86] von Recum, A. F. (1999). Handbook of biomaterials evaluation scientific, technical, and clinical testing of implant materials. 2nd. Taylor & Francis, Philadelphia, PA.

[87] Gell, P. G. H. and Coombs, R. R. A. (1968). Clinical aspects of immunology. Blackwell, Oxford, UK.

[88] Yang, J. and Merritt, K. (1996). Production of monoclonal antibodies to study corrosion products of Co-Cr biomaterials. J. Biomed. Mater. Res. *31*, 71–80.

[89] Meding, B. (2003). Epidemiology of nickel allergy. J. Environ. Monit. *5*, 188–9.

[90] Rostenberg, A., Jr., and Perkins, A. J. (1951). Nickel and cobalt dermatitis. J. Allergy *22*, 466–74.

[91] Büdinger, L. and Hertl, M. (2000). Immunologic mechanisms in hypersensitivity reactions to metal ions: An overview. Allergy *55*, 108–15.

[92] Meyer, J. M., Craig, R. G., Schmalz, G. and Reclaru, L. (1990). Corrosion resistance and biocompatibility of some low gold dental casting alloys. J. Dent. Res. *18*, 74–75.
[93] Acevedo, F. and Vesterberg, O. (2003). Nickel and cobalt activate complement factor C3 faster than magnesium. Toxicology *185*, 9–16.
[94] Wang, X., Yokoi, I., Liu, J. and Mori, A. (1993). Cobalt(II) and nickel(II) ions as promoters of free radicals in vivo: Detected directly using electron spin resonance spectrometry in circulating blood in rats. Arch. Biochem. Biophys. *306*, 402–6.
[95] Granchi, D., Cenni, E., Ciapetti, G., Savarino, L., Stea, S., Gamberini, S., Gori, A. and Pizzoferrato, A. (1998). Cell death induced by metal ions: Necrosis or apoptosis? J. Mater. Sci. Mater. Med. *9*, 31–37.
[96] Kirkpatrick, C. J., Barth, S., Gerdes, T., Krump-Konvalinkova, V. and Peters, K. (2002). Pathomechanisms of impaired wound healing by metallic corrosion products [German]. Mund. Kiefer. Gesichtschir. *6*, 183–90.
[97] Ban, M., Hettich, D. and Cavelier, C. (1995). Use of Mishell-Dutton culture for the detection of the immunosuppressive effect of iron-containing compounds. Toxicol. Lett. *81*, 183–8.
[98] Zelikoff, J. T., Schermerhorn, K. R., Fang, K., Cohen, M. D. and Schlesinger, R. B. (2002). A role for associated transition metals in the immunotoxicity of inhaled ambient particulate matter. Environ. Health Perspect. *110(Suppl. 5)*, 871–5.
[99] Torno, M. D., Kaminski, M. D., Liu, X. Q., Mertz, C. J., Caviness, P., Taylor, L., Holtzman, S. and Rosengart, A. J. (2006). A comprehensive test battery for the in vitro biocompatibility assessment of nanocarriers for medical applications. Biomaterials submitted.
[100] Muller-Eberhard, H. J. (1984). The membrane attack complex. Springer Semin Immunopathol *7*, 93–141.
[101] Moghimi, S. M., Hamad, I., Andresen, T. L., Jørgensen, K. and Szebeni, J. (2006). Methylation of the phosphate oxygen moiety of phospholipid-methoxy(polyethylene glycol) conjugate prevents PEGylated liposome-mediated complement activation and anaphylatoxin production. FASEB J. *20*, 2591–3.
[102] Toxicology Sub-Group of the Tripartite Sub-Committee on Medical Devices: *Tripartite Biocompatibility Guidance for Medical Devices (G87–1)*. Center for Devices and Radiological Health, U.S. Food and Drug Administration.
[103] Warheit, D. B. (2004). Nanoparticles: health impacts? Mater Today *7*, 32–35.

[104] Thomas, K. (2006). Research strategies for safety evaluation of nanomaterials, Part VIII: international efforts to develop risk-based safety evaluations for nanomaterials. Toxicol. Sci. *92*, 23–32.
[105] Decherchi, P., Cochard, P. and Gauthier, P. (1997). Dual staining assessment of Schwann cell viability within whole peripheral nerves using calcein-AM and ethidium homodimer. J. Neurosci. Methods *71*, 205–13.
[106] Wedin, R. (2006). Is nanotechnology safe? Chemistry Spring, 48–50.
[107] ICRP (1994). International commission on radiological protection publication 66: Human respiratory tract model for radiological protection, published in Annals of the ICRP. Vol. 24. Elsevier, Oxford, UK.
[108] Yin, H., Too, H. P. and Chow, G. M. (2005). The effects of particle size and surface coating on the cytotoxicity of nickel ferrite. Biomaterials *26*, 5818–26.
[109] Gomez-Lopera, S. A., Plaza, R. C. and Delgado, A. V. (2001). Synthesis and characterization of spherical magnetite/biodegradable polymer composite particles. J. Colloid. Interface Sci. *240*, 40–47.
[110] Dick, C. A., Brown, D. M., Donaldson, K. and Stone, V. (2003). The role of free radicals in the toxic and inflammatory effects of four different ultrafine particle types. Inhal. Toxicol. *15*, 39–52.
[111] Hilger, I., Rapp, A., Greulich, K. O. and Kaiser, W. A. (2005). Assessment of DNA damage in target tumor cells after thermoablation in mice. Radiology *237*, 500–6.
[112] Roitt, I. M., Martin, S. J., Delves, P. J. and Burton, D. (2006). Roitt's essential immunology. 11th ed. Blackwell Publishing, Malden MA, USA.
[113] Kirkpatrick, C. J., Barth, S., Gerdes, T., Krump-Konvalinkova, V. and Peters, K. (2002). Pathomechanisms of impaired wound healing by metallic corrosion products [German]. Mund Kiefer Gesichtschir *6*, 183–90.

Laboratory Techniques in Biochemistry and Molecular Biology, Volume 32
Magnetic Cell Separation
M. Zborowski and J. J. Chalmers (Editors)

CHAPTER 8

Analytical magnetic techniques in biology

Jeffrey J. Chalmers[1] and Lee R. Moore[2]

[1]*Department of Chemical and Biomolecular Engineering, University Cell Analysis and Sorting Core, The Ohio State University, OH 43210, USA*
[2]*Department of Biomedical Engineering, Lerner Research Institute, Cleveland Clinic, Cleveland, OH 44195, USA*

8.1. Introduction

With the maturation of any technology, further improvement/optimization typically requires both a fundamental understanding of the process and the ability to quantify the various relevant variables responsible for the system performance. With this understanding, needed improvements can be identified and hopefully implemented. In previous chapters the fundamentals of magnetism, as applied to cell separation, as well as various designs of magnet assemblies and separators were presented. In this chapter, measurement techniques will be presented for cells with magnetic particle labels—or in a few cases—for cells with intrinsic magnetism.

8.2. Measurements of magnetic susceptibility

A number of magnetometer devices can be used to measure the magnetic susceptibility of materials, such as Gouy and Faraday balances, and the superconducting quantum interference device (SQUID). While accurate, these devices only provide bulk measurements of materials and particles.

DOI: 10.1016/S0075-7535(06)32008-6

8.3. Measurements of magnetophoretic mobility: cell-tracking velocimetry (CTV)

In Chapter 5, the concept of magnetophoretic mobility was introduced as a fundamental property of a particle or cell and is equally applicable to materials that are diamagnetic, paramagnetic, or ferromagnetic, depending on the relationship between the applied field and particle susceptibility, $\chi_p(H)$:

$$m = \frac{v}{S_m} = \frac{2}{9}\frac{R^2[\chi_p(H) - \chi_f(H)]}{\eta} \tag{8.1}$$

In addition, this parameter combines the particle radius and properties of the fluid: susceptibility and viscosity. The fundamental concept of magnetophoretic mobility enabled the design of separation systems and provided a tool for the operation and evaluation of the magnetic separation process. Given this importance, the following sections will describe a method to experimentally evaluate the magnetophoretic mobility of cells and particles and will provide representative measurements and mathematical relationships.

An instrument has been developed which can measure the magnetophoretic mobility of cells (or particles) on a cell-by-cell basis, which is in contrast to a SQUID that measures only bulk, average properties of cells or particles. A schematic diagram of this instrument, called cell-tracking velocimetry (CTV), is presented in Fig. 8.1. Key to this instrument is a region of constant force field strength, S_m (which is perpendicular to gravity) where the movement of immunomagnetically labeled cells or particles can be microscopically visualized and recorded to a PC hard drive as image files. These images are subsequently processed such that both the settling velocity and magnetically induced velocity of several hundred cells or particles at a time can be determined. Since the value of the force field strength, S_m [Eq. (5.6)] is constant and

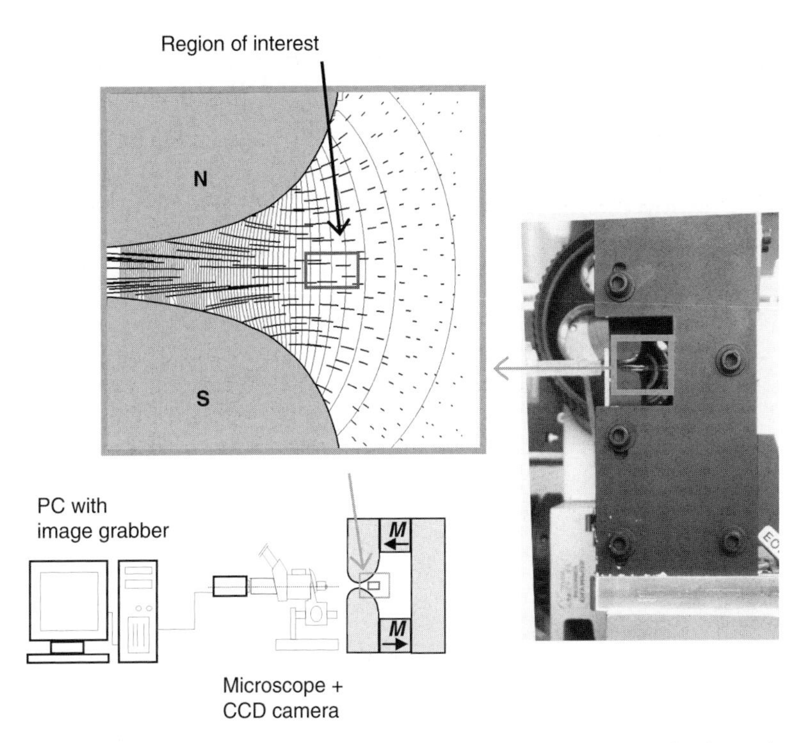

Fig. 8.1. Diagram of the cell tracking velocimetry system. Note isodynamic shape of the pole pieces (reviewed in Chapter 4) and a nearly constant length and direction of magnetic pathlines in the region of interest, in the field of view of the microscope and the camera. The vertical components of pathlines, due to gravitational sedimentation, were omitted for clarity.

accurately measured in the viewing region, the normalization of these measured velocities by S_m allows the magnetophoretic mobility to be determined; see Eq. (8.1). For example, Fig. 8.2(A) is a histogram of human, peripheral blood lymphocytes labeled with an anti-human CD3 antibody conjugated to a magnetic nanoparticle. As a point of comparison, Fig. 8.2 (B) is a flow cytometry histogram

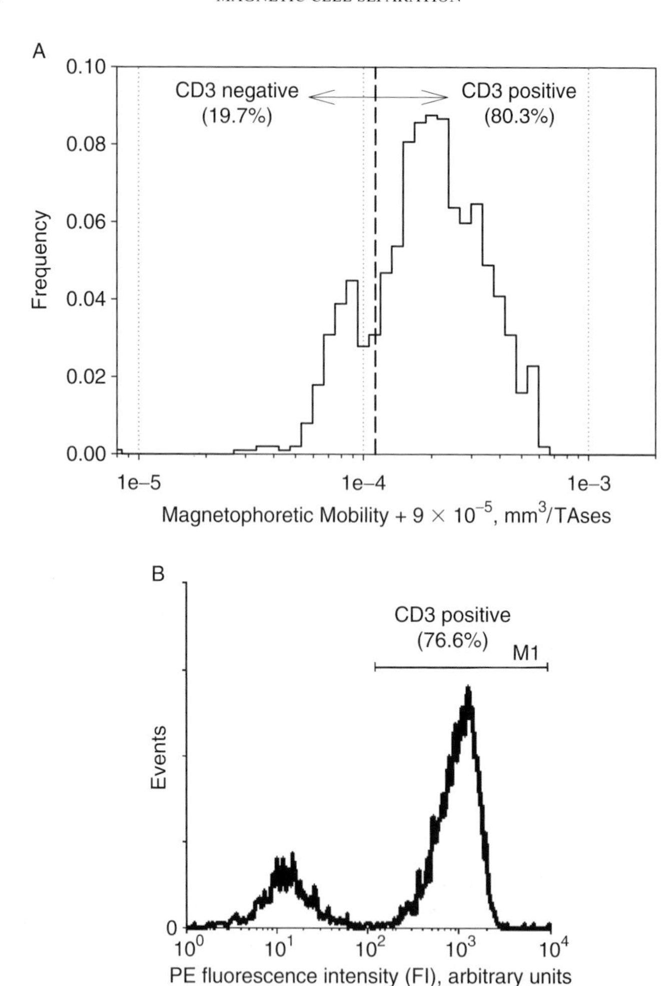

Fig. 8.2. (A) Histogram of peripheral blood lymphocytes labeled with mouse anti-human CD3 antibody conjugated to magnetic nanoparticles. A constant, 9.0e-05, has been added to each mobility so that negative data appear on the plot. A cutoff mobility of 1.1e-04 divides negative from positive mobilities. For comparison, (B) is a flow cytometry analysis histogram of frequency versus fluorescence intensity of the same sample labeled with anti-human CD3 antibody conjugated to the fluorescent probe phycoerythrin (PE).

of the same sample targeted with an anti-CD3 antibody conjugated to phycoerythrin (PE).

The initial version of the CTV instrument used permanent magnets [1]. However, a newer version of the CTV system using electromagnets has been developed, which allows not only the magnetic field to be cycled on an off, but also allows S_m to be varied over a 100-fold range. Figure 8.3 is an image of the trajectories of polystyrene particles tracked by the CTV software. The vertical portion of the lines (which represent the computer tracked trajectories) occurs when the electromagnet is off (the particles are settling) and the horizontal portion corresponds to the magnetically induced velocity. This version of the CTV system allows a number of basic assumptions of

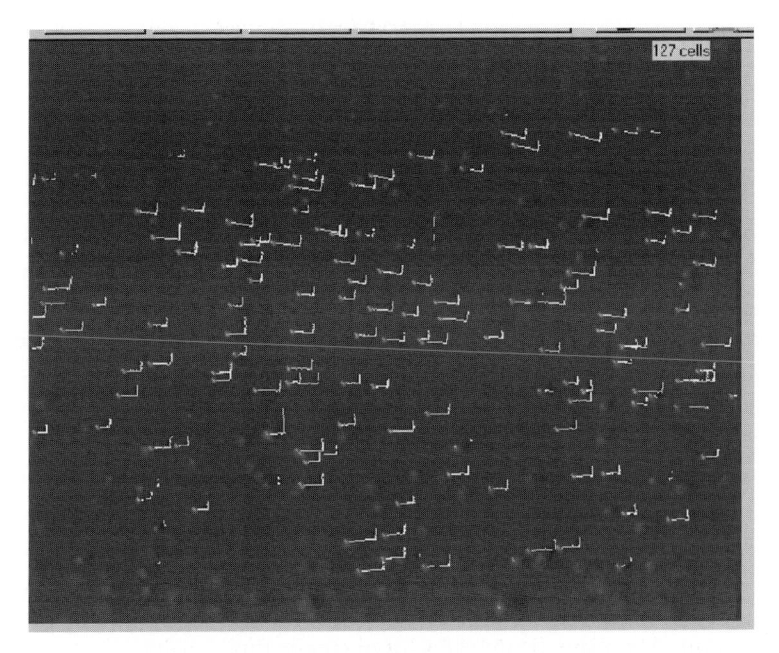

Fig. 8.3. Image of tracked particles obtained from CTV software. The vertical trajectory segments correspond to sedimentation in the absence of the magnetic field, while the horizontal segments correspond to movement in response to a magnetic force.

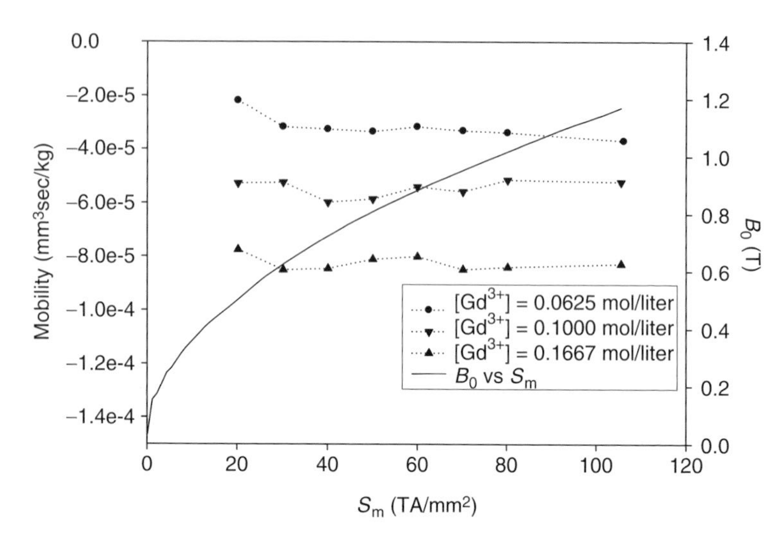

Fig. 8.4. Magnetophoretic mobility (left axis) and magnetic field (right axis) as a function of magnetic energy gradient, S_m, for polystyrene particles suspended in different concentrations of gadolinium.

Eq. (8.1) to be tested. For example, Fig. 8.4 is a plot of the magnetophoretic mobility as a function of the magnetic force field strength, S_m, for 6.7 μm polystyrene particles suspended in three different concentrations of chelated gadolinium ion. In this case, the polystyrene particles are diamagnetic, and the introduction of the paramagnetic gadolinium ions increases the difference in magnetic susceptibility between the particles and the solution. Each data point corresponds to the mean of over 1000 tracked particles. As can be observed, the mobility is constant over a six-fold increase in S_m and a three-fold increase in B_0, the flux density in free space (note the high upper limit of B_0). This result confirms the assumption that the magnetic susceptibilities of diamagnetic polystyrene and paramagnetic gadolinium are constant over a significant range of magnetic flux density.

Recently, it has been discovered that a number of strains of the bacteria *Bacillus*, when sporulated in a specific type of medium, will create spores with a measurable paramagnetism. Elemental analysis

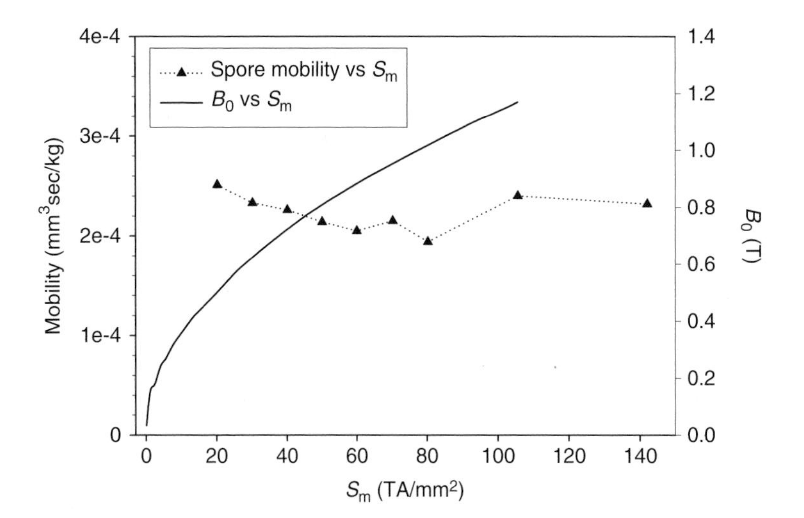

Fig. 8.5. Magnetophoretic mobility (left axis) and magnetic field (right axis) as a function of magnetic energy gradient, S_m, for *Bacillus globigii* spores.

reveals that the most likely element or compound that imparts this paramagnetism to the spores is manganese, or oxides of manganese. Figure 8.5 presents the measured magnetophoretic mobility as a function of S_m, further underscoring the assumption that for specific ranges of the magnetic field, the magnetophoretic mobility of spores containing molecular manganese is independent of the flux density.

However, in contrast to these previous examples of constant mobility—and thereby constant magnetic susceptibility over a significant range of magnetic field strengths — significant examples of the saturation of super paramagnetic material also exist. Most notably, this is encountered with the oxides of iron, and when these crystals are smaller than a single domain, about ten nanometers. In such cases the magnetic dipoles within the material are sufficiently aligned with the field that changes in field strength, within certain limits, produce no discernable effect on the magnetic moment. Figure 8.6 (A) presents the same polystyrene spheres presented in Fig. 8.4, but this time they

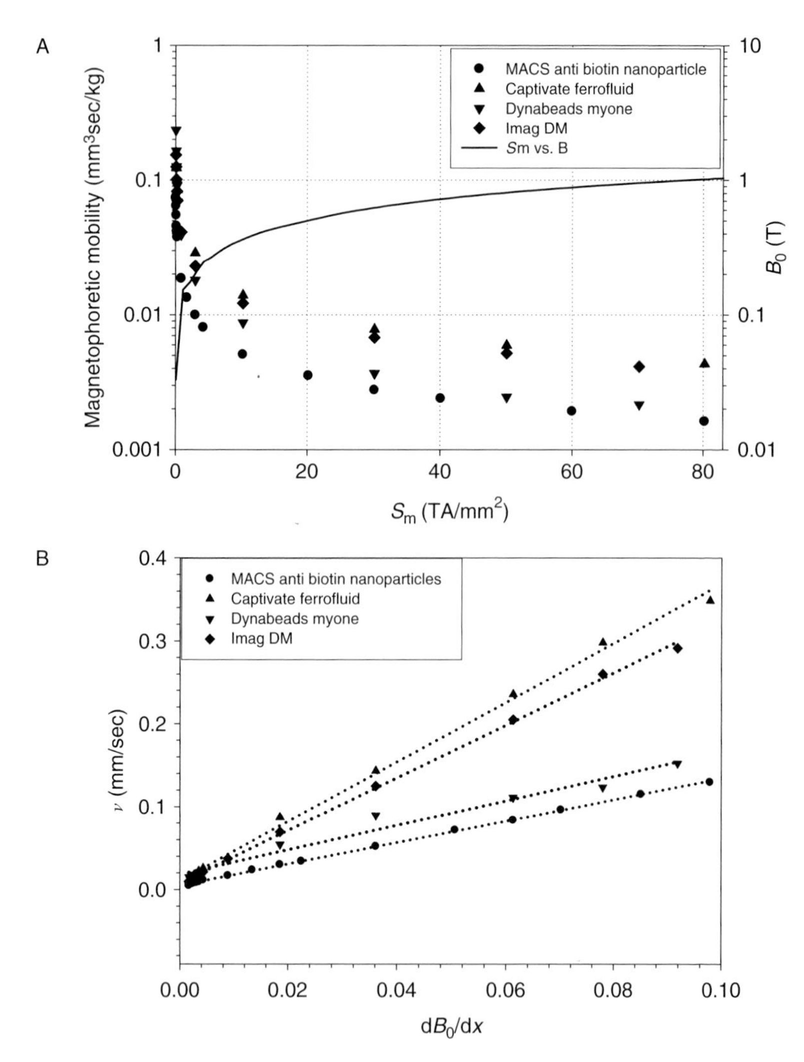

Fig. 8.6. (A) Magnetophoretic mobility (left axis) and magnetic field (right axis) as a function of magnetic energy gradient, S_m, for 6.7-μm polystyrene beads labeled with a number of different, commercial magnetic nanoparticles. (B) Magnetically induced velocity as a function of the magnetic field gradient for the same particles presented in (A).

are immunomagnetically labeled with a number of different, commercially available nanoparticles containing iron oxides. Unlike the previous examples, the magnetophoretic mobility significantly drops as the magnetic force field strength increases. (Note: the solid line corresponds to the measured magnetic flux density.)

Equation (8.1) predicts that when particle susceptibility is independent of the field, mobility is also independent of the external field. Figure 8.6 (A) suggests that this assumption is violated for magnetic nanoparticle labels. It can be shown (Eq. 5.10) that when the iron oxides become saturated, the magnetically induced velocity of the particles is a linear function of ∇B_0:

$$v = \frac{\mu}{6\pi\eta R} \nabla B_0 = \frac{MV}{6\pi\eta R} \nabla B_0 \tag{8.2}$$

where M is the magnetization of the particle, B_0 is the applied magnetic field and V is the volume. Figure 8.6 (B) is a plot of the data presented in Fig. 8.6 (A). However, in this case, the magnetically induced, horizontal velocity component is plotted against the horizontal component of ∇B_0, $\mathrm{d}B_0/\mathrm{d}x$. As can be observed, linear relationships are obtained, implying constant magnetization [Eq. (8.2)] occurring when the iron oxides are saturated. The gradient and flux density are dependent, so that at low gradient, the flux density is also low and the magnetization varies with B_0, and therefore, $\mathrm{d}B_0/\mathrm{d}x$, giving rise to a slight deviation from linearity in the plots at very low values.

8.4. Simultaneous measurements of sedimentation rate and magnetophoretic mobility

In addition to tracking the movement of cells in a magnetic energy gradient, and thereby calculating the magnetophoretic mobility, it is also possible with the CTV system to track the movement of cells (or particles) settling as a result of gravity. Figure 8.7 presents a comparison of the diameter of MCF-7 cancer cells determined with

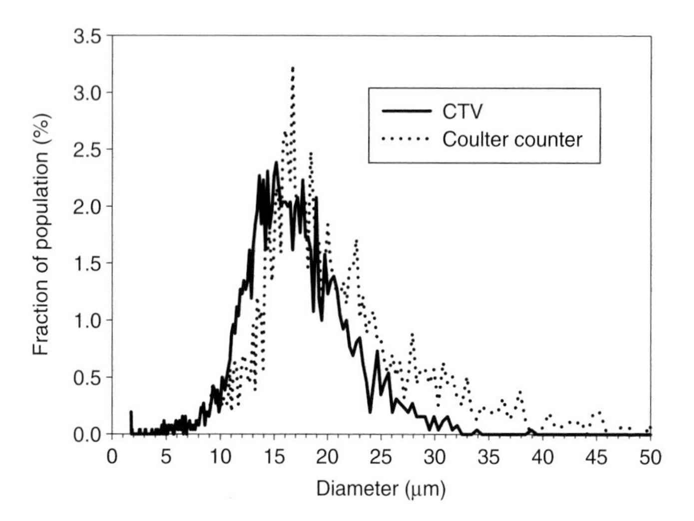

Fig. 8.7. Histograms of cell diameters of MCF-7 cells determined using a Coulter Counter and calculated from settling velocity measurements using the CTV system. An independently measured density of the MCF-7 cells of 1.0657 g/liter was used in the settling velocity calculation.

a Coulter Multisizer II and the CTV system. The latter required an independently measured average cell density 1.0657 g/cm^3, and the following relationship [2]:

$$D_c = \left[\frac{18\eta u_{\text{settling}}}{g(\rho_c - \rho_f)}\right]^{1/2} \tag{8.3}$$

η is the fluid viscosity, u_{settling} is the Stokes regime terminal velocity of settling, and ρ is the density of either the cell or the fluid (c or f).

As discussed previously, the electromagnetic CTV system allows sequential measurements of the settling velocity and magnetically induced velocity on a cell-by-cell, or particle-by-particle basis. Figure 8.8 presents MCF-7 cancer cells, immunomagnetically labeled with anti-HEA-FITC antibodies and anti-FITC-MACS antibody conjugates. The format used is a dot plot in which each "dot" corresponds to a specifically tracked cell.

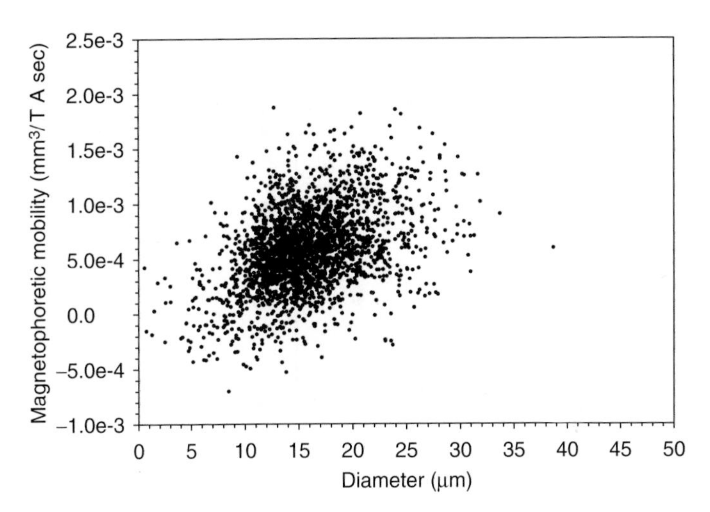

Fig. 8.8. Dot plot of the magnetophoretic mobility and diameter of MCF-7 cells labeled with anti-HEA-FITC and anti-FITC MACS micro beads.

8.5. Magnetophoretic mobility and antibody binding capacity (ABC)

As previously discussed (Chapter 5), magnetophoretic mobility is analogous to the electrophoretic and sedimentation mobilities encountered in electrical- and sedimentation-based separations. Excepting cases of magnetic saturation, the magnetophoretic mobility depends solely on the intrinsic properties of the magnetic carrier and the medium including the viscosity of the medium, the particle size, and magnetic susceptibilities of both the medium and the magnetic carrier, Eq. (8.1). However, when labeling a cell to impart a high magnetic susceptibility, a number of labeling scenarios are possible. Figure 8.9 presents several of these scenarios. For a cell labeled with a two-step process, the magnetic force imparted on the cell is the product of the number of bound nanoparticles and the force per nanoparticle:

$$F_{\mathrm{m}} = (n_1\theta_1\lambda_1)(n_2\theta_2\lambda_2)n_3F_{\mathrm{b}} \tag{8.4}$$

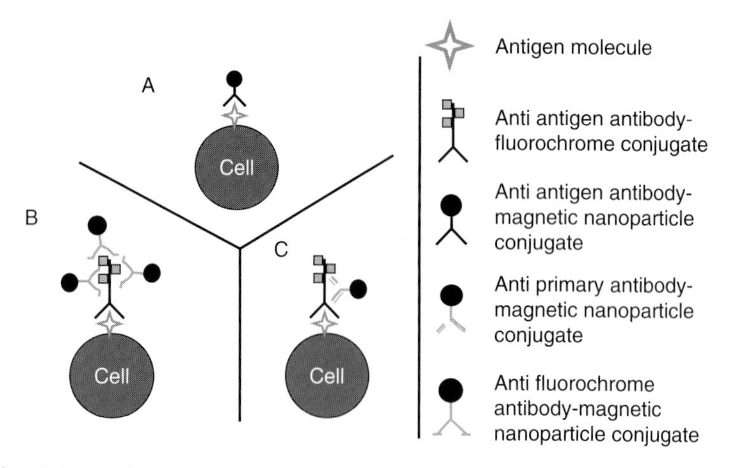

Fig. 8.9. Various immunomagnetic labeling scenarios: one-step (A) and two-step (B and C).

where the subscripts "1" and "2" refer to the primary and secondary labeling antibodies (*Ab*), n_1 is the number of antigen (*Ag*) binding sites per cell, including specific and nonspecific antigen sites ($n_s + n_{ns}$), and θ_1 is the fraction of antigen molecules bound by primary antibody. The parameter, λ_1, represents the valence of the primary antibody binding. The combined term $n_1\theta_1\lambda_1$ is equivalent to the commonly used term "antibody binding capacity" (ABC) of a cell population [3]. Antibody binding capacity is a measure of the number of primary antibodies binding to a cell or microbead. This value includes not only the number of antigen molecules per cell, but also accounts for variables such as the valence of antibody binding, steric hindrance, binding affinities, and nonspecific binding.

The same sequence of parameters is then repeated for the binding of the secondary antibody to sites on the primary antibody. In this case, n_2 is the number of binding sites on the primary antibody recognized by the secondary antibody, θ_2 is the fraction of binding sites on the primary antibodies that are bound by secondary antibodies, and λ_2 represents the valence of the secondary antibody binding. These terms $n_2\theta_2\lambda_2$ can then be combined into one overall term, ψ, representing the antibody amplification due to the ratio of secondary antibodies per primary antibody. The parameter n_3 represents the number of magnetic nanoparticles conjugated to the antibody, in this example, the secondary antibody. Combining parameters $ABC\psi n_3$ into one overall term, N_{mp}, gives a value that represents the number of magnetic nanoparticles bound to each cell or microbead, and is therefore referred to as the "magnetic particle binding capacity" of a cell or microbead. The parameter F_b is the magnitude of magnetic force acting on a single magnetic nanoparticle in the direction of the magnetic energy gradient

$$\mathbf{F}_b = \frac{1}{2\mu_0}\Delta\chi V_m \nabla B_0^2 \tag{8.5}$$

where μ_0 is the magnetic permeability of free space; $\Delta\chi$ is the difference in magnetic susceptibility between the magnetic nanoparticle, χ_b, and the surrounding medium, χ_f; V_m is the volume of magnetic material per nanoparticle; and B_0 is the applied magnetic flux density. Note that the product, $\Delta\chi V_m$, can also be represented by the term, ϕ, the magnetic field interaction parameter [4].

In terms of magnetophoretic mobility, one can use this concept of ABC to write the mobility for one-step labeling, as

$$m = \frac{ABC\psi n_3\phi}{3\pi D_c\eta} = \frac{(n_1\theta_1\lambda_1)n_3\phi}{3\pi D_c\eta} \tag{8.6}$$

And for a two-step labeling, the mobility is

$$m = \frac{ABC\psi n_3\phi}{3\pi D_c\eta} = \frac{(n_1\theta_1\lambda_1)(n_2\theta_2\lambda_2)n_3\phi}{3\pi D_c\eta} \tag{8.7}$$

These equations differ in that in Eq. (8.6), ψ is just equal to unity. From Eq. (8.7), one can observe that the magnetophoretic mobility, m, of a two-step immunomagnetically labeled cell is a function of the parameters ABC, ψ, n_3, $\Delta\chi V_m$, as well as D_c and η. The viscosity of the carrier solution, η, varies in a highly predictable manner over the temperature range of interest (between room temperature and 0°C). Also, the functional number of magnetic carriers conjugated per secondary binding antibody, n_3, is usually 1 (or possibly less than 1 in some cases). Therefore, for constant temperature and carrier fluid composition, the most influential parameters on magnetophoretic mobility are *ABC*, ψ, $\Delta\chi V_m$, and D_c. The magnetophoretic mobility has been reported in units of mm^3/T A s; however, it can also be reported in the more fundamental units of mm^3 s/kg, the magnitude of both units being equivalent.

8.6. Antibody-magnetic nanoparticle binding to cells

The concept of ABC, Eqs. (8.6) and (8.7), and the known mechanisms by which antibodies bind to antigens suggest that a "saturation binding mechanism" exists. If one assumes monovalent binding, at equilibrium the interaction of a receptor–ligand complex can be expressed by

$$R + L \underset{k_d}{\overset{k_a}{\rightleftharpoons}} R \bullet L \tag{8.8}$$

where k_a and k_d are the rate constants for the association and dissociation reactions, and $[R]$, $[L]$, and $[R \bullet L]$ are the concentrations of free receptors, free ligands, and receptor–ligand complexes at equilibrium, respectively. Two constants are widely used to characterize the strength of this interaction: an equilibrium dissociation constant, K_D, and an equilibrium association constant, K_A, given by

$$K_D = \frac{k_d}{k_a} = \frac{[R][L]}{[R \bullet L]} = \frac{1}{K_A} \tag{8.9}$$

It is generally reported that K_D is a quantitative indicator of the stability of receptor–ligand interactions, with low values representing stable (strong) interactions and high values representing weak interactions [5]. Published values of K_D can range from 10^{-15} for avidin–biotin to 10^{-7} to 10^{-11} for antibody–antigen binding [6]. It should be noted that the dissociation constant, K_D, is based on the concept of thermodynamic equilibrium for species suspended in a homogeneous solution [7].

In contrast, when an antibody, or the ligand for a specific antibody, is bound to a solid surface or support (such as a cell or an ELISA plate), the assumption of freely suspended antibodies, ligands, and antibody–ligand conjugates is not valid. Consequently, techniques such as the classical Scatchard analysis are not necessarily valid. This observation is well documented [8, 9]. Given the added complexity of such heterogeneous phase systems, such as the binding of an antibody to the antigen on a cell surface or on an ELISA plate, a typical approach is to use the term "apparent binding constant," which may not have values close to the true binding constants obtained in suspension. This apparent binding constant takes into consideration the potential of antibody valence, steric hindrance, or other nonideal effects [7]. In order to determine this apparent binding constant, the following model was constructed [10].

Taking the antibody–antigen interaction as an example, when an antibody–conjugate binds to a cell, at least five scenarios can occur: (A) monovalent binding, (B) homogeneous bivalent binding, (C) multiple antibodies binding to a single antigen, (D) heterogeneous bivalent binding, and (E) crosslinked binding. Figure 8.10 presents examples of each of these five cases. This complexity is partially the result of the possibility of multiple binding epitopes per surface antigen, as summarized in the literature [11], and discussed more thoroughly as follows.

Of the five scenarios presented in Fig. 8.10, in the following discussion we will only consider Scenarios A, B, and C. The dissociation constant for Scenario A is given by

$$K_{D1} = \frac{[Ag][Ab]}{[Ag \bullet Ab]} \tag{8.10}$$

and potentially for Scenarios B and C:

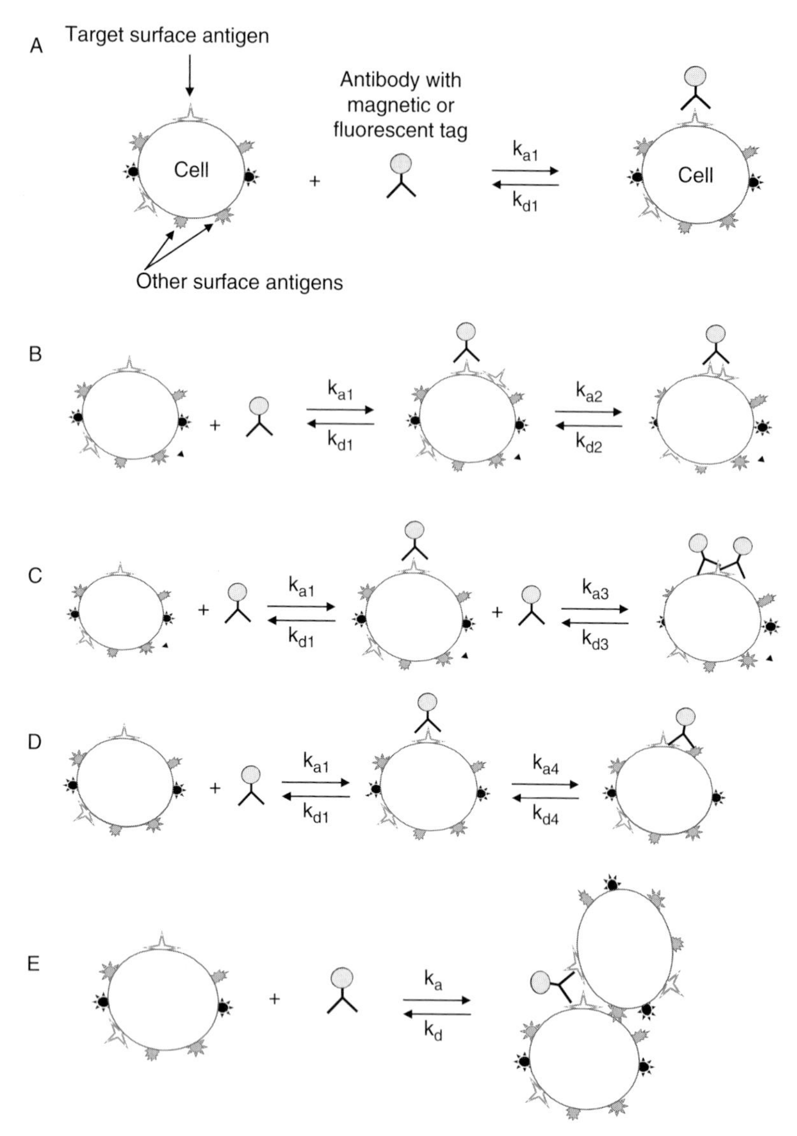

Fig. 8.10. Various antibody receptor interactions on a cell surface: (A) monovalent binding, (B) homogeneous bivalent binding, (C) multiple antibodies binding to a single antigen, (D) heterogeneous bivalent binding, and (E) crosslinking [10].

$$K_{D1} = \frac{[Ag][Ab]}{[Ag \bullet Ab]} \quad K_{D2} = \frac{[Ag][Ag \bullet Ab]}{[Ag \bullet Ab \bullet Ag]} \quad K_{D3} = \frac{[Ab][Ag \bullet Ab]}{[Ab \bullet Ag \bullet Ab]} \tag{8.11}$$

It should be noted that $[Ag]$, $[Ab]$, and $[Ag \bullet Ab]$ represent the concentrations of free antigen, free antibody, and bound antibody–antigen complexes, respectively. A mass balance can be written for Scenario A for the antigen:

$$[Ag]_{\text{total}} = [Ag] + [Ag \bullet Ab] \tag{8.12}$$

If one were to substitute Eq. (8.12) into Eq. (8.10), and solve for $[Ag \bullet Ab]$, one would obtain:

$$[Ag \bullet Ab] = \frac{[Ab][Ag]_{\text{total}}}{K_{D1} + [Ab]} \tag{8.13}$$

Finally, dividing both sides of Eq. (8.13) by $[\text{Ag}]_{\text{total}}$ one obtains the classical form of the Langmuir Isotherm:

$$\theta_1 = \frac{[Ab]}{K_{D1} + [Ab]} \tag{8.14}$$

where θ_1 corresponds to the fraction of the total surface antigen sites bound with antibody, as described above. For Scenarios B and C, the situation becomes more complex. For Scenario B:

$$[Ag]_{\text{total}} = [Ag] + [Ag \bullet Ab] + 2[Ag \bullet Ab \bullet Ag] \tag{8.15}$$

and for Scenario C:

$$[Ag]_{\text{total}} = [Ag] + [Ag \bullet Ab] + [Ab \bullet Ag \bullet Ab] \tag{8.16}$$

When one uses either a flow cytometer (FCM), to quantify the fluorescence intensity of a cell labeled with an antibody–fluorochrome conjugate, or a CTV, to quantify the magnetophoretic mobility of a cell labeled with an antibody–magnetic nanoparticle conjugate, one is not able to distinguish between $[Ag \bullet Ab]$ or $[Ag \bullet Ab \bullet Ag]$. In contrast, in the case of $[Ab \bullet Ag \bullet Ab]$, if the number of Ag is known, in an ideal case, the maximum number of antibodies determined by FCM or CTV would be double that of Ag. Consequently, the term $[complex]$ is introduced to evaluate the concentration of the bound antibody, for Scenario B:

$$[complex] = [Ag \bullet Ab] + [Ag \bullet Ab \bullet Ag] \tag{8.17}$$

and for Scenario C:

$$[complex] = [Ag \bullet Ab] + 2[Ab \bullet Ag \bullet Ab] \tag{8.18}$$

Next, the concept of the antibody binding valence, λ, is introduced as the ratio of concentrations of bound antibody to bound antigen. For Scenario B we have

$$\lambda = \frac{[Ag \bullet Ab] + [Ag \bullet Ab \bullet Ag]}{[Ag \bullet Ab] + 2[Ag \bullet Ab \bullet Ag]} = \frac{[complex]}{[Ag \bullet Ab] + 2[Ag \bullet Ab \bullet Ag]} \tag{8.19}$$

and for Scenario C:

$$\lambda = \frac{[Ag \bullet Ab] + 2[Ab \bullet Ag \bullet Ab]}{[Ag \bullet Ab] + [Ab \bullet Ag \bullet Ab]} = \frac{[complex]}{[Ag \bullet Ab] + [Ab \bullet Ag \bullet Ab]} \tag{8.20}$$

Inspection of Eqs.(8.19) and (8.20) indicates that if all the antibodies bind to the cell in a monovalent nature (Scenario A), the concentrations of $[Ag \bullet Ab \bullet Ag]$ and $[Ab \bullet Ag \bullet Ab]$ are zero and the valence for Scenarios B and C is 1. In contrast, if all the antibody binding is of the homogeneous, bivalent nature (Scenario B), $[Ag \bullet Ab]$ is zero and the valence is 1/2; while for Scenario C, if all of the antigen binding is bivalent the valence is 2, as it should be.

Experimental evidence shows variability of the valence and dissociation constant, presumably due to a variety of factors, which cannot be predicted from first principles. Consequently, it may be shown that Eq. (8.13) should be modified for the case represented in Scenario B:

$$[Ag \bullet Ab] + 2[Ag \bullet Ab \bullet Ag] = \frac{\lambda[Ab][Ag]_{\text{total}}}{\alpha K_{\text{D1}} + [Ab]} \tag{8.21}$$

An equation similar to that of Eq. (8.21) may be written for Scenario C:

$$[Ag \bullet Ab] + [Ab \bullet Ag \bullet Ab] = \frac{\lambda[Ab][Ag]_{\text{total}}}{\alpha K_{\text{D1}} + [Ab]} \tag{8.22}$$

Again, normalizing by dividing out the total antigen concentration yields for scenarios B and C:

$$\theta_1 = \frac{\lambda[Ab]}{\alpha K_{\text{D1}} + [Ab]} \tag{8.23}$$

where λ, as defined above, is the valance of the antibody binding and the term α is an experimentally determined constant. It should be noted that most likely, α is a function of λ as well as the availability of the specific epitopes in cell membrane and membrane fluidity. Given this derivation, and the case for the use of an antibody with a valence of 1, one can rewrite Eq. (8.6) to obtain

$$m = \frac{(n_1\theta_1\lambda_1)n_3\phi}{3\pi D_c\eta} = \left[\frac{[Ab]}{\alpha K_{\text{D1}} + [Ab]}\right]\frac{n_1 n_3\phi}{3\pi D_c\eta} \tag{8.24}$$

8.7. Effect of antibody labeling concentration (antibody titration) on magnetophoretic mobility

Equation (8.24) states that the magnetophoretic mobility of labeled cells is a function of the antibody concentration used to label the cells. Figure 8.11 is a plot of the number of PE molecules bound to a human lymphocyte cell as a function of the concentration of the anti-CD3-PE conjugate in the solution surrounding the cell [12]. Of special note is the observation that the highest data points on the graph correspond to the use of antibodies directly out of the bottle with no dilution in buffer. In addition to antibody–fluorochrome saturation, as one might imagine, the saturation of antibody–magnetic nanoparticle conjugates also is observed. Figure 8.12 is a plot of multiple histograms of MCF-7 breast cancer cells labeled with the same concentration of a primary, anti epithelial surface

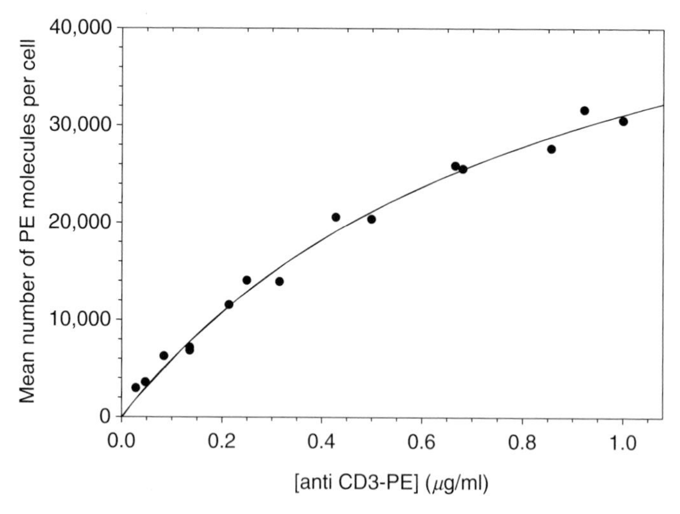

Fig. 8.11. Mean number of PE molecules bound to a cell as a function of the concentration of the anti-CD3-PE antibody in equilibrium with the antibodies bound. The mean number of PE molecules was determined using a calibration curve obtained by the use of BD QuinteBrite beads [10].

marker–FITC conjugate, and various concentrations of anti-FITC magnetic nanoparticle conjugate. An almost two order-of-magnitude increase in mobility can be observed as the concentration of antibody is increased. Figure 8.13 is a plot of mobility versus concentration of secondary antibody for a number of different labeling concentrations on MCF-7 cells [13].

A further implication of this "saturation phenomena" is the importance of the binding affinity/avidities of the antibody conjugate for the surface receptor. As presented above, this affinity/avidity of an antibody is typically quantified using the value of the equilibrium dissociation constant, K_D. An implication of the saturation relationship, Eq. (8.24) is that smaller values of K_D result in lower concentrations of antibody needed to achieve the same level of receptor saturation. Since magnetic separation systems are typically more efficient when the targeted cells have the highest magnetophoretic mobility, it is desirable from an economic

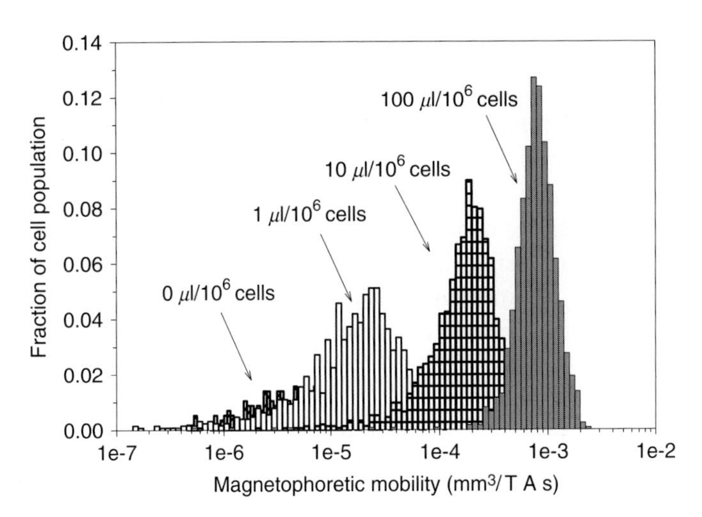

Fig. 8.12. Histograms of the magnetophoretic mobility of MCF-7 cancer cells labeled with a primary antibody of anti-ESA-FITC conjugate and various concentrations of anti-FITC MACS beads [10].

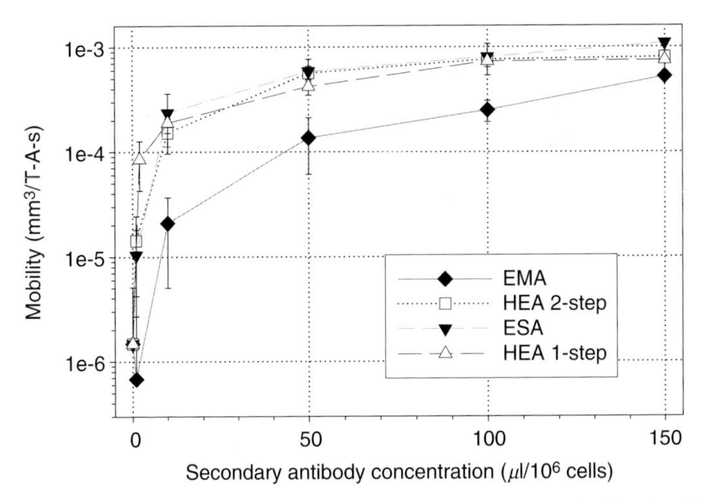

Fig. 8.13. Saturation curve of the magnetophoretic mobility of MCF-7 cells as a function of the antibody concentration for a number of various antibody labeling combinations [10].

point of view (especially when cell separations are scaled-up) to have antibody–magnetic conjugates with the lowest values of K_D. Unfortunately, it has recently been reported that the conjugation of increasingly larger fluorescent molecules results in an increase in the value of K_D by a factor of 2 and a further increase by a factor of 6, when a magnetic nanoparticle is conjugated to the antibody [10]. Practically, this results in an increase by the same factor in the concentration of antibody needed to achieve the same level of bound cell surface marker concentration.

References

[1] Reddy, S., Moore, L. R., Sun, L., Zborowski, M. and Chalmers, J. J. (1996). Determination of the magnetic susceptibility of labeled particles by video imaging. Chem. Eng. Sci. *51*, 947–56.

[2] Nakamura, M., Lasky, L., Zborowski, M. and Chalmers, J. J. (2000). Theoretical and experimental analysis of the accuracy of cell tracking velocimetry. Exp. Fluids *30*, 371–380.

[3] McCloskey, K. E., Chalmers, J. J. and Zborowski, M. (2003). Magnetic cell separation: Characterization of magnetophoretic mobility. Anal. Chem. *75*, 6868–74.

[4] McCloskey, K. E., Comella, K., Chalmers, J. J., Margel, S. and Zborowski, M. (2001). Mobility measurements of immunomagnetically labeled cells allow quantitation of secondary antibody binding amplification. Biotechnol. Bioeng. *75*, 642–55.

[5] Midelfort, K. S., Hernandez, H. H., Lippow, S. M., Tidor, B., Drennan, C. L. and Wittrup, K. D. (2004). Substantial energetic improvement with minimal structural perturbation in a high affinity mutant antibody. J. Mol. Biol. *343*, 685–701.

[6] Garcia, A. A., Bonen, M. R., Ramirez-Vick, J., Sadaka, M. and Vuppu, A. (1999). Bioseparation process science. Blackwell Science Inc., Malden, Massachusetts.

[7] Goldberg, M. E. and Djavadi-Ohaniance, L. (1993). Methods for measurement of antibody/antigen affinity based on ELISA and RIA. Curr. Opin. Immunol. *5*, 278–81.

[8] Underwood, P. A. (1993). Problems and pitfalls with measurement of antibody affinity using solid phase binding in the ELISA. J. Immunol. Methods *164*, 119–30.

[9] Bobrovnik, S. A. (2003). Determination of antibody affinity by ELISA. Theory. J. Biochem. Biophys. Methods *57*, 213–36.

[10] Zhang, H., Williams, P. S., Zborowski, M. and Chalmers, J. J. (2006). Reduction of binding affinities/avidities of antibody-antigen and streptavidin-biotin: Quantification and scale-up implications. Biotechnol. Bioeng. *95*, 812–29.

[11] Davis, K. A., Abrams, B., Iyer, S. B., Hoffman, R. A. and Bishop, J. E. (1998). Determination of CD4 antigen density on cells: Role of antibody valency, avidity, clones, and conjugation. Cytometry *33*, 197–205.

[12] Zhang, H., Moore, L. R., Zborowski, M., Williams, P. S., Margel, S. and Chalmers, J. J. (2005). Establishment and implications of a characterization method for magnetic nanoparticle using cell tracking velocimetry and magnetic susceptibility modified solutions. Analyst *130*, 514–27.

[13] Chosy, E. J., Nakamura, M., Melnik, K., Comella, K., Lasky, L. C., Zborowski, M. and Chalmers, J. J. (2003). Characterization of antibody binding to three cancer-related antigens using flow cytometry and cell tracking velocimetry. Biotechnol. Bioeng. *82*, 340–51.

Laboratory Techniques in Biochemistry and Molecular Biology, Volume 32
Magnetic Cell Separation
M. Zborowski and J. J. Chalmers (Editors)

CHAPTER 9

Preparative applications of magnetic separation in biology and medicine

Jeffrey J. Chalmers,[1] Xiaodong Tong,[1] Oscar Lara,[1] and Lee R. Moore[2]

[1]Department of Chemical and Biomolecular Engineering, University Cell Analysis and Sorting Core, The Ohio State University, OH 43210, USA,
[2]Department of Biomedical Engineering, Lerner Research Institute, Cleveland Clinic, Cleveland, OH 44195, USA

9.1. Introduction

A complete review of the use of magnetic cell separation technology for preparation of cell samples for biological or clinical use is beyond the scope of this chapter. For example, a Pub Med search on the phrase "magnetic cell separation" in August 2006 listed over 1200 "hits." However, what can be reviewed are examples of separations using specific mature methodologies. In addition, the types of entities on cells that are immunomagentically targeted will also be reviewed.

9.1.1. Types of methodologies used in magnetic cell separation

There are a number of methodologies that practioners of magnetic cell separation technologies use: (1) positive selection of the targeted cell population, (2) depletion of undesired cells from a mixed population by targeting the undesired cell population (e.g., removal of a rare cancer cells from a blood sample), and (3) enrichment of

DOI: 10.1016/S0075-7535(06)32009-8

the targeted population by removal of at least one population of undesired cells (e.g., immunomagnetically targeting CD45-positive lymphocytes in a peripheral blood sample to enrich for progenitor cells).

9.1.2. *Types of entities targeted for magnetic cell separation*

The versatility and relative ease in the creation, production, and usage of antibodies makes them the most commonly used "labeling entity" or ligands for magnetic cell separation. However, a number of entities, in addition to antibodies, are currently used to make targeted cells magnetic for subsequent separation. Examples include streptavidin–biotin, nonspecific or specific endocytosis of magnetic nanoparticles, specific receptor ligands conjugated to magnetic nanoparticles, and DNA or RNA oligonucleotides.

9.1.3. *Measures of performance*

While there are no universal parameters that are used to evaluate the performance of magnetic cell separation, or cell separation, in general, a number of reasonably common terms can be used. These terms include the initial and final *purities* of the sample, the *depletion* (usually given in a $\log_{10}$ format) of an undesirable cell population, the *recovery* of the targeted cell population after a separation, and the *enrichment* of the targeted cell population (also usually given in a $\log_{10}$ format). Mathematically, these terms are defined as:

$$purity = P_t = \frac{N_t}{N_t + N_{nt}} \tag{9.1}$$

$$recovery = \frac{N_{t,\,final}}{N_{t,\,initial}} = \left(\frac{N_{final} \cdot P_{t,\,final}}{N_{initial} \cdot P_{t,\,initial}}\right) \tag{9.2}$$

$$\log_{10} \textit{depletion} = \log\left(\frac{N_{\text{initial}} \cdot P_{\text{nt, initial}}}{N_{\text{final}} \cdot P_{\text{nt, final}}}\right) = -\log(\textit{recovery}_{\text{nt}}) \quad (9.3)$$

$$\textit{enrichment} = \left(\frac{P_{\text{t, final}}}{P_{\text{t, initial}}}\right) \quad (9.4)$$

where N refers to the total number of cells, the subscripts "t" and "nt" refer to targeted and nontargeted cells, "initial" refers to the initial cell population (preseparation), and "final" refers to the final cell population (postseparation).

9.2. *Examples of positive selection*

The range of cells that can be positively targeted for magnetic separation is only limited by the number of distinct affinity labels that are available; consequently, a complete review is effectively impossible. However, if one were to group positive magnetic selection into types of cells commonly separated, these would include rare cancer cells in peripheral blood and bone marrow, progenitor cells in peripheral blood, bone marrow and neonatal umbilical cord blood, contaminating bacteria in food, and various lymphocyte subpopulations differentiated by "cluster of differentiation" (CD) markers.

Table 9.1 summarizes some of the more recent studies using positive magnetic cell separation to select for rare cancer cells in peripheral blood and bone marrow. In addition to listing the type of cancer, blood sample source, magnetic cell separation technology, and surface marker targeted, if provided, the performance of the separation is summarized. Table 9.2 summarizes some of the more recent studies using positive magnetic cell separation to select for hematopoetic stem cells from human blood and bone marrow.

9.3. *Depletion of undesirable cells*

As with positive selection, the potential applications for depletion of undesired cells are only limited by the availability of affinity labels. However, clinical demand significantly reduces the number

TABLE 9.1

Examples of immunomagnetic separation of cancer cells from blood and bone marrow using positive selection methods

Target cell	Cell sample	Antigen for IMS	Detection method	Initial purity	Recovery of target cells	Enrichment	Sensitivity	References
MACS (Miltenyi Biotec, Germany)								
Renal carcinoma CL	Caki-1in PBL	CK	ICC	$1/10–1/10^7$	73.6%	–	–	[7]
Breast cancer Patients	PB, BM, PBPC	HEA-125	ICC	$6/10^6$	24.4%	6.8	–	[8]
Melanoma CL	SK-Mel-30 in PB	MCSP	ICC	50–1000 per 50 ml PB	11.4%	–	–	[9]
Breast cancer CL	BT474 in PBL	CK-8	ICC, FC	$4/10^8–4/10^5$	57.7%	10477 (two IMS)	–	[10]
Dynabeads (Dynal, Norway)								
Cancer CL	4 CLs in PBL	Ber-EP4	ICC	?	9–45%	–	–	[11]
Urological malignant CL	5 CLs in PBL	Ber-EP4	ICC	$4/10^5–4/10^8$	61.2%	15.3	–	[12]
Head and neck SCC	PB, BM	Ber-EP4	ICC	$1–25/10^6$	22–32%	–	–	[13]
Breast cancer	PB, BM	MUC-1	RT-PCR	$1/10^3–1/10^8$	–	–	$1/1 \times 10^8$ MNC	[14]

Colorectal cancer CL	SW-480 in PBL	Ber-EP4	RT-PCR	$1/10^4$–$1/10^6$	–	–	$1/1 \times 10^6$ MNC	[15]
Colon caner CL	SNUC4 in PB	CEA	RT-PCR	1–10^6/1ml PB	–	–	10/1ml Blood	[16]
Breast cancer	BM	EpCAM	ICC	–	–	29	$10/5 \times 10^7$ MNC	[17]
Immunicon (USA)								
Breast cancer CL	SKBr3 in PB	EpCAM	ICC, FC	50–4500 per 5 ml PB	76%	–	1/1 ml Blood	[18]
Prostate cancer CL	PC3 in PB	EpCAM	FC	25–800 per 7 ml PB	73.7%	–	–	[19]
Breast cancer CL	Colo-205 in PB	EpCAM	ICC, FC	1–200 per 5 ml PB	60–90%	–	–	[20]
Prostate cancer	PB	EpCAM	FC	–	–	–	1/7.5 ml Blood	[21]
Magnetic deposition microscopy (CCF, USA)								
Breast cancer CL	MCF-7 in PBL	EMA	ICC, FC	$1/10^6$–$1/10^8$	20–60%	–	$1/1 \times 10^7$ MNC	[22]
QMS (OSU, USA)								
Breast cancer CL	HCC1954 in PBL	HER-2/ neu	FC	$1/10^2$	66–89%	5–50	$1/1 \times 10^7$ MNC	[22]

TABLE 9.2
Examples of magnetic separation of hematopoietic stem cells from blood and bone marrow using positive selection methods. In each case, the detection method was flow cytometry analysis

Selection method	Type of blood sample	Antigen targeted	Final progenitor cell purity (%)	Recovery of HSC (%)	References
Isolex 300i	Fresh leukapheresis	CD34	97	57	[23]
Isolex 300i	Cryopreserved leukapheresis	CD34	94	47.2	[24]
Isolex 300SA	Peripheral mononuclear cells	CD34	92.8	69.8	[25]
Isolex 300i	Fresh leukapheresis	CD34	84.3	51.4	[26]
Isolex 300i	Cryopreserved leukapheresis	CD34	98.0	51.0	[27]
MiniMACS	Bone marrow	CD34	78.4	68.9	[28]
SuperMACS-SuperMACS	Fresh leukapheresis	CD34	97	98	[29]
CliniMACS	Fresh leukapheresis	CD133	90	80.6	[4]
CliniMACS	Fresh leukapheresis	CD34	90	77.3	[4]
CliniMACS	Fresh leukapheresis	CD34	97.5	46.2	[30]
CliniMACS	Fresh leukapheresis	CD34	96.1	52.3	[31]
CliniMACS	Fresh leukapheresis	CD34	85.1	74.2	[32]
Dynabeads	Bone marrow	CD34	33.9	34.5	[28]
QMS	Fresh leukapheresis	CD34	90 ($n = 11$)	39 ($n = 11$)	[33]
QMS	Cryopreserved leukapheresis	CD34	77 ($n = 11$)	30 ($n = 11$)	[33]

of prevalent examples of this type of separation methodology. One of the more common separations is the depletion of T-cells for potentially mismatched bone marrow transplants. Table 9.3 presents examples of immunomagnetic approaches to remove unwanted T-cells.

9.4. *Enrichment of rare cells by depletion of normal cells*

The rise in the use of molecular analysis technology including RT-PCR and microarrays has led to the desire to detect specific markers in blood. Like other highly specific markers, the potential targets for RT-PCR are only limited by the availability of specific primers. However, it has also been reported that the illegitimate expression of the targeted molecule by normal cells hinders this approach, and it is impossible for the molecular approaches to conclusively identify the targeted molecule (typically contained in a targeted cell) when its concentration is very low in a cell population.

For example, numerous studies have shown that conventional RT-PCR can be used to detect circulating tumor cells in body fluids of patients with various carcinomas, and that the sensitivity is one cancer cell per 10^5 or 10^6 mononuclear cells. However, the detection sensitivity strongly depends on the type of tumor, the choice of the amplified gene, the operative skill of the researcher, and the efficiency of the experiment. Variation in the degree of RNA degradation, the efficiency of RNA extraction, the amount of RNA used in RT-PCR, and the PCR efficiency may lead to false negatives, even though a sufficient amount of mRNA template for detection may be present. Additionally, the expression level of the target gene is often up-regulated or down-regulated in cultured cells [1–3].

To overcome the aforementioned problems of the present techniques, it is highly desirable to apply an enrichment step prior to the actual detection procedure. In the recent literature, immunomagnetic cell separation has been used as one of the most

TABLE 9.3

Removal of unwanted T-cells from blood samples using either a positive selection method, a depletion method, or a combined method of positive selection of HSC followed by a negative depletion of T-cells

Selection method	Antigen targeted	Detection method	Initial T-cell purity (%)	Final T-cell purity (%)	$\log_{10}$ T-cell depletion[a]	Recovery of HSC (%)	References
Positive selection of HSC							
Isolex 300i cell selection system	CD34	FC	34.3	0.04	5.1	57	[23]
CEPRATE LC	CD34	FC, LDA	23.5 ± 6.0	~7.8	2–3	57 ± 13	[34]
			N/A	0.3 ± 0.3	4[b]	46 ± 24	
Two step MACS systems	CD34	FC	N/A	0.1 ± 0.19	5[c]	84.2 ± 20.5	[29]
			N/A	0.12 ± 0.19	5[d]	84 ± 20	
CliniMACS	CD133	FC	28.4	0.09	3.75	80.6	[35]
CliniMACS	CD34	FC	32.4	0.06	4.1	77.3	[35]
CliniMACS	CD34	FC	9.4–9.66	0.05	4	46.2	[30]

Depletion of T-cells							
CliniMACS	CD3	FC	35.8	0.15	3.4	82	[36]
Super MACS	CD3	FC	16.2 ($n = 10$)	0.06 ± 0.04 ($n = 10$)	2.89 ($n = 10$)	63 ± 8.5%	[4]
QMS	CD3	FC	22 ($n = 16$)	0.03 ± 0.04 ($n = 16$)	3.98 ($n = 16$)	57.9 ± 8.5% ($n = 16$)	[4]
Positive selection of HSC followed by depletion of T-cells							
Isolex 300i cell selection and MaxSep device	CD34 CD3	FC	30.5	0.18	4.6	46	[23]
CEPRATE SC	CD34 CD2	FC	97.41	1.9	4.2[e]	70.5	[10]
Isolex 300i cell selection system	CD34 CD3	FC	N/A	N/A	4.72	47.2	[24]

[a]Unless otherwise stated, no information available on the log depletion calculation.
[b]Depletion in two steps using CellPro and VarioMACS.
[c]Depletion in two steps using SuperMACS and VarioMACS.
[d]Depletion in two steps using SuperMACS and SuperMACS.
[e]Log depletion calculation based on cell number pre- and post-depletion.

promising techniques because of its specificity, rapidity, and high efficiency.

Batch, commercial systems for the immunomagnetic enrichment of suspensions containing rare cells are readily available [i.e., magnetically activated cell sorter (MACS) system]. However, concerns about their limited load capacity and potential, irreversible and nonspecific entrapment of cells have led us to develop and optimize "flow-through" cell separation devices, one of which is referred to as the quadrupole magnetic cell sorter (QMS), for rare cell enrichment [4].

While positive selection has proved effective for enrichment and isolation of rare—and not so rare—cells from various sources (see Table 9.1), there are significant limitations. One of these major limitations is that one frequently lacks information about the phenotype of the rare targeted cell. Moreover, the expression level and frequency of target antigen on the rare cell is potentially diverse. For example, in the case of a rare, circulating cancer cell, one is assuming that the antibody conjugate can label the cancer cell specifically and sufficiently to allow an acceptable separation. Recently, a number of studies have been published which demonstrate a significant antigen expression level diversity for a number of commercial antibodies targeting antigens associated with cancer cells. These studies also discuss the implications of this diversity on separator performance [5, 6].

Another potential limitation of the positive selection mode of operation is that after separation, the targeted cells contain antibodies or other labeling agents bound to surface antigens. This can, in some cases, cause illicit physiological responses within the cell, or limit further labeling for analysis, because the primary antibodies sterically hinder access to secondary sites.

To remove these potential limitations, a number of studies are published in which a preenrichment procedure is conducted prior to the molecular detection by using a negative depletion of unwanted, normally occurring cells. Table 9.4 lists some of those studies. Table 9.5 defines the abbreviations used in the previous tables.

TABLE 9.4

Preenrichment of cancer cells by negative immunomagnetic depletion of unwanted, normally occurring cells

Target cell	Cell sample	Antigen for IMS	Detection method	Initial purity	Recovery of target cells	Enrichment	References
Renal carcinoma CL	Caki-1in PBL	CD45	ICC	$1/10–1/10^7$	84.3%	–	[7]
Gastric cancer CL	MKN-45 in PBL	CD45	FC	$1/10–1/10^6$	–	9	[37]
Colorectal cancer CL	SW-480 in PBL	CD45	RT-PCR	$1/10–1/10^6$	–	–	[37]
Urological malignant CL	5 CLs in PBL	CD45	ICC	$4/10^5–4/10^8$	57.3%	13	[12]
Head and neck SCC	PB, BM	CD45	ICC	$1–25/10^6$	36–62%	–	[13]
Breast cancer	PB, BM	CD45	ICC	–	–	10	[38]
Head and neck SCC	PB, BM	CD45	ICC	–	–	–	[39]
Spiked breast cancer CL	PB	CD45	ICC	1.4×10^8	46	5.1 $\log_{10}$	[4]

TABLE 9.5
List of abbreviations

BM	Bone marrow
CL	Cell line
FC	Flow cytometry
HSC	Hematopoietic stem cells
ICC	Immunocytochemistry
IMS	Immunomagnetic separation
LDA	Limited dilution assay
MNC	Mononuclear cells
PB	Peripheral blood
PBL	Peripheral blood lymphocytes
PBPC	Peripheral blood progenitor cells
QMS	Quadrupole magnetic flow sorter
RT-PCR	Reverse transcription polymerase chain reaction
SCC	Squamous cell carcinoma

References

[1] Davids, V., Kidson, S. H. and Hanekom, G. S. (2003). Accurate molecula detection of melanoma nodal metastases: an assessment of multimarke assay specificity, sensitivity, and detection rate. Mol. Pathol. *56*, 43–51.

[2] Gradilone, A., Gazzaniga, P., Silvestri, I., Gandini, O., Trasatti, L. Lauro, S., Frati, L. and Agliano, A. M. (2003). Detection of CK19 CK20 and EGFR mRNAs in peripheral blood of carcinoma patients Correlation with clinical stage of disease. Oncol. Rep. *10*, 217–22.

[3] Ko, Y., Grunewald, E., Totzke, G., Klinz, M., Fronhoffs, S., Gouni Berthold, I., Sachinidis, A. and Vetter, H. (2000). High percentage c false-positive results of cytokeratin 19 RT-PCR in blood: A model for th analysis of illegitimate gene expression. Oncology *59*, 81–88.

[4] Lara, O., Tong, X., Zborowski, M. and Chalmers, J. J. (2004). Enrich ment of rare cancer cells through depletion of normal cells using densit and flow-through, immunomagnetic cell separation. Exp. Hematol. *32* 891–904.

[5] McCloskey, K. E., Chalmers, J. J. and Zborowski, M. (2003). Magneti cell separation: characterization of magnetophoretic mobility. Anal Chem. *75*, 6868–74.

[6] Chosy, E. J., Nakamura, M., Melnik, K., Comella, K., Lasky, L. C., Zborowski, M. and Chalmers, J. J. (2003). Characterization of antibody binding to three cancer-related antigens using flow cytometry and cell tracking velocimetry. Biotechnol. Bioeng. *82*, 340–51.

[7] Bilkenroth, U., Taubert, H., Riemann, D., Rebmann, U., Heynemann, H. and Meye, A. (2001). Detection and enrichment of disseminated renal carcinoma cells from peripheral blood by immunomagnetic cell separation. Int. J. Cancer *92*, 577–82.

[8] Kruger, W., Datta, C., Badbaran, A., Togel, F., Gutensohn, K., Carrero, I., Kroger, N., Janicke, F. and Zander, A. R. (2000). Immunomagnetic tumor cell selection—implications for the detection of disseminated cancer cells. Transfusion *40*, 1489–93.

[9] Benez, A., Geiselhart, A., Handgretinger, R., Schiebel, U. and Fierlbeck, G. (1999). Detection of circulating melanoma cells by immunomagnetic cell sorting. J. Clin. Lab. Anal. *13*, 229–33.

[10] Martin-Hernandez, M. P., Arrieta, R., Martinez, A., Garcia, P., Jimenez-Yuste, V. and Hernandez-Navarro, F. (1997). Haploidentical peripheral blood stem cell transplantation with a combination of CD34 selection and T cell depletion as graft-versus-host disease prophylaxis in a patient with severe combined immunodeficiency. Bone Marrow Transplant. *20*, 797–9.

[11] Sieben, S., Bergemann, C. and Lübbe, A. (2001). Comparison of different particles and methods for magnetic isolation of circulating tumor cells. J. Magn. Magn. Mater. *225*, 175–9.

[12] Zigeuner, R. E., Riesenberg, R. and Pohla, H. (2000). Immunomagnetic cell enrichment detects more disseminated cancer cells than immunocytochemistry in vitro. Urology *164*, 1834–7.

[13] Partridge, M., Phillips, E., Francis, R. and Li, S. R. (1999). Immunomagnetic separation for enrichment and sensitive detection of disseminated tumour cells in patients with head and neck SCC. J. Pathol. *189*, 368–77.

[14] Zhong, X. Y., Kaul, S., Lin, Y. S., Eichler, A. and Bastert, G. (2000). Sensitive detection of micrometastases in bone marrow from patients with breast cancer using immunomagnetic isolation of tumor cells in combination with reverse transcriptase/polymerase chain reaction for cytokeratin-19. J. Cancer Res. Clin. Oncol. *126*, 212–8.

[15] Hardingham, J., Kotasek, D. and Farmer, B. (1993). Immunobead-PCR: a technique for the detection of circulating tumor cells using immunomagnetic beads and the polymerase chain reaction. Cancer Res. *53*, 3455–8.

[16] Park, S., Lee, B., Kim, I., Choi, I., Hong, K., Ryu, Y., Rhim, J., Shin, J., Park, S. C., Chung, H. and Chung, J. (2001). Immunobead RT-PCR

versus regular RT-PCR amplification of CEA mRNA in periphera
blood. J. Cancer Res. Clin. Oncol. *127*, 489–94.

[17] Choesmel, V., Anract, P., Hoifodt, H., Thiery, J. P. and Blin, N. (2004
A relevant immunomagnetic assay to detect and characterize epitheli
cell adhesion molecule-positive cells in bone marrow from patients wit
breast carcinoma: Immunomagnetic purification of micrometastase
Cancer *101*, 693–703.

[18] Racila, E., Euhus, D., Weiss, A. J., Rao, C., McConnell, J
Terstappen, L. W. and Uhr, J. W. (1998). Detection and characterizatio
of carcinoma cells in the blood. Proc. Natl. Acad. Sci. USA *95*, 4589–94

[19] Moreno, J. G., O'Hara, S. M., Gross, S., Doyle, G., Fritsche, H
Gomella, L. G. and Terstappen, L. W. (2001). Changes in circulatin
carcinoma cells in patients with metastatic prostate cancer correlate wit
disease status. Urology *58*, 386–92.

[20] Liberti, P. A., Rao, C. G. and Terstappen, L. W. M. M. (2001). Optimi
zation of ferrofluids and protocols for the enrichment of breast tumor
cells in blood. J. Magn. Magn. Mater. *225*, 301–7.

[21] O'Hara, S. M., Moreno, J. G., Zweitzig, D. R., Gross, S., Gomella, L. G
and Terstappen, L. W. (2004). Multigene reverse transcription-PCR
profiling of circulating tumor cells in hormone-refractory prostate cancer
Clin. Chem. *50*, 826–35.

[22] Fang, B., Zborowski, M. and Moore, L. R. (1999). Detection of rar
MCF-7 breast carcinoma cells from mixtures of human peripheral leuko
cytes by magnetic deposition analysis. Cytometry *36*, 294–302.

[23] Martin-Henao, G. A., Picon, M., Amill, B., Querol, S., Ferra, C
Granena, A. and Garcia, J. (2001). Combined positive and negative ce
selection from allogeneic peripheral blood progenitor cells (PBPC) by us
of immunomagnetic methods. Bone Marrow Transplant. *27*, 683–7.

[24] Debelak, J., Shlomchik, M. J., Snyder, E. L., Cooper, D., Seropian, S
McGuirk, J., Smith, B. and Krause, D. S. (2000). Isolation and flov
cytometric analysis of T-cell-depleted CD34 + PBPCs. Transfusion *40*
1475–81.

[25] Hoppe, B., Mohr, M., Roots-Weiss, A., Kienast, J. and Berdel, W. E
(1999). Improvement of tumor cell depletion by combining immunomag
netic positive selection of CD34-positive hematopoietic stem cells an
negative selection (purging) of tumor cells. Bone Marrow Transplant
23, 809–17.

[26] Hildebrandt, M., Serke, S., Meyer, O., Ebell, W. and Salama, A. (2000)
Immunomagnetic selection of CD34 + cells: Factors influencing compo
nent purity and yield. Transfusion *40*, 507–12.

[27] Mohr, M., Dalmis, F., Hilgenfeld, E., Oelmann, E., Zuhlsdorf, M., Kratz-Albers, K., Nolte, A., Schmitmann, C., Onaldi-Mohr, D., Cassens, U., Serve, H. Sibrowski, W. et al. (2001). Simultaneous immunomagnetic CD34 + cell selection and B-cell depletion in peripheral blood progenitor cell samples of patients suffering from B-cell non-Hodgkin's lymphoma. Clin. Cancer Res. *7*, 51–57.

[28] de Wynter, E. A., Ryder, D., Lanza, F., Nadali, G., Johnsen, H., Denning-Kendall, P., Thing-Mortensen, B., Silvestri, F. and Testa, N. G. (1999). Multicentre European study comparing selection techniques for the isolation of CD34 + cells. Bone Marrow Transplant. *23*, 1191–6.

[29] Lang, P., Schumm, M., Taylor, G., Klingebiel, T., Neu, S., Geiselhart, A., Kuci, S., Niethammer, D. and Handgretinger, R. (1999). Clinical scale isolation of highly purified peripheral CD34 + progenitors for autologous and allogeneic transplantation in children. Bone Marrow Transplant. *24*, 583–9.

[30] Gaipa, G., Dassi, M., Perseghin, P., Venturi, N., Corti, P., Bonanomi, S., Balduzzi, A., Longoni, D., Uderzo, C., Biondi, A., Masera, G. Parini, R. et al. (2003). Allogeneic bone marrow stem cell transplantation following CD34 + immunomagnetic enrichment in patients with inherited metabolic storage diseases. Bone Marrow Transplant. *31*, 857–60.

[31] Richel, D. J., Johnsen, H. E., Canon, J., Guillaume, T., Schaafsma, M. R., Schenkeveld, C., Hansen, S. W., McNiece, I., Gringeri, A. J., Briddell, R., Ewen, C. Davies, R. et al. (2000). Highly purified CD34 + cells isolated using magnetically activated cell selection provide rapid engraftment following high-dose chemotherapy in breast cancer patients. Bone Marrow Transplant. *25*, 243–9.

[32] Chou, T., Sano, M., Ogura, M., Morishima, Y., Itagaki, H. and Tokuda, Y. (2005). Isolation and transplantation of highly purified autologous peripheral CD34 + progenitor cells: Purging efficacy, hematopoietic reconstitution following high dose chemotherapy in patients with breast cancer: Results of a feasibility study in Japan. Breast Cancer *12*, 178–88.

[33] Jing Y, Chalmers J. J., Zborowski M. (2007). Blood progenitor cell separation from clinical leukapheresis product by magnetic nanoparticle binding and magnetophoresis. Biotechnol. Bioeng. *96*, 1139–1154.

[34] Clarke, E., Potter, M. N., Oakhill, A., Cornish, J. M., Steward, C. G. and Pamphilon, D. H. (1997). A laboratory comparison of T cell depletion by CD34 + cell immunoaffinity selection and in vitro Campath-1M treatment: clinical implications for bone marrow transplantation and donor leukocyte therapy. Bone Marrow Transplant. *20*, 599–605.

[35] Lang, P., Bader, P., Schumm, M., Feuchtinger, T., Einsele, H., Fuhrer, M., Weinstock, C., Handgretinger, R., Kuci, S., Martin, D., Niethammer, D. and Greil, J. (2004). Transplantation of a combination of CD133 + and CD34 + selected progenitor cells from alternative donors. Br. J. Haematol. *124*, 72–79.

[36] Gordon, P. R., Leimig, T., Mueller, I., Babarin-Dorner, A., Holladay, M. A., Houston, J., Kerst, G., Geiger, T. and Handgretinger, R. (2002). A large-scale method for T cell depletion: Towards graft engineering of mobilized peripheral blood stem cells. Bone Marrow Transplant. *30*, 69–74.

[37] Linuma, H., Okinaga, K. and Adachi, M. (2000). Detection of tumor cells in blood using CD45 magnetic cell separation followed by nested mutant allele-specific amplification of p53 and K-ras genes in patients with colorectal cancer. Int. J. Cancer *89*, 337–44.

[38] Naume, B., Borgen, E., Nesland, J. M., Beiske, K., Gilen, E., Renolen, A., Ravnas, G., Qvist, H., Karesen, R. and Kvalheim, G. (1998). Increased sensitivity for detection of micrometastases in bone-marrow/peripheral-blood stem-cell products from breast-cancer patients by negative immunomagnetic separation. Int. J. Cancer *78*, 556–60.

[39] Partridge, M., Brakenhoff, R., Phillips, E., Ali, K., Francis, R., Hooper, R., Lavery, K., Brown, A. and Langdon, J. (2003). Detection of rare disseminated tumor cells identifies head and neck cancer patients at risk of treatment failure. Clin. Cancer Res. *9*, 5287–94.

Laboratory Techniques in Biochemistry and Molecular Biology, Volume 32
Magnetic Cell Separation
M. Zborowski and J. J. Chalmers (Editors)

CHAPTER 10

Commercial magnetic cell separation instruments and reagents

Maciej Zborowski

Department of Biomedical Engineering, Lerner Research Institute, Cleveland Clinic, Cleveland, OH 44195, USA

The selection of magnetic separators presented in this chapter was made to illustrate differences in approaches to magnetic cell separation. As discussed in Chapter 3, the magnetic force exerted on the cell-magnetic label complex is a product of the magnetic dipole moment of the complex and the field gradient (for saturated dipole) or the magnetic susceptibility of the cell-label complex and the field energy density gradient (for unsaturated dipole). Therefore, the engineering design of the magnetic cell separators depends on the type of the magnetic particle, available for cell labeling. For relatively large, micrometer-size magnetic particles that do not require high fields for saturation, relatively weak magnets and open-gradient magnetic field configurations are sufficient. The advantages of the strong magnetic particle–weak magnetic field combination are the simplicity of the magnetic configuration and its relatively low price, and the ability to use it with the standard laboratory tubes. In other words, such separators do not require special types of columns for the cell-magnetic label capture from the solution. The cell-label complex is captured directly on the tube walls, and the unlabeled cells are separated from the labeled ones by decanting the solution from the tube. Such separators require a high quality magnetic label (a microbead),

DOI: 10.1016/S0075-7535(06)32010-4

however, because as little as one bead is sufficient for removing th
attached cell from solution, and therefore one relies on the specificit
of the bead binding to the cell for the best separation.

A highly successful open-gradient type of separators wa
designed around micrometer-sized Dynabead^{TM} magnetic particle
(from Invitrogen, Inc.) [1, 2]. Dynabeads were the first magneti
particles used for clinical cell separation applications, and thei
use has been extended to applications in molecular biology [3
An example of the Dynabead-Invitrogen separator is described i
Section 10.1. The disadvantages of the micrometer-size magneti
include slow reaction kinetics and precipitation from suspension
(related to their relatively large size), potential for increased non
specific cell entrapment due to the bead–bead interactions an
aggregation, and the potential for cell damage due to large shea
forces involved in the bead pulling the attached cell through th
solution (related to the bead large magnetic moment) [4]. Th
preferred applications of the Dynabeads are to negative cells selec
tion, following which the cells targeted by the magnetic label ar
discarded, and only the cells staying in suspensions are used.

Magnetic cell capture from suspensions directly in the field o
view of a microscope offers unique opportunities for rare cell detec
tion. A specialized system based on bioferrofluids, that is submicro
magnetic particles conjugated to targeting antibodies, has bee
developed by Immunicon Corp., as described in Section 10.2. Th
system requires a special type of chamber and magnet configuratio
to maximize magnetic forces acting in a relatively weakly magneti
bioferrofluid particle [5, 6]. The system has been successfully teste
for detection of rare circulating cancer cells in patients undergoin
therapy [7]. The Immunicon Corp. offers other products an
reagents for magnetic cell separation, as noted in Section 10.5.

The use of colloidal magnetic particles as magnetic labelin
reagents offers advantages of their forming stable suspensions and
fast cell binding reaction kinetics. In this respect, they are no
much different from other cytological staining reagents, such a

immunofluorescence labels, and therefore require similar types of cell-labeling protocols. The small size of the particles, in the range of 10–100 nm, comes with the cost of their low magnetic moment, requiring high fields and gradients, however. These could be reasonably obtained by soft steel alloy inserts inside the cell suspension container to produce local high gradient magnetic fields (as illustrated in Chapter 4). A particularly successful design has been developed and commercialized by Miltenyi Biotec GmbH, Bergisch Gladbach, Germany, around the magnetic colloid of a trade name MACS MicroBead™, as described in Section 10.3 [8]. The company offers a full line of products, from small, laboratory use MiniMACS columns to large, automated separators currently undergoing clinical trials, CliniMACS [9]. The disadvantage of magnetic colloidal reagents is their low magnetic moment that requires close contact of the cell suspension with the magnetized surfaces, and therefore the use of specialized, magnetic affinity-type columns, with the associated potential complications of column overloading and column cost.

In search for an ideal combination of a colloidal magnetic reagent with an open-gradient magnetic field configuration (not requiring steel alloy solid support inserts and dedicated columns), a number of designs have been proposed and now are available commercially. One of these designs is based on a concentric, quadrupole magnetic field (as illustrated in Chapter 4) that fits naturally the cylindrical shape of the laboratory tube, used in combination with magnetic dextran colloid, available from StemCell Technologies, Inc., Vancouver, Canada, as described in Section 10.4. The other design uses a rectangular magnet assembly that is a source of high field gradients, used in combination of a laboratory tube stand and a customized magnetic colloid, from BD Immunosciences, San Jose, CA, as described in Section 10.6. The recent development of these open-gradient magnetic field separators is made possible by the increasing availability of high energy product permanent magnets (illustrated in Chapter 2) and the advances in the material science as related to synthesis and characterization of the magnetic colloids.

10.1. Invitrogen DYNAL Magnetic Particle Concentrator

The description provided in this chapter has been adapted from the company website [10]. (In this and in the later examples, the interested reader is advised to check company websites for the most current updates.) The Dynabeads® magnetic separation technology from Invitrogen (formerly Dynal Biotech) is a simple, tube-based system for cell isolation. With the use of Dynabeads® and a Dynal magnet, any type of cell can be isolated from any starting sample included whole blood. Dynabeads® are superparamagnetic, mono-sized polymer beads (Fig. 10.1). Each bead has an even dispersion of superparamagnetic material coated with a thin polymer shell to encase the magnetic material. Dynabeads® are made of undegradable material and designed with an inert coating to eliminate the risk that any unwanted material is left in the sample or adversely affect cells during isolation. Dynabeads® provide three different separation

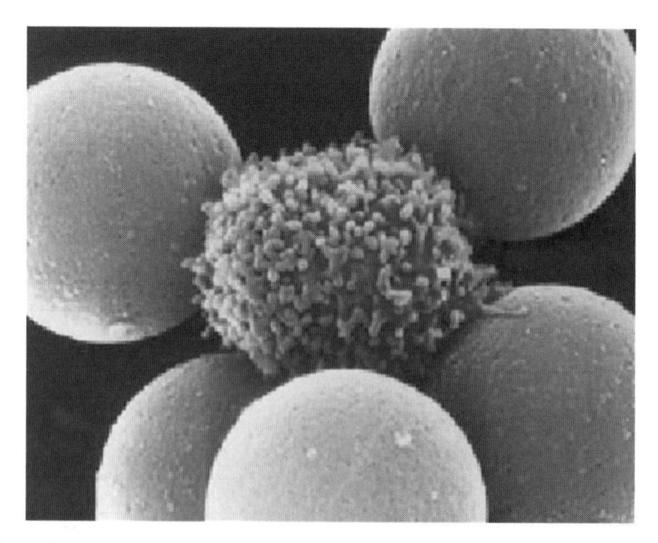

Fig. 10.1. SEM microphotograph of a mononuclear cell (MNC) rosetted with Dynabead M-450. The diameter of the M-450 microsphere is 4.5 μm [10].

strategies for isolation of cells of various types; positive isolation, depletion and negative isolation.

1. In positive isolation, Dynabeads® coated with specific antibody are mixed with sample and bind to target cells and isolate them from rest of sample with help of a Dynal magnet (Fig. 10.2). The supernatant is then discarded. This method is recommended when cells of highest purity are required. If cells need to be detached after positive isolation for downstream applications, three strategies can be adopted: (i) with the DETACHaBEAD® solution, a polyclonal Fab antibody specific for several of the primary antibodies on Dynabeads® competes with antibody/antigen binding at the cell surface, releasing the antibody to leave viable, untouched bead-free cells. (ii) The CELLection™ Positive Isolation

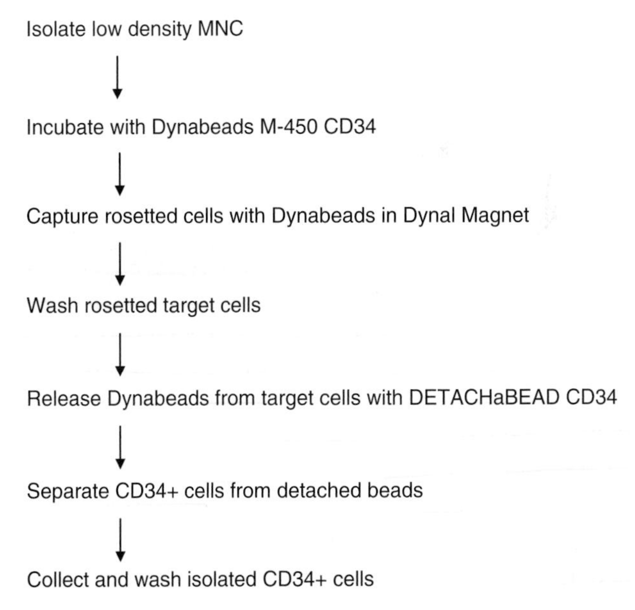

Fig. 10.2. Example of a positive cell isolation flow chart using Dynabeads® and Dynal Magnet for hematopoietic progenitor cell (CD34+ cell) capture (rosetting) [10].

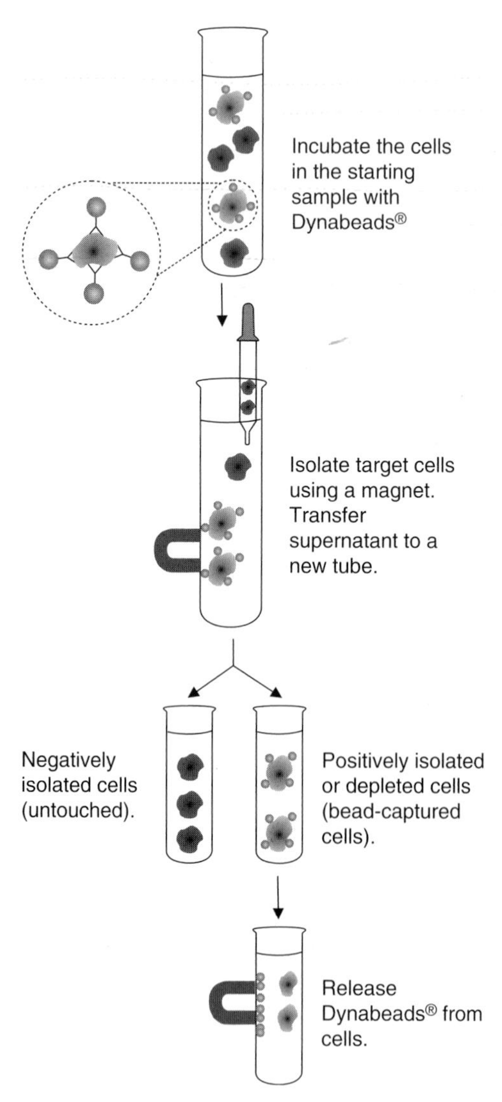

Fig. 10.3. Dynabead® applications to positive and negative cell isolation [10].

with Universal Detachment method is designed for positive isolation of any cell type. Antibodies or streptavidin are attached to the surface of the CELLection™ Dynabeads® via a DNA linker. After isolation cells can be detached from the beads by cleaving the DNA linker with DNase. (iii) Dynabeads FlowComp™ is the new magnetic cell isolation technology leaving the cells bead-free after isolation. Cells isolated with Dynabeads® FlowComp™ keep their functional characteristics and are suitable for any downstream cell-based assay such as flow cytometry. For downstream applications including molecular studies and some cell cultures (e.g. endothelial cells), Dynabeads do not need to be released from the cells.

2. In depletion unwanted cell types are targeted with Dynabeads® and then easily removed them from a mixed cell sample with high efficiency (>99%, according to the Company information) with help of a Dynal magnet. The depleted supernatant - still containing the wanted cells - is then recovered for downstream analyses and assays.

3. In negative isolation the target cell is isolated by removing unwanted cell types present in a mixed starting sample. First the sample is mixed with an antibody cocktail containing antibodies against all the unwanted cells. After short incubation the cells are washed to remove any unbound free antibodies. The cells are resuspended back in the buffer and mixed with Dynabeads coated with secondary antibody that is specific toward the primary antibodies on all the unwanted cells. After short incubation again, the unwanted cells are captured by the beads and removed from the rest of sample using magnet. The supernatant contains now untouched, bead-free target cells for downstream analyses and assays. All three cell isolation strategies are summarized in Fig. 10.3.

10.2. R&D Systems Immunicon MagCellect™ Ferrofluid

The description of the product and the applications is adapted from the company website [11]. R&D Systems, in association with Immunicon Corporation, offers MagCellect™ ferrofluid conjugates of a variety goat anti-mouse IgG antibody, goat anti-rabbit Ig antibody and streptavidin, in combination with DSB-X biotin co jugates. Ferrofluids are superparamagnetic particles ~200 nm i diameter that respond to a magnetic field but demagnetize whe the field is removed. The key feature of the MagCellect^{TM} ferroflui is its small and relatively uniform particle size, which results in rapi diffusion of the ferrofluid conjugate and fast binding reaction kine ics. The ferrofluid conjugates exhibit significantly higher ligan binding capacities per mass, as compared with larger-diamete superparamagnetic particles [12].

The unique particle size of the MagCellect ferrofluid permi fluorescence microscope-based cell sorting, imaging, and analytic options. To accomplish this, the cell suspension or blood samp is incubated with the appropriate fluorescent probes and the mixed with the MagCellect ferrofluid conjugate. The capture cells can then be imaged and analyzed, or recovered free fro unselected cells.

R&D Systems also has available Captivate magnetic separato for both microplates and microtubes for use with the MagCellect^{T} ferrofluid products. The microplate separator is compatible with mo 96-well microplates, whereas the microtube separator can accomm date six 1.5 ml microcentrifuge tubes (Fig. 10.4). Both separators pu magnetic particles to one side, allowing easier removal of supe natants. The DSB-X biotin technology permits reversible labelin of DSB-X biotin derivatives by avidin and streptavidin conjugate DSB-X biotin has moderate affinity for avidin and streptavidi however, its binding is rapidly reversed by addition of excess D-bioti or D-desthiobiotin. The applications include selective isolation of cel with a combination of MagCellect^{TM} ferrofluid streptavidin, DSB-X biotin-labeled secondary antibody, and a cell-selective prima ry antibody, followed by release into fresh medium at neutral p and physiological temperatures with D-biotin. Cells that have bee isolated with the Captivate ferrofluid streptavidin by the abo method retain the DSB-X biotin immunolabel; consequently, the

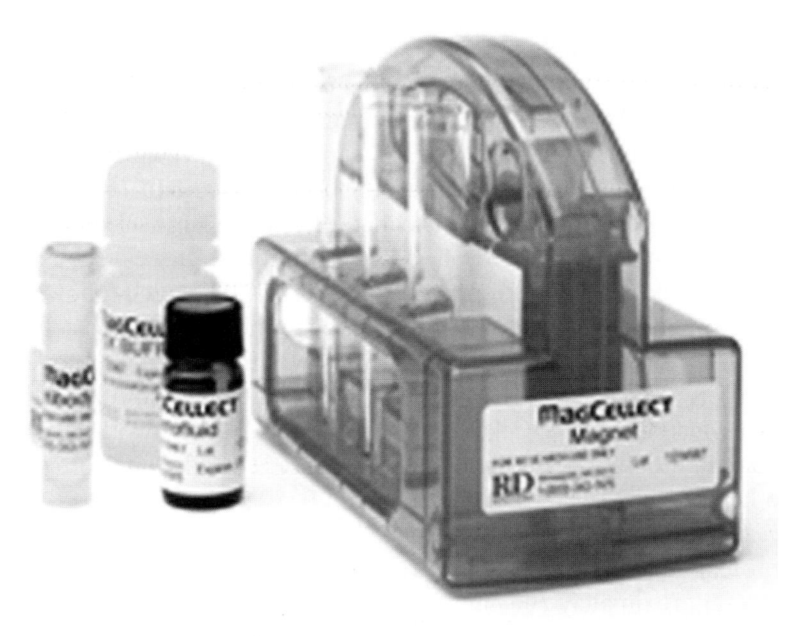

Fig. 10.4. MagCellect^{TM} magnetic separators for both microplates and laboratory tubes for use with the MagCellect ferrofluid products [11].

can be detected and subsequently analyzed with any of our wide array of avidin and streptavidin conjugates.

10.3. Miltenyi Biotec GmbH

The description of the product and the applications is adapted from the company website [13]. MACS® Technology is based on MACS MicroBeads, manual or automated MACS Separators, and MACS Columns. When MACS Columns are placed in a MACS Separator, the MACS Column matrix provides a magnetic field strong enough to retain cells labeled with colloidal MACS MicrBeads. The magnetically labeled cells are retained on the column, while unlabeled cells pass through. These cells can be collected as the unlabeled fraction. The retained cells are eluted from the MACS Column after removal from the magnet. The viable and functionally active cells can be separated in less than 30 min. High purity and recovery of both labeled and unlabeled cells have been reported [14]. The method and the large selection of reagents and columns offer flexible separation strategies, such as positive selection, depletion, and subset sorting (depletion followed by positive selection or MACS MultiSort applications). It is compatible with downstream applications: flow cytometry, microscopic analysis, molecular biology experiments, cell culture, and in vivo experiments. MACS Cell Separation Technology is available for human, mouse, rat, and nonhuman primate cell isolation. For any other cell type Indirect MicroBeads can be used.

MACS MicroBeads are coupled to highly specific antibodies or proteins. MACS MicroBeads are small, superparamagnetic particles, approximately 100 nm in diameter [15, 16]. They are made of biodegradable dextran matrix; therefore, there is no need to remove them after the separation process, and the cells can be transferred directly to experiment. The company reports that they have no known effect on structure, function, or activity status of labeled cells and do not interfere with subsequent experiments.

Direct labeling is the fastest way of magnetic labeling. It requires only one labeling step since the specific antibody is directly coupled to the magnetic particle. Direct labeling limits the number of washing steps and therefore avoids unnecessary cell loss. The company offers highly specific monoclonal antibodies, thus reducing background and optimizing the separation. For this purpose a variety of antibody conjugated MicroBeads targeting many human, mouse, rat, and nonhuman primate cell surface markers are available. Fluorescent staining can simultaneously be performed for subsequent analysis of the separated fractions by flow cytometry or microscopy.

Indirect labeling is performed when no direct MicroBeads are available. Almost any monoclonal or polyclonal antibody targeting any cell type from any species can be used for indirect labeling.

Cells are labeled with a primary antibody that is unconjugated, biotinylated, or fluorochrome-conjugated. In a second step, magnetic labeling is performed by using Anti-Immunoglobulin, Anti-Biotin, Streptavidin or Anti-Fluorochrome MicroBeads, respectively. A cocktail of antibodies can also be used for isolating or depleting a number of cell types concurrently. Indirect labeling may be the method of choice if dimly expressed markers are targeted for magnetic separation because it amplifies the binding of the magnetic label [17].

Positive selection means that the wanted target cells are magnetically labeled and isolated directly as the positive cell fraction. The company advises considering a positive selection strategy for best purity, especially for rare cell enrichment, best recovery, and fast procedure. They consider positive selection as the most direct and specific way to isolate the target cells from a heterogenous cell suspension.

Depletion (or negative selection) means that the unwanted cells are magnetically labeled and eliminated from the cell mixture, that is, the nonmagnetic, unlabeled cell fraction are the desired cells. The company advises that the depletion should be considered in the following instances: for the removal of unwanted cells; if no specific antibody is available for target cells; if binding of the antibody to the target cells is not desired; for the subsequent isolation of a cell subset by means of the positive selection. A single depletion procedure can remove up to 99.99% (equivalent to 4 logs) of the magnetically labeled cells, leaving a highly pure fraction of unlabeled cells. This can be achieved by combining strong magnetic labeling with MACS columns. Miltenyi Biotec offers optimized MACS kits for many cell types that contain pretitrated cocktails of antibodies directed against nontarget cells.

The selection of reagents and columns makes possible a combination of depletion followed by positive selection: cell subsets can be isolated by first depleting the nontarget cells and then magnetically labeling and positively selecting the subset cells. If the isolation of extremely rare cells is required (the number frequency of 1:10,000 or lower), the strategy can be useful for first depleting the

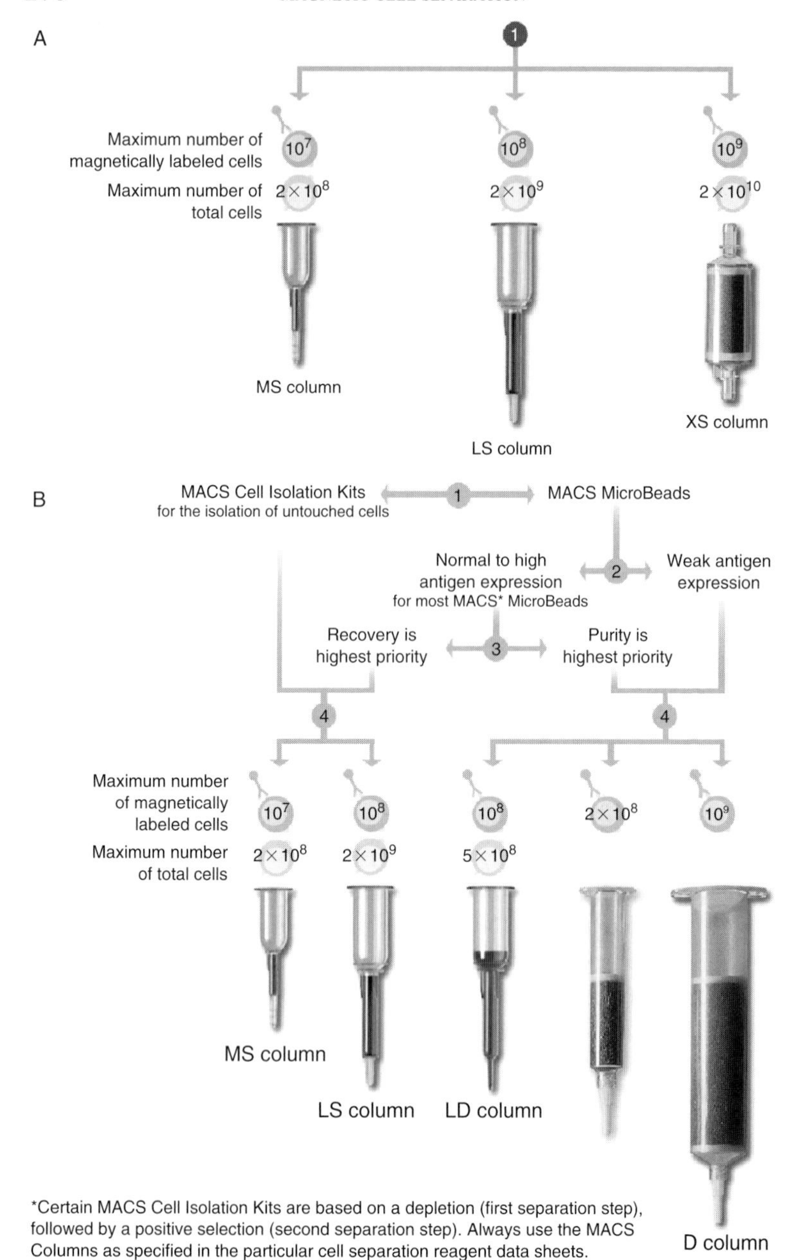

Fig. 10.5. (A) HGMS column size and (B) column selection criteria from Miltenyi Biotec [13].

nontarget cells from the suspension (so-called "debulking" of the cell suspension). Positive selection can then be done on the enriched, low volume fraction to obtain very pure cells. Multiparameter sorting with MACS MultiSort Kits allows sequential positive selections of cells. The target cells are first labeled with MACS MultiSort MicroBeads and positively selected for the first parameter. Then the cells are incubated with the MultiSort Release Reagent which enzymatically removes the MicroBeads from the antibodies. In the next step these cells are magnetically labeled with a second antibody-conjugated MicroBead directed against a subset marker. After labeling, the cells are again magnetically separated. See Fig. 10.5 for additional column selection criteria and Table 10.1 for MACS Column data.

MS Columns, LS Columns, and XS Columns are optimized for positive selection of cells. They can also be used for depletion if the magnetic labeling is strong. Large cell columns were designed for positive selection of large human or animal cells (e.g., megakaryocytes) and have a larger pore size compared to MS Columns. LD Columns, as well as CS Columns and D Columns are recommended for depletion of unwanted cells. Highest depletion efficiency is obtained with these column types, even if the magnetic labeling of the cells—due to dim antigen expression—is weak. The autoMACS Columns are designed for use with the autoMACS™ Separator.

TABLE 10.1
MACS column data

Column	Maximum number of labeled cells	Maximum number of total cells	Separator
MS	10^7	2×10^8	MiniMACS, OctoMACS, VarioMACS, SuperMACS
LS	10^8	2×10^9	MidiMACS, QuadroMACS, VarioMACS, SuperMACS
XS	10^9	2×10^{10}	SuperMACS
autoMACS	2×10^8	4×10^9	autoMACS

The company offers a large selection of separators, designed for different classes of applications.

1. MiniMACS™ Separator: small-scale cell separations in combination with MS Columns (Fig. 10.6A)
2. OctoMACS™ Separator: multiple separations with MS Columns
3. MidiMACS™ Separator: separation of larger samples using LS Columns or LD Columns
4. QuadroMACS™ Separator: multiple separations with LS Columns or LD Columns
5. VarioMACS™ Separator: can be equipped with MS Columns, LS Columns, LD Columns, and CS Columns
6. SuperMACS™II Separator: large-scale separations with up to 10^{11} cells
7. autoMACS™ Separator: bench-top automated magnetic cell sorter
8. MACSiMAG Separator (Fig. 10.6B): the MACSiMAG™ Separator is designed for the removal of MACSiBead™ particles from cell suspensions. Cells that have been activated or expanded by means of antibody-loaded MACSiBead particles, for example by using the Human T Cell Activation/Expansion Kit, or cells magnetically labeled with Anti-Biotin MACSiBead particles for depletion are placed in the extremely strong magnetic field. Within a few minutes, the 3.5-μm-sized MACSiBead particles adhere to the tube wall and the supernatant containing the MACSiBead-depleted cells is subsequently removed. The MACSiMAG Separator holds up to three 50-ml tubes and four 15-ml tubes. The tube rack supplied with the separator can be equipped with eight tubes from 0.5 ml to 5 ml in size.
9. AutoMACS® Technology for automated cell separations, compatible with all MACS MicroBeads for cell sorting, designed for applications to any cell type from any species. The company reports that it is fast–high-speed sorting of more than 10 million cells per second, for up to 4×10^9 cells or up to 3 ml of whole blood; simple to

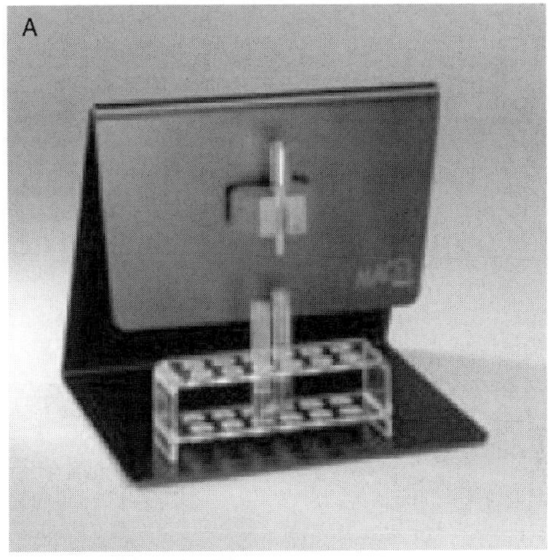

Fig. 10.6. (A) The MiniMACS™ and (B) the MACSiMAG™ magnetic cell separators from Miltenyi Biotec GmbH, Bergisch Gladbach, Germany [13].

operate–intuitive interface and preset programs, ready-to-use MACS Buffers for easy operation; and compatible with flow cytometry as no MicroBead removal is required for further experiments.

10.4. StemCell Technologies, Inc.

The description of the product and the applications is adapted from the company website [18]. EasySep® is a column-free immunomagnetic cell selection system (Fig. 10.7A and B) that operates on the

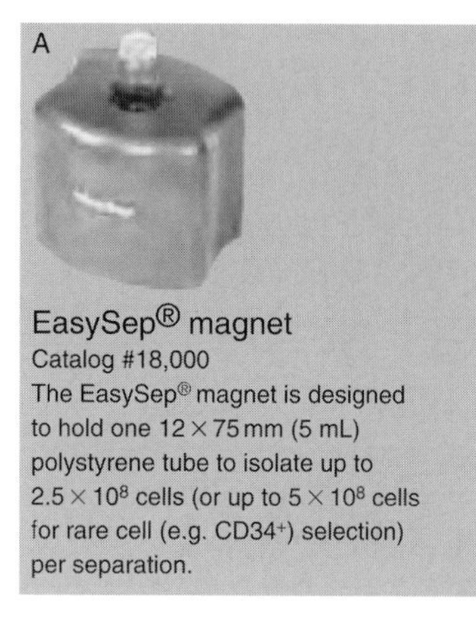

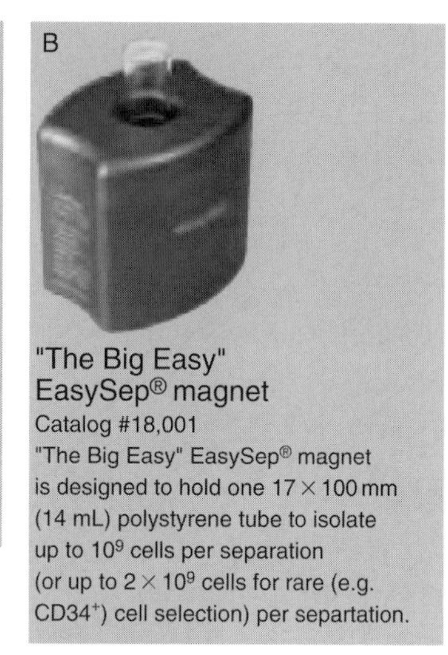

Fig. 10.7. (A) EasySep®, (B) "The Big Easy" EasySep®, and (C) the automated RoboSep® line of magnetic separators operating on the open gradient principle, from StemCell Technologies, Inc., Vancouver, British Columbia, Canada [18].

open gradient principle (discussed in Chapter 4). Cells can be isolated manually using handheld magnets, or using the RoboSep® automated system (Fig. 10.7C). Cells are targeted for selection or depletion (positive or negative selection) using monoclonal antibodies directed against specific cell surface antigens (Fig. 10.8A). These targeted cells are then cross-linked to EasySep® magnetic nanoparticles in a standard polystyrene tube. The tube is placed directly into a unique EasySep magnet, which generates a high gradient magnetic field in the interior cavity (similar to that shown in Chapter 4 as a quadrupole field). The field is strong enough to separate cells labeled with EasySep nanoparticles without the additional magnetic field gradients provided by a column matrix. After a 5- to 10-min incubation, the cells that are not bound to the magnetic nanoparticles are removed by pouring off (manually) or pipetting (RoboSep) the supernatant. Unlike larger microparticles used with other column-free systems, the EasySep nanoparticles do not interfere with subsequent flow cytometric analysis and do not need to be removed, a consideration when cells of interest are isolated by positive selection. If negative selection is preferred, cells of interest are left untouched, and are not labeled with antibodies or magnetic particles.

Tetrameric antibody complexes are composed of two mouse IgG1 monoclonal antibodies held in tetrameric array by two rat antimouse IgG1 monoclonal antibody molecules (shown in Fig. 10.8A). One mouse antibody recognizes the specific cell surface antigen while the other recognizes dextran on the EasySep® magnetic nanoparticle.

10.4.1. Positive cell selection

Highly purified cells from any species can be obtained by targeting the cells of interest with monoclonal antibodies to a specific cell surface antigen. Biotinylated, PE-, APC- or FITC-conjugated antibody, or any mouse IgG1 antibody directed against a cell surface antigen on cell of interest are used. The labeled cell suspension is

Schematic drawing of EasySep® magnetic labeling of human cells

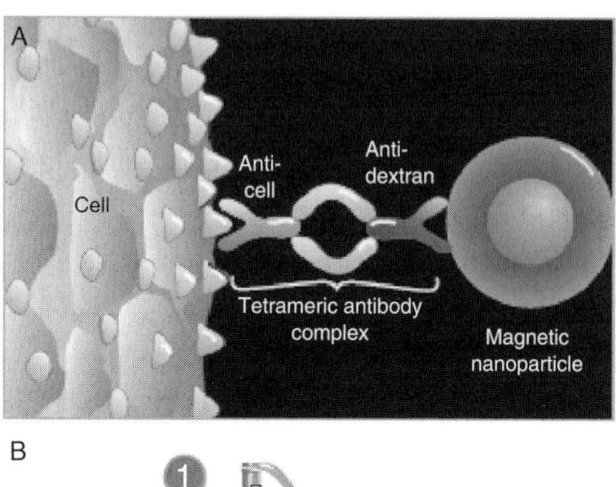

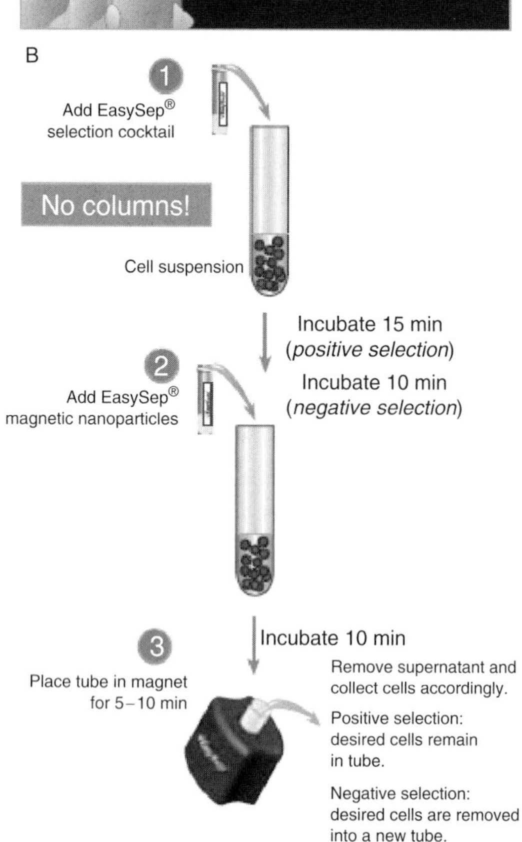

Fig. 10.8. (A) Schematic illustration of EasySep® tetrameric antibody complex (TAC) labeling of human cells and (B) cell separation procedure [18].

then mixed with the appropriate EasySep selection reagent and magnetic nanoparticles in a standard polystyrene tube. The tube is placed directly into an EasySep magnet. After a 5-min incubation, the cells that are not bound to the magnetic nanoparticles are removed, leaving the magnetically labeled cells of interest in the tube (Fig. 10.8). The cells remaining in the tube are easily collected by removing the tube from the magnet. The EasySep magnetic nanoparticles do not interfere with subsequent flow cytometric analysis, and the cells are ready for immediate use.

10.4.2. Negative selection

Unwanted cells are labeled with a cocktail of biotinylated antibodies, and then cross-linked to magnetic nanoparticles. The sample is placed in the EasySep magnet, and after incubation, the desired cells are poured or pipetted into a new tube, while the magnetically labeled (unwanted) cells remain in the original tube (Fig. 10.8). Desired cells are not labeled with antibody.

The company describes EasySep as a flexible system, suitable for the separation of many cell types from any species and tissue, and provides literature references describing applications, such as enrichment circulating human tumor cells from whole mouse blood in a model system of human metastatic breast cancer [19]. Depletion of the mouse leukocytes using a PE-conjugated anti-CD45 antibody increased the sensitivity of tumor cell detection approximately tenfold, enabling detection of one tumor cell in 100,000 mouse leukocytes. Depletion of human leukocytes using the Human CD45 Depletion Kit has been used to highly enrich mesenchymal progenitor cells from human bone marrow [20]. Human CD34 Positive Selection Kit has been used in several studies to select progenitor cells from cord blood [21–23]. This kit has also been used to isolate CD34+ cells from collagenase-treated adipose tissue [24, 25]. The selected cells were functional in endothelial cell differentiation assays, and were demonstrated to enhance revascularization after ischemic limb injury in mice [24]. Problematic (e.g., clumpy, adherent) tissue samples can be processed using EasySep since the problems often encountered using columns (such as clogging) are avoided. The cell depletion can be accomplished using EasySep positive selection cocktails to deplete unwanted cells from various samples. For example, EasySep has been used to deplete mouse T cells from spleen tissue [26] and human CD14+ monocytes from progenitor cell bulk cultures [27]. The company offers help in designing a custom selection kit. For example, a recent study used a Custom EasySep Selection Kit to select a CD27+ subset of B cells [28]. Multiple functional and biochemical assays were performed on the purified cells to assess the expression of K^+ channels during B cell development. In addition to the standard EasySep products for the isolation of T cells, CD4+ T cells, and CD8+ T cells, EasySep kits are available for the isolation of: Naive CD4+ T cells, Memory CD4+ T cells, Regulatory CD4+ CD25+ T cells. In addition, the company plans on introducing EasySep and RoboSep human positive selection kits for direct use with whole blood. The protocols are designed to minimize sample handling by lyzing the red blood cells with the buffer in the same tube used for cell separation.

10.5. Immunicon Corporation

The description has been adapted from the company website [29]. Immunicon Corporation is developing and commercializing proprietary cell-based research and diagnostic products with an initial focus on cancer. Immunicon's platform technologies can identify, count, and characterize a small number of tumor cells present in the blood of a patient. Immunicon's collaborator, Veridex, LLC, a Johnson & Johnson company, received 510(k) clearance from the US FDA in January 2004 for use of the CellSearch™ Epithelial Cell Kit, which incorporates Immunicon reagent technology, in the management of metastatic breast cancer [7, 30, 31]. This product was launched for in vitro diagnostic use in August 2004. The technology may have other applications in cancer diagnostics, in the clinical development of cancer drugs, and in cancer research. In addition, its proprietary technologies may have applications in other fields of medicine, such as cardiovascular and infectious diseases.

The company develops technologies that identify, count, and characterize a small number of circulating tumor cells (CTC) and other rare cells present in a blood sample of a patient. The company offers CellSave™ Preservative Tube, an evacuated blood collection tube with preservative; CellTracks™ AutoPrep System, an automated instrument to capture and label cells from 7.5-ml blood samples; CellTracks MagNest™ Cell, a presentation device, which presents magnetically labeled cells for analysis (Fig. 10.9) [5, 6]; CellSpotter™ Analyzer; CellTracks Analyzer II, which is a semi-automated fluorescence microscope used to count and characterize

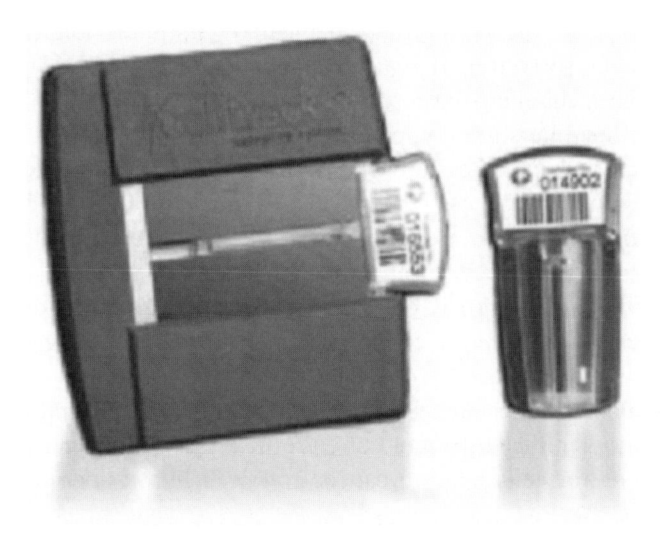

Fig. 10.9. CellTracks MagNest Cell, a presentation device, which presents magnetically labeled cells for microscopic analysis (Immunicon Corp., Huntingdon Valley, PA) [29].

cells; the CellSearch™ Circulating Tumor Cell Kit, which is used for counting CTCs in blood; CellSearch Profile Kit, which is used to isolate CTCs for subsequent molecular or cellular analysis; CellTracks Endothelial Cell Kit that is used in isolating and counting endothelial cells; CellTracks Bone Marrow Tumor Kit; Molecular Diagnostic Kits; and Tumor Phenotyping Reagents, which are research reagents for cell characterization.

The MagNest® device is a fixture of two magnets held together by steel (Fig. 10.9). The strong magnetic field causes the magnetically labeled target cells to move to the surface of the cartridge and form a monolayer on one focal plane inside the reaction cartridge [5, 6]. This technology enables essentially all of the cells from the starting samples (7.5 ml) to be analyzed with minimal loss. In contrast, manual enrichment methods and cytospin slide preparation typically result in up to 50% cell loss. The CellTracks® Analyzer II is a semiautomated fluorescence microscope used to enumerate fluorescently labeled cells that are immunomagnetically selected and aligned (with a dedicated computer loaded with CellTracks software). It scans the entire surface of the chamber four times, changing fluorescence filters between each scan. Images of the four filters are compiled and presented in a gallery format for final cell classification by the user. Rare cell isolation kits are designed to be used with the CellTracks AutoPrep System. It is an automated, walk-away sample preparation system for immunomagnetic cell capture and fluorescence staining of rare cells. Standardization is critical for optimal recovery and reproducible results. Several configurations are available to meet the needs of the clinical lab or scientists. Cellular analysis kits and associated marker reagents for counting and characterizing circulating epithelial (tumor) or endothelial cells. Rare cell enrichment kits for cell isolation can be followed by off-line molecular analysis.

Magnetic nanoparticles, called ferrofluids, are at the heart of rare cell isolation. The company reagent kits include ferrofluid conjugated to antibodies directed against rare cells, including circulating tumor cells and circulating endothelial cells. After the rare cells are enriched from the patient sample, they are fluorescently labeled. Ferrofluid consists of a magnetic core coated with bovine serum albumin (BSA) conjugated with antibodies for capturing cells. Ferrofluid particles are colloidal, which permits long incubations without the ferrofluid settling. Specifications: average diameter = 145 nm; composition: >85% of oxide is Fe_3O_4, ~80% w/w magnetite, magnetic susceptibility = 125 emu/g; ~4 × 10^{11} particles per milligram of iron; ~3 fg Fe_3O_4 per particle; ~1 fg BSA per particle.

Binding capacity: ~15,000 small biotinylated molecules per particle; ~5000 large biotinylated molecules per particle; ~50- to 150-μg monobiotinylated monoclonal antibody (MAb) per milligram of iron; ~10 nanomoles biotin-FITC per milligram of iron.

10.6. BD Biosciences

The description has been adapted from the company website [32].

BD Biosciences Pharmingen offers antibody-labeled magnetic particles for enrichment or depletion of leukocyte subpopulations. The BD IMag™ Cell Separation System is compatible with BD Pharmingen antibodies. BD IMag particles range in size between 0.1 and 0.45 μm and are coated with BD Pharmingen monoclonal antibodies. These particles are optimized for positive or negative selection of leukocyte subpopulations using the BD IMagnet™ direct magnet (Fig. 10.10). BD IMag particles coated with specific monoclonal antibodies are added to a cell suspension. The BD IMag particles will specifically bind to the subpopulation of interest. The labeled cell suspension can then be placed in the magnetic field of the BD IMagnet direct magnet (BD IMag-DM™, Fig. 10.11), or alternatively, the cells can be run over a separation column that has been

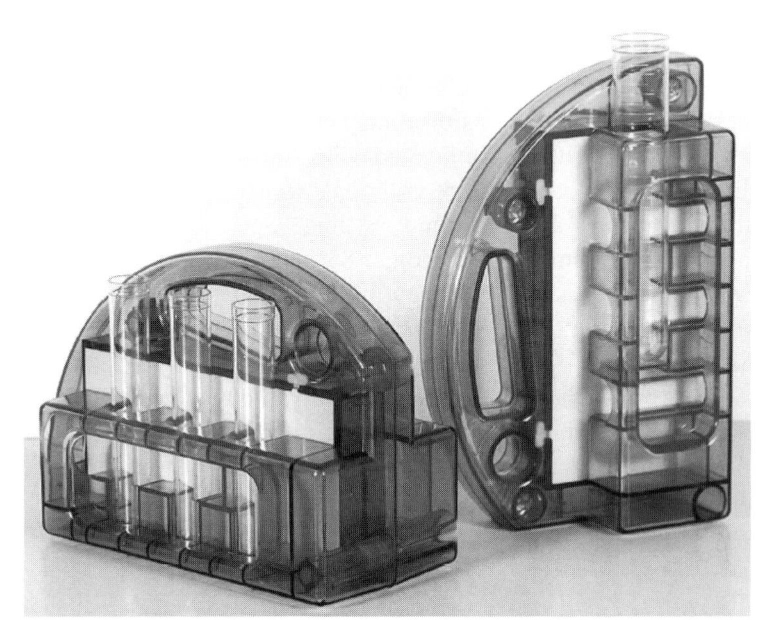

Fig. 10.10. BD IMagnet direct magnet (BD Biosciences, San Jose, CA) [32].

placed in a magnetic field. Captured cells can be run on a flow cytometer with the BD IMag particles intact. In addition to particles labeled with monoclonal antibodies to leukocyte subsets, the company offers particles coated with streptavidin, which can be used with the biotinylated antibody of choice.

BD IMagnet is a direct magnet that is designed for use with BD IMag Particles-DM^{TM}. The BD IMagnet is a neodymium iron boron permanent magnet that can hold either six 12 × 75 mm test tubes or two 17 × 100 mm test tubes (Fig. 10.10). Yield and purity of cell enrichment using BD IMag Particles-DM with the BD IMagnet are similar to that observed with magnetic separation

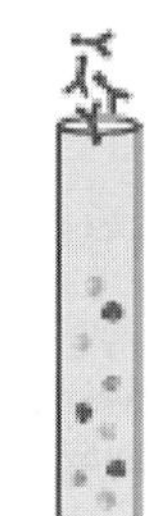

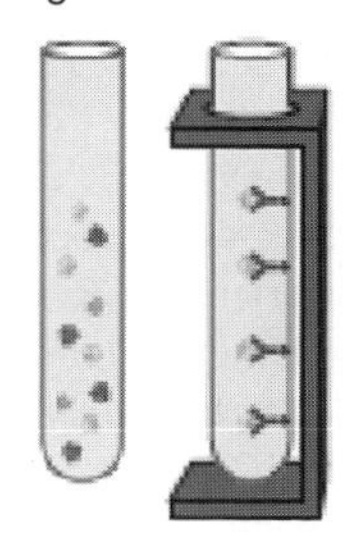

Fig. 10.11. The BD Imag cell separation procedure using the BD IMagnet direct magnet and BD IMag Particles-DM (BD Biosciences, San Jose, CA) [32].

columns. Therefore, the BD IMagnet offers an economical option for cell separations, such as pre-enrichment for flow cytometric sorting.

References

[1] Ugelstad, J., Stenstad, P., Kilaas, L., Prestvik, W. S., Herje, R., Berge, A. and Hornes, E. (1993). Monodisperse magnetic polymer particles. New biochemical and biomedical applications. Blood Purif. *11*, 347–69.

[2] Fonnum, G., Johansson, C., Molteberg, A., Morup, S. and Aksnes, E. (2005). Characterisation of Dynabeads(R) by magnetization measurements and Moessbauer spectroscopy. J. Magn. Magn. Mater. *293*, 41–47.

[3] Kemshead, J. T. and Ugelstad, J. (1985). Magnetic separation techniques: Their application to medicine. Mol. Cell Biochem. *67*, 11–18.

[4] Manyonda, I. T., Soltys, A. J. and Hay, F. C. (1992). A critical evaluation of the magnetic cell sorter and its use in the positive and negative selection of CD45RO+ cells. J. Immunol. Methods *149*, 1–10.

[5] Tibbe, A. G. J., de Grooth, B. G., Greve, J., Dolan, G. J., Rao, C. and Terstappen, L. W. M. M. (2002). Magnetic field design for selecting and aligning immunomagnetic labeled cells. Cytometry *47*, 163–72.

[6] Tibbe, A. G. J., de Grooth, B. G., Greve, J., Liberti, P. A., Dolan, G. J. and Terstappen, L. W. M. M. (1999). Optical tracking and detection of immunomagnetically selected and aligned cells. Nat. Biotechnol. *17*, 1210–3.

[7] Budd, G. T., Cristofanilli, M., Ellis, M. J., Stopeck, A., Borden, E., Miller, M. C., Matera, J., Repollet, M., Doyle, G. V., Terstappen, L. W. and Hayes, D. F. (2006). Circulating tumor cells versus imaging–predicting overall survival in metastatic breast cancer. Clin. Cancer Res. *12*, 6403–9.

[8] Miltenyi, S., Müller, W., Weichel, W. and Radbruch, A. (1990). High gradient magnetic cell separation with MACS. Cytometry *11*, 231–8.

[9] Ringhoffer, M., Wiesneth, M., Harsdorf, S., Schlenk, R. F., Schmitt, A., Reinhardt, P. P., Moessner, M., Grimminger, W., Mertens, T., Reske, S. N., Dohner, H. and Bunjes, D. (2004). CD34 cell selection of peripheral blood progenitor cells using the CliniMACS device for allogeneic transplantation: Clinical results in 102 patients. Br. J. Haematol. *126*, 527–35.

[10] Dynal: http://www.invitrogen.com/content.cfm?pageid=11049.

[11] R&D Systems: http://www.rndsystems.com/product_detail_objectname_cell_selection_kits_product.aspx

[12] Liberti, P. A., Rao, C. G. and Terstappen, L. W. M. M. (2001). Optimization of ferrofluids and protocols for the enrichment of breast tumor cells in blood. J. Magn. Magn. Mater. *301*, 301–7.

[13] MACS: http://www.miltenyibiotec.com/nn.21,3437a19c0cef71f755ca1145348d27d9,navigation.html

[14] Kantor, A. B., Gibbons, I., Miltenyi, S. and Schmitz, J. (1998). Magnetic cell sorting with colloidal superparamagnetic particles. In: Cell Separation

Methods and Applications (Recktenwald, D. and Radbruch, A., eds.). Marcel Dekker, Inc., New York, pp. 153–73.
[15] Williams, S. K. R., Lee, H. and Turner, M. M. (1999). Size characterization of magnetic cell sorting microbeads using flow field-flow fractionation and photon correlation spectroscopy. J. Magn. Magn. Mater. *194*, 248–53.
[16] Zhang, H., Moore, L. R., Zborowski, M., Williams, P. S., Margel, S. and Chalmers, J. J. (2005). Establishment and implications of a characterization method for magnetic nanoparticle using cell tracking velocimetry and magnetic susceptibility modified solutions. Analyst *130*, 514–27.
[17] McCloskey, K. E., Chalmers, J. J. and Zborowski, M. (2003). Magnetic cell separation: Characterization of magnetophoretic mobility. Anal. Chem. *75*, 6868–74.
[18] StemCell: http://www.stemcell.com/technical/product_reference.asp
[19] Allan, A. L., Vantyghem, S. A., Tuck, A. B., Chambers, A. F., Chin-Yee, I. H. and Keeney, M. (2005). Detection and quantification of circulating tumor cells in mouse models of human breast cancer using immunomagnetic enrichment and multiparameter flow cytometry. Cytometry A *65*, 4–14.
[20] Baksh, D., Davies, J. E. and Zandstra, P. W. (2003). Adult human bone marrow-derived mesenchymal progenitor cells are capable of adhesion-independent survival and expansion. Exp. Hematol. *31*, 723–32.
[21] Hidalgo, A. and Frenette, P. S. (2005). Enforced fucosylation of neonatal CD34+ cells generates selectin ligands that enhance the initial interactions with microvessels but not homing to bone marrow. Blood *105*, 567–75.
[22] Imren, S., Fabry, M. E., Westerman, K. A., Pawliuk, R., Tang, P., Rosten, P. M., Nagel, R. L., Leboulch, P., Eaves, C. J. and Humphries, R. K. (2004). High-level beta-globin expression and preferred intragenic integration after lentiviral transduction of human cord blood stem cells. J. Clin. Invest. *114*, 953–62.
[23] Sundstrom, J. B., Little, D. M., Villinger, F., Ellis, J. E. and Ansari, A. A. (2004). Signaling through Toll-like receptors triggers HIV-1 replication in latently infected mast cells. J. Immunol. *172*, 4391–401.
[24] Miranville, A., Heeschen, C., Sengenes, C., Curat, C. A., Busse, R. and Bouloumie, A. (2004). Improvement of postnatal neovascularization by human adipose tissue-derived stem cells. Circulation *110*, 349–55.
[25] Curat, C. A., Miranville, A., Sengenes, C., Diehl, M., Tonus, C., Busse, R. and Bouloumie, A. (2004). From blood monocytes to adipose tissue-resident macrophages: Induction of diapedesis by human mature adipocytes. Diabetes *53*, 1285–92.
[26] Storck, S., Delbos, F., Stadler, N., Thirion-Delalande, C., Bernex, F., Verthuy, C., Ferrier, P., Weill, J. C. and Reynaud, C. A. (2005). Normal

immune system development in mice lacking the Deltex-1 RING finger domain. Mol. Cell Biol. *25*, 1437–45.

[27] Chen, W., Antonenko, S., Sederstrom, J. M., Liang, X., Chan, A. S., Kanzler, H., Blom, B., Blazar, B. R. and Liu, Y. J. (2004). Thrombopoietin cooperates with FLT3-ligand in the generation of plasmacytoid dendritic cell precursors from human hematopoietic progenitors. Blood *103*, 2547–53.

[28] Wulff, H., Knaus, H. G., Pennington, M. and Chandy, K. G. (2004). K+ channel expression during B cell differentiation: Implications for immunomodulation and autoimmunity. J. Immunol. *173*, 776–86.

[29] Immunicon: http://www.immunicon.com/products/products.html

[30] Racila, E., Euhus, D., Weiss, A. J., Rao, C., McConnell, J., Terstappen, L. W. and Uhr, J. W. (1998). Detection and characterization of carcinoma cells in the blood. Proc. Natl. Acad. Sci. USA *95*, 4589–94.

[31] Cristofanilli, M., Budd, G. T., Ellis, M. J., Stopeck, A., Matera, J., Miller, M. C., Reuben, J. M., Doyle, G. V., Allard, W. J., Terstappen, L. W. and Hayes, D. F. (2004). Circulating tumor cells, disease progression, and survival in metastatic breast cancer. N. Engl. J. Med. *351*, 781–91.

[32] BDBiosciences: http://www.bdbiosciences.com/index_us.shtml

Laboratory Techniques in Biochemistry and Molecular Biology, Volume 32
Magnetic Cell Separation
M. Zborowski and J. J. Chalmers (Editors)

CHAPTER 11

Worked examples of cell sample preparation and magnetic separation procedures

Maciej Zborowski,[1] Lee R. Moore,[1] Kristie Melnik[2] and Ying Jing[1]

[1]*Department of Biomedical Engineering, Lerner Research Institute, Cleveland Clinic, Cleveland, OH 44195, USA*
[2]*Columbus NanoWorks Inc., Columbus, OH 43212, USA*

11.1. Dynal® T Cell Negative Isolation Kit

The information provided in this section is adapted from company website [1]. The kit depletes activated T cells. Intended use: isolate untouched human T cells by depleting non-T cells (B cells, NK cells, monocytes, platelets, dendritic cells, granulocytes, erythrocytes) and activated T cells from peripheral blood. Isolated T cells are bead- and antibody-free and are suitable for any downstream application. This kit depletes activated T cells (HLA class II positive cells). Isolated T cells can be used in applications such as cell culture, flow cytometry, functional assays, and molecular studies. The kit will process up to 5×10^8 cells. The kit is intended for research use only.

11.1.1. Principle of isolation

Add a mixture of monoclonal antibodies against the non-T cells to the starting sample. Add Dynabeads® to bind to the non-T cells

DOI: 10.1016/S0075-7535(06)32011-6

during a short incubation. Separate the bead-bound cells by a magnet. Discard the bead-bound cells and use the remaining, untouched human T cells for any application. Figures 10.2 and 10.3 illustrate the process.

11.1.2. Description of materials

Dynabeads are uniform, superparamagnetic polystyrene beads (4.5-μm diameter) coated with a monoclonal human anti-mouse IgG antibody. The antibody coated onto Dynabeads recognizes all mouse IgG subclasses and is Fc-specific. The antibody on the Dynabeads is a human IgG4 anti-mouse IgG. The source of the human monoclonal antibody is free of *Human immunodeficiency virus* (HIV), *Hepatitis B virus* (HBV), and *Hepatitis C virus* (HCV).

11.1.2.1. Materials supplied

1. Dynal T Cell Negative Isolation Kit contains Depletion Dynabeads and Antibody Mix.
2. 5-ml Depletion Dynabeads. Supplied in phosphate buffered saline (PBS), pH 7.4, containing 0.1% bovine serum albumin (BSA), and 0.02% sodium azide (NaN_3).
3. 1-ml Antibody Mix. The Antibody Mix contains mouse IgG antibodies for CD14, CD16 (specific for CD16a and CD16b), HLA class II DR/DP, CD56, and CD235a (Glycophorin A). Supplied in PBS and 0.02% (NaN_3).

11.1.2.2. Additional materials required

1. Magnet (Dynal Magnetic Particle Concentrator): MPC-l for 1- to 5-ml samples, MPC-15 for 1- to 15-ml samples, and MPC-50 for 15- to 50-ml samples. (The interested reader is directed to Company website for information update.)
2. Mixer allowing both tilting and rotation.
3. Heat-inactivated fetal calf serum (FCS).

4. Buffer 1: PBS (without Ca^{2+} and Mg^{2+}) with 0.1% BSA and 2-mM EDTA, pH 7.4.
5. Buffer 2: PBS (without Ca^{2+} and Mg^{2+}), pH 7.4.
6. Lymphoprep™ for MNC preparation [Axis Shield Point-of-care (PoC), Norway, www.axis-shield-poc.com].

Important notes:

1. BSA can be replaced by human serum albumin (HSA) or FCS. EDTA can be replaced by sodium citrate. PBS containing Ca^{2+} or Mg^{2+} is not recommended.
2. Keep the buffers cold!

11.1.3. Protocols

11.1.3.1. Dynabeads washing procedure

Dynabeads should be washed before use.

1. Resuspend the Dynabeads in the vial to a homogenous suspension.
2. Transfer the desired volume of Dynabeads to a tube.
3. Add the same volume of Buffer 1, or at least 1 ml, and mix.
4. Place the tube in a magnet for 2 min and discard the supernatant.
5. Remove the tube from the magnet and resuspend the washed Dynabeads in the same volume of Buffer 1 as the initial volume of Dynabeads.

11.1.3.2. Preparation of MNC from buffy coat to obtain low platelet numbers

1. Dilute 10- to 18-ml buffy coat with Buffer 2 to a total volume of 35 ml at 18–25 °C (room temperature, RT).
2. Add the diluted buffy coat on top of 15 ml of Lymphoprep.

3. Centrifuge at $160 \times g$ for 20 min at RT. Allow to decelerate without brakes.

4. Remove 20 ml of supernatant to eliminate platelets.

5. Centrifuge at $350 \times g$ for 20 min at RT. Allow to decelerate without brakes.

6. Recover MNC from the plasma/Lymphoprep interface and transfer the cells to a 50-ml tube.

7. Wash MNC once with Buffer 1 by centrifugation at $400 \times g$ for 8 min at 2–8 °C.

8. Wash MNC twice with Buffer 1 by centrifugation at $225 \times g$ for 8 min at 2–8 °C and resuspend the MNC at 1×10^8 MNC per milliliter in Buffer 1. For other recommended sample preparation procedures, visit https://catalog.invitrogen.com/

11.1.3.3. Critical steps for cell isolation

1. Use a mixer that provides tilting and rotation of the tubes to ensure Dynabeads do not settle at the bottom of the tube.
2. Never use less than 100-μl Dynabeads per 1×10^7 MNC sample.
3. It is critical to follow the magnet recommendations to ensure a successful isolation.

11.1.3.4. Isolation of human T cells from MNC

This protocol is based on 1×10^7 MNC. It is scalable from 1×10^7 to 5×10^8 cells.

1. Transfer 100-μl (1×10^7) MNC in Buffer 1 to a tube.
2. Add 20-μl heat-inactivated FCS.
3. Add 20 μl of Antibody Mix.
4. Mix well and incubate for 20 min at 2–8 °C.
5. Wash the cells by adding 2-ml Buffer 1. Mix well by tilting the tube several times and centrifuge at $300 \times g$ for 8 min at 2–8 °C. Discard the supernatant.
6. Resuspend the cells in 900-μl Buffer 1.

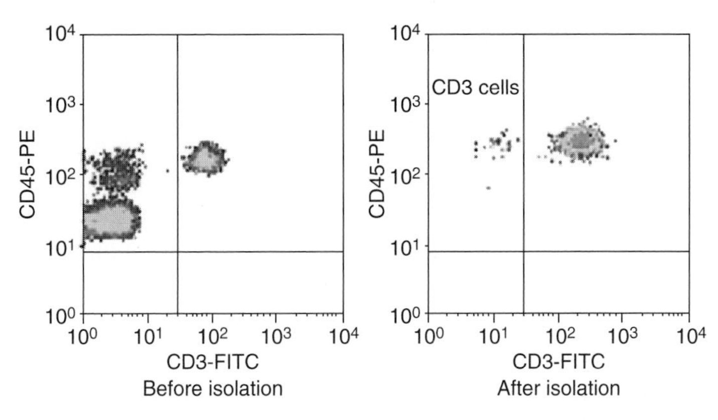

Fig. 11.1. Untouched CD3+ T cells negatively isolated from MNC with the Dynal T Cell Negative Isolation Kit. The figure shows the CD3+ population before and after negative isolation (from the Company website [1], used with permission). (See Color Insert.)

7. Add 100-μl prewashed Depletion Dynabeads.
8. Incubate for 15 min at 18–25 °C with gentle tilting and rotation.
9. Resuspend the bead-bound cells by gently pipetting five times using a pipette with a narrow tip opening, (e.g., a 1000-μl pipette tip or a 5-ml serological pipette).
10. Add 1-ml Buffer 1.
11. Place the tube in the magnet for 2 min.
12. Transfer the supernatant to a new tube.

Repeat steps 10–12. The supernatant contains the negatively isolated human T cells. Figure 11.1 shows typical flow cytometry (FCM) results of CD3+ T cells negatively isolated with the Dynal T Cell Negative Isolation Kit.

11.2. Dynal CD34 Progenitor Cell Selection System

The information provided in this section is adapted from company websites [1, 2].

11.2.1. Principle of isolation

Invitrogen Dynal offers DYNAL Magnetic Particle Concentrators such as the DYNAL MPC-1, DYNAL MPC-2, or DYNAL MPC-6. (The interested reader is directed to Company website for information update.) The optimal concentration of Dynabeads required may vary for different samples. The cell:bead mixture should contain a concentration of 4×10^7 Dynabeads M-450 CD34/ml of MNC suspension. Invitrogen Dynal recommends an initial MNC concentration in the range of 4×10^7 to 4×10^8 cells/ml. During rosetting the incubation volume should be at least 1 ml. Depletion will become slightly more efficient in the lower end of this range. Detachment will become slightly more efficient in the higher end of this range (when the bead concentration is kept constant), because lower bead to cell ratio favor detachment. Overall, a cell concentration in the range of 4×10^7 to 4×10^8 MNC/ml will result in the same final percent recovery of CD34+ target cells. The Company offers local Invitrogen Dynal Technical Service for details on isolating CD34+ cells from rhesus monkey bone marrow (BM) and cord blood (CB) samples. The protocol presented below is for positive isolation of CD34+ cells from human BM, CB, or PB.

Figure 11.2 shows typical FCM results of CD34+ progenitor cells positively isolated with the Dynal Progenitor Cell Selection System.

11.2.2. Description of materials

11.2.2.1. Required buffers and media

All reagents should be analytical grade and filter sterilized before use.

1. Lymphocyte separation medium (e.g., Ficoll®).

2. Isolation buffer: per 100-ml isolation buffer add 70 ml of double-distilled water (ddH_2O) to 10 ml 10× PBS, pH 7.4 without Ca^{2+} and Mg^{2+} (supplier e.g., GIBCOBRL). While gently stirring add 10 ml 20% BSA/HSA and 0.6-g trisodium citrate dihydrate (supplier e.g., Sigma), stirring until sodium citrate is completely dissolved. Add 2-ml PS, bring volume to 100 ml with ddH_2O, and

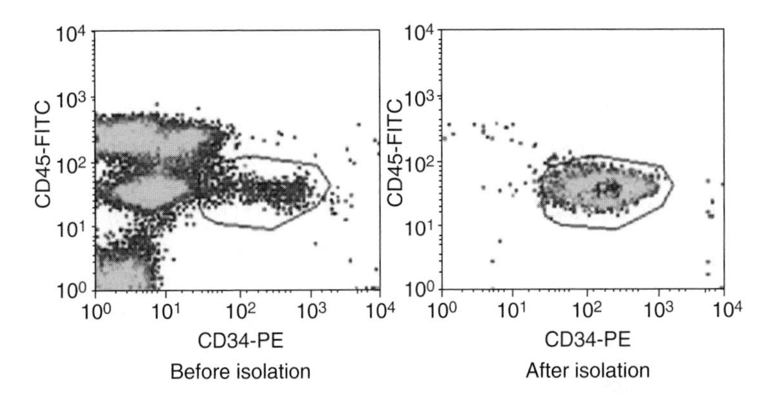

Fig. 11.2. Isolation results: CD34+ cells positively isolated from bone marrow-derived MNC with the Dynal® CD34 Progenitor Cell Selection System (from the Company website [2], used with permission). (See Color Insert.)

filter sterilize. The final solution should contain 1× PBS, 2% BSA/HSA, 0.6% citrate, and 100 IU/ml PS.

3. Culture medium with 10% FCS, 100 IU/ml PS: Iscove's Modified Dulbecco's Medium (IMDM) (supplier e.g., GIBCOBRL) or RPMI 1640 (supplier e.g., GIBCOBRL). Supplement with heat-inactivated FCS (to a final concentration of 10%) (supplier e.g., GIBCOBRL) and 100 IU/ml PS (see below). FCS should be tested for cytotoxity before use. The final culture medium should include 10% heat-inactivated FCS and 100 IU/ml PS.

4. Stock solutions:

a. 10× PBS: pH 7.4 without Ca^{2+} and Mg^{2+}, $NaH_2PO_4{\cdot}H_2O$ 1.57 g, $Na_2HPO_4{\cdot}12H_2O$ 19.80 g, NaCl 81.00 g, ddH_2O to 1liter. Filter sterilize before use. Or:

b. 10× Dulbecco's Phosphate Buffer Saline (D-PBS) without Ca^{2+} and Mg^{2+} (supplier e.g., GIBCOBRL)

c. 20% BSA: Dissolve 20-g BSA, fraction V (supplier e.g., GIBCOBRL) in 70-ml ddH_2O with gentle stirring. Bring volume to 100 ml with ddH_2O and filter sterilize.

d. 20% HSA: from a supplier, for example Pharmacia

e. Penicillin-streptomycin solution (PS): 5000 IU/ml penicillin, 5000 IU/ml streptomycin (supplier e.g., GIBCOBRL)

11.2.3. Protocols

11.2.3.1. Prewash of Dynabeads M-450 CD34

Calculate the volume of Dynabeads M-450 CD34 required for your experiment. Dynal Biotech recommends a concentration of 4×10^7 Dynabeads (100 μl) per milliliter of initial MNC. Thoroughly resuspend Dynabeads M-450 CD34 and transfer required volume to a round bottom tube. Wash the Dynabeads M-450 CD34 by placing the tube on a DYNAL MPC for 1 min and then removing the supernatant. Remove the tube from the DYNAL MPC and resuspend Dynabeads M-450 CD34 in 1 ml of isolation buffer. Repeat. Place the tube on DYNAL MPC for 1 min after last buffer addition. Pipette off buffer and remove the tube from the DYNAL MPC before cells are added.

11.2.3.2. Preparation of cells

1. Collect BM, PB, or CB in presence of anticoagulant. Note: This protocol recommends using fresh sample collected the day of CD34+ cell selection. If bone marrow is collected the day before selection MNC should be isolated the day of harvest and stored overnight at +4°C at a concentration $\leq 5 \times 10^6$ MNC/ml in appropriate medium such as IMDM or RPMI supplemented with 10% FCS. During isolation the isolation buffer should be added 10-μg DNase/ml. If isolating CD34+ cells from BM or PB, dilute cells in an equal amount of isolation buffer. CB samples should be diluted 1:4 in isolation buffer. PB apheresis material should be collected the same day of CD34+ cell selection is performed and diluted to a concentration of $\leq 1 \times 10^7$cells/ml in isolation buffer.
2. Transfer 35 ml of cell suspension onto 15-ml Ficoll® and centrifuge 20 min at 800 × *g*.
3. Collect interface, suspend cells in an equal volume of isolation buffer, and centrifuge 20 min at 500 × *g*.
4. Resuspend pellet in isolation buffer again and centrifuge 20 min at 300 × *g*.

5. Resuspend pellet to a concentration in the range of 4×10^7 to 4×10^8 cells/ml in cold isolation buffer. The total volume should be at least 1 ml. Maintain at +4 °C.

11.2.3.3. Positive selection of CD34+ cells: Mixing of beads and cells
Optimal concentration of Dynabeads M-450 CD34 is 4×10^7 beads/ml. It is critical to use round bottom tubes to obtain optimal mixing during incubation. Resuspend cells thoroughly with a pipette with a narrow tip to prevent cells from aggregating before adding them to the beads. Vortex cell-bead mixture 2–3 sec to mix. Incubate the cell-bead suspension for 30 min at +4 °C with gentle tilt rotation at 10–20 rpm (e.g., DYNAL Sample Mixer). Note: During incubation and separation it is important to keep the cell suspension and buffers cold (+4 °C) to prevent nonspecific attachment of phagocytic cells to Dynabeads. Make sure that the Dynabeads M-450 CD34 are kept in suspension during incubation.

11.2.3.4. Positive selection of CD34+ cells: Magnetic separation of rosetted cells
After incubation, add isolation buffer to the height of the magnet. Resuspend the cell-bead complexes. You may vortex for 2–3 sec, if desired. Place tube in a DYNAL MPC for 2 min to separate Dynabeads M-450 CD34-rosetted cells from nontarget cells. Aspirate supernatant containing nonrosetted cells while the tube is still exposed to the magnet. Repeat three times. After the final wash, resuspend rosettes in 100 μl of isolation buffer per 4×10^7 beads used, in a volume of at least 100 μl. Proceed with detachment as follows.

11.2.3.5. Detachment of beads from purified cells

1. Add 100-μl DETACHaBEAD CD34 per 4×10^7 Dynabeads M-450 CD34 in a volume of at least 100 μl (i.e., use the same volume of DETACHaBEAD as used for resuspension of the rosettes).
2. Vortex 2–3 sec to mix.

a. Incubate at room temperature with gentle, tilt rotation (e.g., DYNAL Sample Mixer) for 45 min. After incubation proceed with step 3 below.

or

b. Incubate for 15 min at 37 °C with gentle tilt rotation. If rotation is not possible it is essential to shake the tube by hand every 5 min to keep rosettes in suspension. Note: Detachment will become less efficient if the incubation temperature is below 20 °C. Due to relatively small sample volume care should be taken that the cells remain near the bottom of the tube to avoid loss of cells.

3. After incubation, add 2 ml of isolation buffer and vortex 2–3 sec to enhance detachment of Dynabeads M-450 CD34 from the cells.

4. Place the tube on the DYNAL MPC and allow the Dynabeads M-450 CD34 to accumulate on the tube wall for 2 min.

5. Transfer released cells to a new tube. Repeat steps 3 and 4 three times, pooling released cells into a single tube.

6. Place the tube with the pooled isolated cells on the DYNAL MPC for 2 min to remove any residual beads.

7. Transfer supernatant containing released cells to a fresh, conical tube.

11.2.3.6. Washing positively selected CD34+ cells

1. Wash the cells twice in 10 ml or more of isolation buffer, centrifuging 10 min at 400 × *g* to pellet cells.

2. To avoid cell loss, leave 100-μl fluid over the pellet when removing the supernatant. Avoid decanting the supernatant as it may lead to loss of cells.

3. After final centrifugation, resuspend pellet in 1-ml isolation buffer.

4. Purified CD34+ cells can now be stained, sorted, cultured, and counted as desired.

5. If FACS analysis is to be performed, we recommend the rigorous gating settings as described by Sutherland et al. [3]. If storing cells overnight, store at +4 °C at a concentration $\leq 5 \times 10^6$ cells/ml in appropriate medium such as IMDM or RPMI 1640 supplemented with 10% FCS, 100 IU/ml PS.

11.2.3.7. Purity and recovery of selected CD34+ cells

Percent recovery as determined by flow cytometry is defined as: the total number of CD34+ cell selected multiplied by hundred divided by the total number of CD34+ cells in the initial MNC population. See Fig. 11.1.

11.3. Miltenyi Biotec MACS® T Cell Isolation Kit II (human)

The information provided in this section is adapted from company website [4]. Components (for 10^9 total cells, up to 100 separations):

1. 1-ml Pan T Cell Biotin-Antibody Cocktail: Cocktail of biotin-conjugated monoclonal antibodies against CD14, CD16, CD19, CD36, CD56, CD123, and Glycophorin A. The Biotin-Antibody Cocktail is supplied in a solution containing stabilizer and 0.05% (NaN_3).
2. 2-ml Anti-Biotin MicroBeads: MicroBeads conjugated to a monoclonal anti-biotin antibody (isotype: mouse IgG1). The Anti-Biotin MicroBeads are supplied as a suspension containing stabilizer and 0.05% (NaN_3).

11.3.1. Principle of MACS® separation

Using the Pan T Cell Isolation Kit II, human T cells are isolated by depletion of non-T cells (negative selection). Non-T cells are indirectly magnetically labeled with a cocktail of biotin-conjugated monoclonal antibodies, as primary labeling reagent, and anti-biotin monoclonal antibodies conjugated to MicroBeads, as secondary labeling reagent. In between the two labeling steps no washing

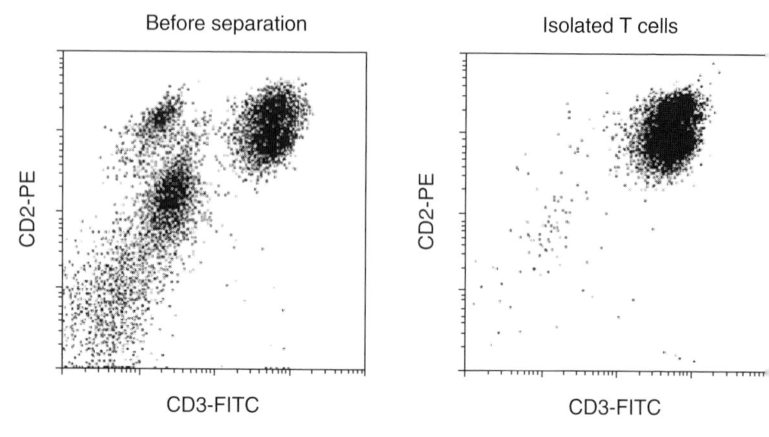

Fig. 11.3. Example of a separation using Miltenyi Biotec Pan T Cell Isolation Kit II and an LS Column. Cells are fluorescently stained with CD3-FITC and CD2-PE. Cell debris and dead cells were excluded from the analysis based on scatter signals and PI fluorescence (from the Company website [4]).

steps are required. The magnetically labeled non-T cells are depleted by retaining them on a MACS Column in the magnetic field of a MACS Separator, while the unlabeled T cells pass through the column. The selection of MACS columns is described in Chapter 10. Figure 11.3 illustrates FCM evaluation following enrichment of T cells with this method.

11.3.2. Background and product applications

The Pan T Cell Isolation Kit II is an indirect magnetic labeling system for the isolation of untouched T cells from human peripheral blood mononuclear cells (PBMCs). Non-T cells, that is B cells, NK cells, dendritic cells, monocytes, granulocytes, and erythroid cells, are indirectly magnetically labeled by using a cocktail of biotin-conjugated antibodies against CD14, CD16, CD19, CD36, CD56, CD123, and Glycophorin A, and Anti-Biotin

MicroBeads. Isolation of highly pure T cells is achieved by depletion of magnetically labeled cells.

Examples of applications:

1. Functional studies on T cells in which effects due to antibody cross-linking of cell surface proteins should be avoided.
2. Studies on signal requirements for T-cell activation, induction of T-cell proliferation, induction of T-cell anergy, and others.
3. Studies on signal transduction in T cells.
4. Studies on regulation of T cell cytokine expression.

11.3.3. Reagent and instrument requirements

1. Buffer (degassed): Prepare a solution containing PBS (phosphate buffered saline) pH 7.2, 0.5% BSA, and 2-mM EDTA by diluting MACS BSA Stock Solution 1:20 with autoMACS™ Rinsing Solution. Keep buffer cold (4–8 °C).

Note: EDTA can be replaced by other supplements such as anticoagulant citrate dextrose formula-A (ACD-A) or citrate phosphate dextrose (CPD). BSA can be replaced by other proteins such as human serum albumin, human serum or fetal calf serum. Buffers or media containing Ca^{2+} or Mg^{2+} are not recommended for use.

2. MACS Columns and MACS Separators: Choose the appropriate MACS Separator and MACS Columns according to the number of labeled cells and to the number of total cells (as shown in Chapter 10).

Note: Column adapters are required to insert certain columns into VarioMACS™ Separator or SuperMACS™ Separator. For details, see MACS Separator data sheets.

3. (Optional) Fluorochrome-conjugated antibodies (e.g., CD3-FITC, CD2-PE, Anti-Biotin-PE, Anti-Biotin-APC).

4. (Optional) Pre-Separation Filters to remove cell clumps.

5. (Optional) PI (propidium iodide) or 7-AAD for the flow cytometric exclusion of dead cells.

11.3.4. Protocols

11.3.4.1. Sample preparation

When working with anticoagulated peripheral blood or buffy coat, peripheral blood mononuclear cells (PBMCs) should be isolated by density gradient centrifugation (e.g., Ficoll-Paque®). Note: Remove platelets after density gradient separation: resuspend cell pellet in buffer and centrifuge at 200 × *g* for 10–15 min at 20 °C. Carefully remove supernatant. Repeat washing step and carefully remove supernatant.

When working with tissues, prepare a single-cell suspension by a standard preparation method. Note: Dead cells may bind nonspecifically to MACS MicroBeads. To remove dead cells, we recommend using density gradient centrifugation or the Dead Cell Removal Kit (available from the Company).

11.3.4.2. Magnetic labeling

Work fast, keep cells cold, use precooled solutions. This will prevent capping of antibodies on the cell surface and nonspecific cell labeling. Volumes for magnetic labeling given below are for up to 10^7 total cells. When working with fewer than 10^7 cells, use same volumes as indicated. When working with higher cell numbers, scale up all reagent volumes and total volumes, accordingly (e.g., for 2×10^7 total cells, use twice the volume of all indicated reagent volumes and total volumes). For optimal performance it is important to obtain a single cell suspension before magnetic separation. Pass cells through 30-μm nylon mesh (Pre-Separation Filters, available from the Company) to remove cell clumps that may clog the columns. Working on ice may require increased incubation times. Higher temperatures and/or longer incubation times lead to nonspecific cell labeling.

1. Determine cell number.
2. Centrifuge cell suspension at 300 × *g* for 10 min. Pipette off supernatant completely.
3. Resuspend cell pellet in 40 μl of buffer per 10^7 total cells.

4. Add 10 μl of Biotin-Antibody Cocktail per 10^7 total cells.
5. Mix well and incubate for 10 min at 4–8 °C.
6. Add 30 μl of buffer per 10^7 total cells.
7. Add 20 μl of Anti-Biotin MicroBeads per 10^7 total cells.
8. Mix well and incubate for an additional 15 min at 4–8 °C.
9. Wash cells with buffer by adding 10–20× labeling volume and centrifuge at 300 × *g* for 10 min. Pipette off supernatant completely.
10. Resuspend up to 10^8 cells in 500 μl of buffer. Note: For higher cell numbers, scale up buffer volume accordingly.
11. Proceed to magnetic separation.

11.3.4.3. Magnetic separation

Choose an appropriate MACS column and MACS separator according to the number of labeled cells and to the number of total cells (as illustrated in Chapter 10).

11.3.4.3.1. Magnetic separation with MS and LS columns

1. Place column in the magnetic field of a suitable MACS Separator.

2. Prepare column by rinsing with appropriate amount of buffer: MS: 500 μl, LS: 3 ml.

3. Apply cell suspension onto the column.

4. Allow the cells to pass through and collect effluent as fraction with unlabeled cells, representing the enriched T-cell fraction.

5. Wash column with appropriate amount of buffer. Perform washing steps by adding buffer three times, each time once the column reservoir is empty (MS: 3 × 500 μl, LS: 3 × 3 ml). Collect entire effluent in the same tube as effluent of step 4. This fraction represents the enriched T cells.

6. (Optional) Elute retained cells outside of the magnetic field. This fraction represents the magnetically labeled non-T cells.

Note: Magnetic separation with XS Columns: for instructions on the column assembly and the separation, refer to the "XS Column data sheet" (available from the Company).

11.3.4.3.2. Magnetic separation with the autoMACS™ Separator
Refer to the "autoMACS™ User Manual" for instructions on how to use the autoMACS Separator.

1. Prepare and prime the autoMACS Separator.
2. Place tube containing the magnetically labeled cells in the autoMACS Separator. Choose program "Deplete."
3. Collect the negative fraction (outlet port "neg1"). This fraction represents the enriched T cells.
4. (Optional) Collect positive fraction (outlet port "pos1"). This fraction represents the magnetically labeled non-T cells.
5. (Optional) Evaluation of T-cell purity. The purity of the enriched T cells can be evaluated by flow cytometry or fluorescence microscopy. Stain aliquots of the cell fractions with a fluorochrome-conjugated antibody against a T-cell marker, for example, CD3-FITC (available from the Company), as recommended in the respective data sheet. Analyze cells by flow cytometry or fluorescence microscopy. Labeling of non-T cells with the Biotin-Antibody Cocktail can be visualized by counter-staining with fluorochrome-conjugated anti-biotin antibodies, for example Anti-Biotin-PE or Anti-Biotin-APC. Staining with fluorochrome-conjugated streptavidin is not recommended.

11.4. Miltenyi Biotec Direct CD34 Progenitor Cell Isolation Kit

The Direct CD34 Progenitor Cell Isolation Kit contains MicroBeads directly conjugated to CD34 antibodies for magnetic labeling of CD34 expressing hematopoietic progenitor cells from peripheral blood, cord blood, bone marrow, or apheresis harvest. Hematopoietic progenitor cells, present at a frequency of about 0.05–0.2% in peripheral blood, 0.1–0.5% in cord blood and 0.5–3% in bone marrow. The Company provides instrumentations and reagents for enrichment to a purity of about 85–98%.

11.4.1. Research applications of CD34 progenitor cells

1. Characterization of hematopoietic progenitor cells and their developmental pathways.
2. Studies on stimulation of proliferation and maturation by cytokines.

11.4.2. Isolation strategy

The isolation of hematopoietic progenitor cells is performed by positive selection of CD34 expressing cells. Mononuclear cells from peripheral blood (PBMC), cord blood, bone marrow, or apheresis harvest are obtained by density gradient centrifugation over Ficoll Paque®. Apheresis harvest can also be used as sample without further density gradient centrifugation. For MACS separation, CD34+ hematopoietic progenitor cells are magnetically labeled using MACS CD34 MicroBeads. The magnetically labeled cells are enriched on positive selection columns in the magnetic field of the MiniMACS, MidiMACS, VarioMACS, or SuperMACS (as described in Chapter 10).

Figure 11.4 illustrates an example of flow cytometry dot plots following direct progenitor cell enrichment with this kit.

11.4.3. Components

1. 2-ml MACS CD34 MicroBeads: Colloidal superparamagnetic MicroBeads conjugated to monoclonal mouse anti-human CD34 antibody (isotype: mouse IgG1, clone: QBEND/10).
2. 2-ml FcR Blocking Reagent. Human IgG.
3. (Optional) Fluorochrome conjugated CD34 antibody (e.g., CD34-PE) and fluorochrome conjugated CD45 antibody (e.g., CD45-FITC, available from the Company) for control of CD34 progenitor cell isolation.

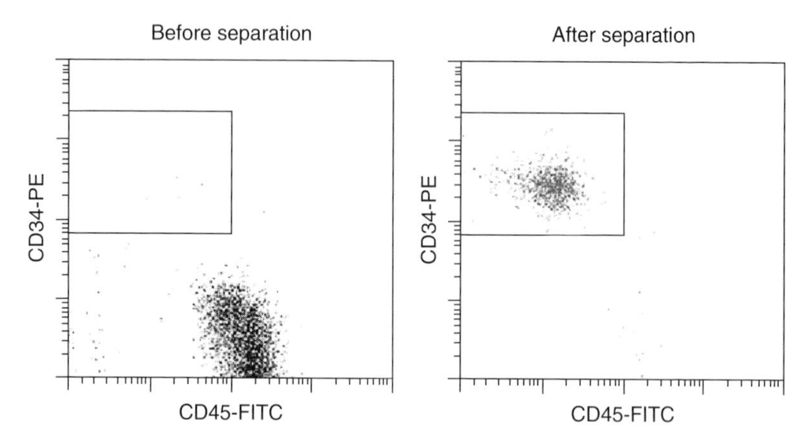

Fig. 11.4. Isolation of CD3+ cells using Direct CD34 Progenitor Cell Isolation Kit and MiniMACS. Cells were stained with CD34-PE and CD45-FITC (from the Company website [4]).

11.4.4. Equipment required

The description of the available magnetic separation columns from Miltenyi Biotec is provided in Chapter 10.

11.4.5. Protocols

The description of the available magnetic separation columns from Miltenyi Biotec is provided in Chapter 10.

11.4.5.1. Preparation of peripheral blood mononuclear cells

Start with fresh human blood treated with an anticoagulant [e.g. heparin, citrate, acid citrate dextrose-anticoagulant (ACD-A) or citrate phosphate dextrose (CPD)] or leukocyte-rich buffy coat not older than 8 hours.

1. Dilute cells with 2–4 volumes of PBS containing 2-mM EDTA or 0.6% ACD-A.
2. Carefully layer 35 ml of diluted cell suspension over 15-ml Ficoll Paque® (1.077 density) in a 50-ml conical tube and centrifuge

at 400 × *g* for 30–40 min at 20 °C in a swinging-bucket rotor without brake.

3. Aspirate the upper layer leaving the mononuclear cell layer undisturbed at the interphase.

4. Carefully transfer the interphase cells (lymphocytes and monocytes) to a new 50-ml conical tube.

5. Fill the conical tube with PBS containing 2-mM EDTA or 0.6% ACD-A, mix and centrifuge at 300 × *g* for 10 min at 20 °C. Carefully remove the supernatant completely.

6. Resuspend the cell pellet in 50 ml of PBS containing 2-mM EDTA or 0.6% ACD-A and centrifuge at 200 × *g* for 10–15 min at 20 °C. Carefully remove the supernatant completely.

7. Resuspend the cell pellet in 50 ml of buffer and centrifuge at 200 × *g* for 10–15 min at 20 °C. Carefully remove the supernatant completely.

8. Resuspend cell pellet in a final volume of 300 μl per 10^8 total cells (PBMC of about 100-ml blood). For less than 10^8 total cells, use 300 μl. Proceed to magnetic labeling.

Notes: The peripheral blood or buffy coat should not be older than 8 hours and supplemented with anticoagulants. PBMC may be stored in refrigerator overnight in PBS containing 0.5% BSA supplemented with autologous serum after the last washing step.

11.4.5.2. Preparation of cord blood cells

1. Dilute anticoagulated cord blood 1:4 with PBS containing 2-mM EDTA or 0.6% ACD-A and carefully layer 35 ml of diluted cell suspension over 15 ml of Ficoll-Paque.

2. Centrifuge for 35 min at 400 × g at 20 °C in a swinging-bucket rotor (without brake).

3. Aspirate the upper layer leaving the mononuclear cell layer undisturbed at the interphase.

4. Carefully collect interphase cells and wash twice in PBS containing 2-mM EDTA or 0.6% ACD-A. Centrifuge for 10 min at 200 × g at 20 °C.

5. Resuspend cell pellet in a final volume of 300 μl of buffer per 10^8 total cells. For less than 10^8 total cells, use 300 μl. Proceed to magnetic labeling.

Notes: Do not use cord blood older than 4 hours. The cord blood should be drawn directly into a 50-ml tube containing 5 ml of PBS supplemented with 2-mM EDTA or 0.6% ACD-A or 200-U/ml heparin. The cord blood should be stored at 4 °C prior to separation.

11.4.5.3. Preparation of bone marrow cells

1. Collect bone marrow in 50-ml tubes containing 5-ml PBS containing 2-mM EDTA or 0.6% ACD-A or 200-U/ml heparin and store at 4 °C if the cells cannot be processed immediately.
2. For release of the cells, dilute in 10× excess of RPMI 1640 containing 0.02% collagenase B and 100-U/ml DNAse and shake gently at room temperature for 45 min.
3. Pass cells through 30-μm nylon mesh or filter (available from the Company). Wet filter with buffer before use.
4. Carefully layer 35 ml of diluted cell suspension over 15 ml of Ficoll-Paque®.
5. Centrifuge for 35 min at 400 × g at 20 °C in a swinging-bucket rotor (without brake).
6. Aspirate the upper layer leaving the mononuclear cell layer undisturbed at the interphase.
7. Carefully collect interphase cells and wash twice in PBS containing 2-mM EDTA or 0.6% ACD-A. Centrifuge for 10 min at 300 × g at 20 °C.
8. Resuspend cell pellet in a final volume of 300 μl of buffer per 10^8 total cells. For less than 10^8 total cells, use 300 μl. Proceed to magnetic labeling.

Notes: If cells cannot be separated on the day of harvest, store cells at 4 °C. Remove all cell clumps by passing cells through 30-μm nylon mesh or Pre-Separation Filter (available from the Company) during the cell preparation. Wet filter with buffer before use.

11.4.5.4. Preparation of cells from leukapheresis material
Filter apheresis harvest through 30-μm nylon mesh, or Pre-Separation Filter (available from the Company or other sources, e.g., "Cell Strainers" from Becton Dickinson), in order to remove clumps, wash cells once with buffer and resuspend in a final volume of 300 μl of buffer per 10^8 cells. For less than 10^8 total cells, use 300 μl. Proceed to magnetic labeling.

11.4.5.5. Magnetic labeling of CD34+ progenitor cells

1. Add 100 μl FcR blocking reagent per 10^8 total cells to the cell suspension to inhibit unspecific or Fc-receptor mediated binding of CD34 MicroBeads to nontarget cells.
2. Label cells by adding 100-μl CD34 MicroBeads per 10^8 total cells, mix well, and incubate for 30 min in the refrigerator at 4–8 °C.
3. (Optional) Add fluorochrome conjugated CD34 antibody recognizing another epitope than QBEND/10 (e.g., CD34-PE, Clone: AC136) and fluorochrome conjugated CD45 antibody (e.g., CD45-FITC, available from the Company) at the titer recommended by manufacturer and incubate for further 10 min in the refrigerator at 4–8 °C.
4. Wash cells carefully and resuspend in appropriate amount of buffer (MS+/RS+ column: 500–1000 μl; LS+/VS+ column: 1–10 ml, maximum 2×10^8 cells per milliliter).
5. Proceed to magnetic separation.

11.4.5.6. Magnetic separation of $< 2 \times 10^9$ mononuclear cells

1. Choose a column type (MS+/RS+ or LS+/VS+) according to the number of total unseparated cells and place it (with column adapter) in the magnetic field of the MACS separator (see Chapter 10). Fill and rinse with buffer (MS+/RS+: 500 μl; LS+/VS+: 3 ml; for details, see "Column and Adapter Data Sheets," available from the Company).
2. Pass cells through 30-μm nylon mesh or Pre-Separation Filter to remove clumps. Wet filter with buffer before use.

3. Apply cells to the column, allow cells to pass through the column, and wash with buffer (MS+/RS+: $3 \times 500\ \mu l$; LS+/VS+: 3×3 ml).

4. Remove column from separator, place column on a suitable tube, and pipette buffer on top of column (MS+/RS+: 1 ml; LS+/VS+: 5 ml). Firmly flush out retained cells with pressure using the plunger supplied with the column.

5. Repeat magnetic separation step: apply the eluted cells to a new prefilled positive selection column (for $<10^7$ CD34+ cells: MS+/RS+; for $<10^8$ CD34+ cells: LS+/VS+), wash, and elute retained cells in buffer (MS+/RS+: 500 μl; LS+/VS+: 2.5 ml).

11.4.5.7. Magnetic separation of 2×10^9 to 2×10^{10} mononuclear cells

1. Assemble XS+ column and place it in the column holder of the SuperMACS using XS+ column adapter (for details, see "XS+ Selection Column and SuperMACS Data Sheets," available from the Company).

2. Turn 3-way-stopcock to position "fill."

3. Fill the column from the bottom with buffer from the syringe until the buffer reaches the syringe cylinder.

4. Turn the 3-way-stopcock to position "run" and rinse column by filling from the top with buffer. Allow buffer to run into the column. Then, add more buffer. Rinse with 50 ml of buffer.

5. Close 3-way-stopcock; leave the syringe attached during separation, except when refilling with buffer.

6. Move column in the magnetic field of the SuperMACS by turning the handle.

7. Pass cells through 30-μm nylon mesh or filter to remove cell clumps.

8. Apply cells into the syringe cylinder that is set up on the XS+ column and turn 3-way-stopcock to position "run." Allow the cells to pass through the column.

9. Remove flow resistor and wash with 4×30-ml buffer.

10. Close 3-way-stopcock and remove column out of the magnetic field of the SuperMACS by turning the XS+ adapter handle

backward (see "Super MACS Starting Kit Data Sheet," available from the Company).

11. Detach syringe from the 3-way-stopcock, fill with buffer, and attach to port A of the XS+ column.

12. Elute retained cells with 20-ml buffer using the syringe.

13. Repeat magnetic separation step: apply the eluted cells to a new prefilled XS+ column or VS+ column, wash, and elute retained cells in buffer.

11.4.5.8. Evaluation of hematopoietic progenitor cell purity (optional)
The purity of the isolated hematopoietic progenitor cells can be evaluated by flow cytometry or fluorescence microscopy. Fluorescent staining of CD34+ cells can be accomplished by direct immunofluorescent staining using an antibody recognizing an epitope different from that recognized by the CD34 monoclonal antibody QBEND/10 (e.g., CD34-PE, Clone: AC136, available from the Company). For optimal discrimination of CD34+ cells from other leukocytes, counterstain cells with an antibody against CD45 (e.g., CD45-FITC). CD34+ cells express CD45 at a lower level as compared to lymphocytes. Use the antibodies in appropriate concentrations recommended by the manufacturers. Typically, staining for 5 min at 4–8 °C should be sufficient. After fluorescence staining, cells should be washed and resuspended in buffer.

11.4.5.9. Important notes

1. Avoid capping of antibodies on the cell surface during labeling by working fast, and keeping cells cold. Use cold solutions only. Attention: Working on ice requires increased incubation times.

2. Increased temperature and prolonged incubation time for labeling may lead to unspecific cell labeling.

3. If progenitor cells are taken into culture, EDTA in the buffer may have a slightly negative effect on cell proliferation. EDTA can be replaced by other supplements such as 0.6% ACD-A or citrate phosphate dextrose (CPD).

4. Use degassed buffer only! Excess of gas in buffer will form bubbles in the matrix of the column during separation. This may lead to clogging of the column and decreases the quality of separation.

5. Contamination of the cell preparation with excessive number of thrombocytes can result in low purities and can also cause cell clumping which may clog the column. Additional washes after density gradient centrifugation over Ficoll Paque® at 200 × *g* for 10 min will reduce the number of thrombocytes in the cell preparation.

11.5. MACS Progenitor Cell Kit performance evaluation

In an independent study, the MACS progenitor kit was evaluated for the enrichment of the CD34+ hematopoietic progenitor cells from human umbilical cord blood [5].

11.5.1. Materials and methods

11.5.1.1. Cells

Umbilical cord blood was used for these studies and was obtained through informed consent as approved by Ohio State University's Institutional Review Board.

The mononuclear cell (MNC) layers were isolated by density gradient centrifugation over Ficoll-Hypaque (1.077 g/cm^3, Accurate Chemicals, Westbury, NY) at 400 × *g* for 30 min at 18 °C. The MNC layers were collected, washed with Ca^{2+}, Mg^{2+}-free Dulbecco's phosphate-buffered saline (PBS) containing 1% HSA. The MNC layers were then resuspended in degassed PBS containing 0.5% HSA and 2-mM EDTA (buffer). Cells were washed by adding 5–10× of the original volume in milliliters of buffer solution and centrifuged at 400 × *g* for 7 min.

11.5.1.2. Cell counts

The concentration of cells in a specific suspension was determined by using a Unopette test (Becton Dickinson, San Jose, CA) and a hemocytometer.

11.5.1.3. Magnetic labeling of CD34+ progenitor cells

The MNC layers were resuspended in buffer to obtain a concentration of 10^8 total cells/300 μl. To this cell suspension was added Fc receptor blocker (human IgG, Miltenyi Biotec, Auburn, CA) in the amount of 100 μl/10^8 total cells. Simultaneously, a hapten conjugated CD34 antibody in the amount of 100 μl/10^8 total cells was added (Reagent A2, Miltenyi Biotec). This suspension was incubated for 15 min at 6–12 °C. After incubation, the cell suspension was washed with buffer and resuspended in buffer to obtain a cell suspension at a concentration of 10^8 total cells/400 μl. To this suspension was added MACS microbeads recognizing the hapten-conjugated CD34 antibody (Reagent B, Miltenyi Biotec) in the amount of 100 μl/10^8 total cells. This suspension was, once again, incubated at 6–12 °C for 15 min. Finally, in preparation of loading into the magnetic column, the cell suspension was washed and resuspended in buffer to obtain a concentration of approximately 10^8 total cells/500 μl. The initial concentrations for each separation run are presented in Table 11.1.

11.5.1.4. Magnetic separation

After the two-step immunomagnetic labeling procedure, the cells were immediately introduced into a prefilled (with degassed buffer) MS+ column (Miltenyi Biotec). The actual process, as recommended by the manufacturer, consisted of introducing the labeled cell suspension into the top of the MS+ column, followed by 1.5 ml of buffer. The added cell suspension was allowed to flow through the column and was collected at the bottom exit. This cell suspension was referred to as the first, negative eluent. Next, the MS+ column was removed from its magnetic housing and 1.0 ml of buffer solution was forced through the column by using a plunger supplied by the manufacturer to elute the retained cells. This collected cell suspension was referred to as the first positive eluent. This first positive eluent was used as a feed to a second prefilled column mounted in the magnetic housing, and the process was

TABLE 11.1
Summary of CD34 cell enrichment from cord blood

	Feed					Second positive eluent		
Run	Volume (μl)	Concentration (cells/μl, $\times 10^{-6}$)	Total number of cells ($\times 10^{-6}$)	CD34 purity (%)	CD34 number ($\times 10^{-6}$)	CD34 recovery (%)	CD34 purity (%)	Fraction total cells recovered (%)[a]
1	300	0.25	75	0.4	0.3	114[d]	76	55
2	300	0.18	53	0.1	0.053	55	29	92
3	300	0.15	44	0.1	0.044	0	0	69
4	600	0.33	200	0.5	1.0	39	86	65
5[c]	300	0.31	93	1.3	1.2	19	57	29
6[c]	600	0.33	200	0.9	1.8	67	76	93
7[c]	600	0.33	940	1.5	3.0	21	65	93
8	600	0.33	1400	0.3	0.6	101	81	93
9	600	0.33	630	0.2	0.4	78	78	56
10[b]	600							
11[c]	600	0.33	200	1.5	3.0	8	53	107
Average		0.29	380	0.068	1.1	50.2	60	75

[a]The sum of the total number of cells in the first and second negative eluent and the second positive eluent divided by the total number of cells in feed.
[b]In this experiment, the FACS results were not available.
[c]Indicates that the column was "pulsed."
[d]Recovery higher than 100% results from random error in cell count.

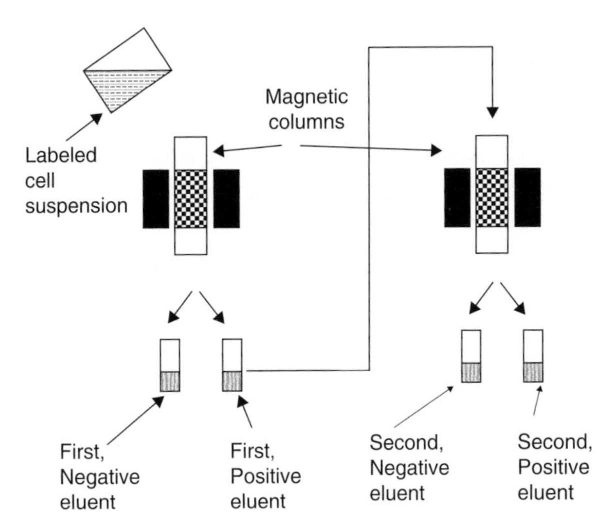

Fig. 11.5. Schematic diagram of the cell separation process [5] (reprinted with permission).

repeated a second time, producing a second negative eluent and a second positive eluent.

Figure 11.5 presents a diagram of the process. In runs 5–7 and 11 the first feed did not flow continuously through the column. Consequently, as suggested by the manufacturer, the plunger was attached and pressure was applied to force the cell suspension through the column. The experimental runs in which this pressure was needed are marked in the table as "pulsed."

11.5.1.5. Flow cytometry

To determine the performance of the separation system, initial aliquots of cord blood MNC were taken prior to immunomagnetic labeling for flow cytometry background IgG and autofluorescence studies and to determine the initial percentage of positive CD34 cells in the sample. In addition, aliquots were removed from all eluents for flow cytometry analysis. These aliquots ranged in size from 1×10^4 to 3×10^5 cells and were resuspended in 80–100 μl of buffer.

For studies in which the percentage of CD34+ cells present in a specific eluent was desired, a sample was removed in which 20 μl/10^6 cells of anti-human CD34+ FITC conjugated antibody (Becton-Dickinson) was added. In all cases the final cell concentration was 10^6 cells/100 μl and the cell suspensions containing the fluorescent antibodies were incubated for 30 min at 6–12 °C. After incubation, the cell suspension was washed in 1–2 ml of buffer, centrifuged, and resuspended in the appropriate amount of buffer. These various cell suspensions were analyzed on a Beckman-Coulter EPICS Elite II ESP flow cytometer. Both autofluorescence and isotype controls were performed to produce the appropriate gates to quantify the positive cell populations.

11.5.2. Results

A total of 11 separations of human cord blood were conducted and the overall performance is presented in Table 11.1. As is observed, a significant range in overall recovery and purity of CD34+ cells in the final second positive fraction was obtained. Total cell recovery was calculated by dividing the sum of the total number of cells in each of the recovered eluents (first and second negative, and second positive) by the total number of cells added to the system. The total number of cells in each eluent was obtained by knowing the volume of the collected eluent in calibrated tubes (also by keeping track of the total amount of fluid added to the column) and multiplying by the cell concentration as determined by a hemacytometer. A calibration test of the reproducibility of the hemacytometer data resulted in a coefficient of variation (CV) of 22% (data not shown). The final recovery of CD34+ cells was calculated using the following relationship:

$$R \equiv \frac{P_{2+e}N_{2+e}}{P_{\mathrm{f}}N_{\mathrm{f}}} \tag{11.1}$$

where P is the fraction of positive cells (purity) as measured by flow cytometry, N is the total number of cells in a given cell sample as measured by hemacytometer and cell suspension volume, subscript $2+e$ refers to the second positive eluent, and subscript f refers to the feed.

In an attempt to understand the causes of the large range in the recovery and final purity of CD34+ cells, both values were plotted as a function of initial, total cell loading to the column and as a function of initial number of CD34+ cells loaded (plots not shown). The only trend that could be observed is that both the purity and recovery appear to reach a peak when the initial, total number of CD34+ cells is in the approximate range of 10^5 to 10^6 cells. Taking the average of the recovery and purity in the range of 9×10^4 to 1×10^6 CD34+ cells, one obtains significantly higher values of purity and recovery of 0.82 and 0.68, respectively. Of the 11 total runs, 4 had sufficiently slow flow through the column such that the plunger had to be applied to the top of the MS column to create a slight pressure to resume flow. These runs, referred to as "pulsed," are marked with a "c" in Table 11.1.

To further study the separation performance, the process was divided into two independent stages. This was possible since flow cytometry analysis was also conducted on the first negative eluent sample and the first positive eluent sample. These data are presented in a plot of the fraction of CD34+ cells in the negative and positive eluents (y-axis) as a function of the fraction of CD34+ in the feed (x-axis) (Fig. 11.6A). Ideally, one would expect that the fraction of CD34+ cells in the positive eluent (open symbols) would not vary and remain near a value of 1.0 and that the fraction of CD34+ cells in the negative eluent (closed symbols) would remain low (below a fraction of 0.001) and also not vary with the fraction of CD34+ cells in the feed. However, in practice this was not observed. Although there is a noticeable increase (as can be observed by the slope of the solid line) in the fraction of CD34+ cells in the positive eluent with increasing fraction of CD34+ in the feed, no trends in the fraction of CD34+ in the negative eluent can be observed. An alternative to Fig. 11.6A is to plot the fraction positive in the positive and negative eluent as a function of the total number of CD34+ loaded onto the column, Fig. 11.6B.

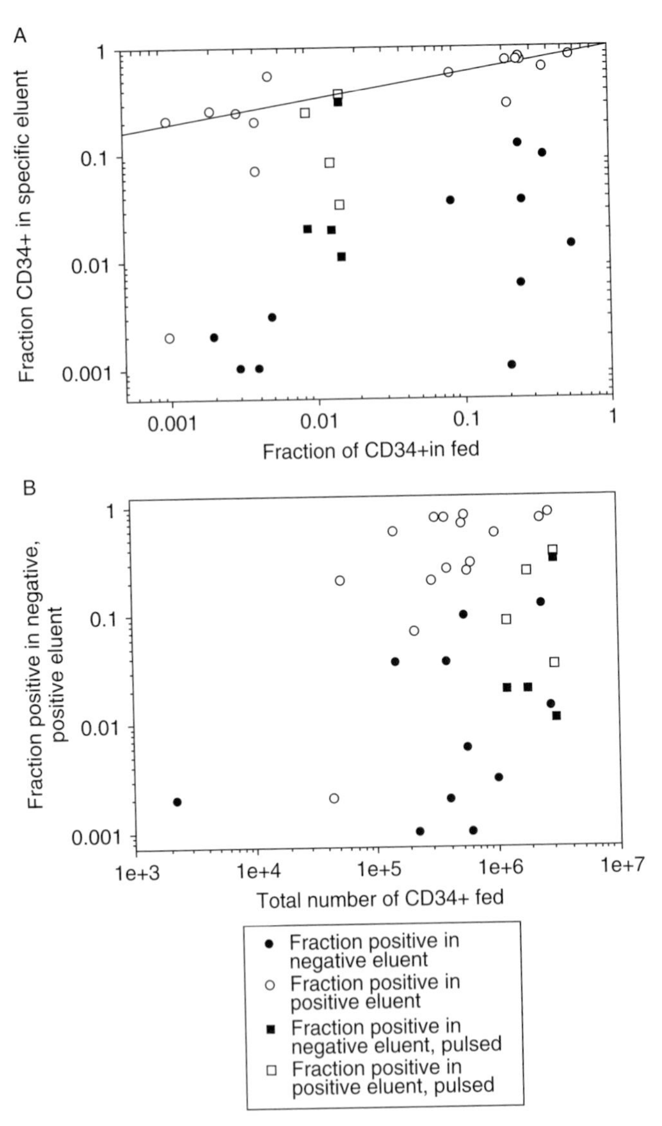

Fig. 11.6. Fraction of CD34+ in the positive and negative eluent as a function of (A) the fraction in the feed and (B) the total number of CD34 positive cells loaded [5] (reprinted with permission).

There appears to be a general increase in the fraction of CD34+ cells in the negative eluent as the number of total CD34+ cells fed to the column increases. While care should be taken in interpreting log-log plots, this trend has analogies to the breakthrough concept in chromatography, namely, that there is a limited number of binding sites for the cell to be retained (bound) and that once those sites are occupied, further positive cells cannot bind and are thus not retained. The analysis was based on total cell counts using a hemacytometer, volumes of cell suspension measured with a pipette, and fractions of a cell population positive for CD34 cells through flow cytometry analysis. These 11 experiments were conducted over a period of approximately one and three-quarters years.

11.6. Post-separation analyses in other magnetic separation systems

11.6.1. Progenitor cell purity analysis by flow cytometry

In an independent study, separate from the one described above, the CD34+ hematopoietic progenitor cell enrichment was evaluated following magnetic flow cell sorting in a quadrupole field (the quadrupole magnetic flow sorter, or QMS, is described in the next Chapter 12) [6, 7]. The separations were performed on samples of leukapheresis product from patients undergoing stem cell transplantation therapy (with consent, approved by the Institutional Review Board, or IRB). Following QMS depletion of unwanted cells from leukapheresis product, the progenitor cell purity in both the original sample O (before separation) and the sorted cell product, P (after separation) were measured by flow cytometry (FCM) according to the protocol developed by the International Society for Hematotherapy and Graft Engineering, or ISHAGE (now International Society for Cellular Therapy, ISCT) [8]. Samples were stained with two immunofluorescent dyes, anti-CD34 PE mAb (8G12) (BD Bioscience, San Jose, CA) and anti-CD45 PerCP mAb (2D1) (BD Bioscience, San Jose, CA). Between 10,000 and 50,000 cells per sample were processed on a BD LSR I flow cytometer (BD Bioscience, San Jose, CA). In the data analysis process, sequential gating strategy was used to identify progenitor cells by their expression of CD34, characteristic low to intermediate CD45 antigen expression, and characteristic low side scatter light properties. A typical flow cytometry result from a leukapheresis sample CD34 analysis before and after enrichment is shown in Fig. 11.7.

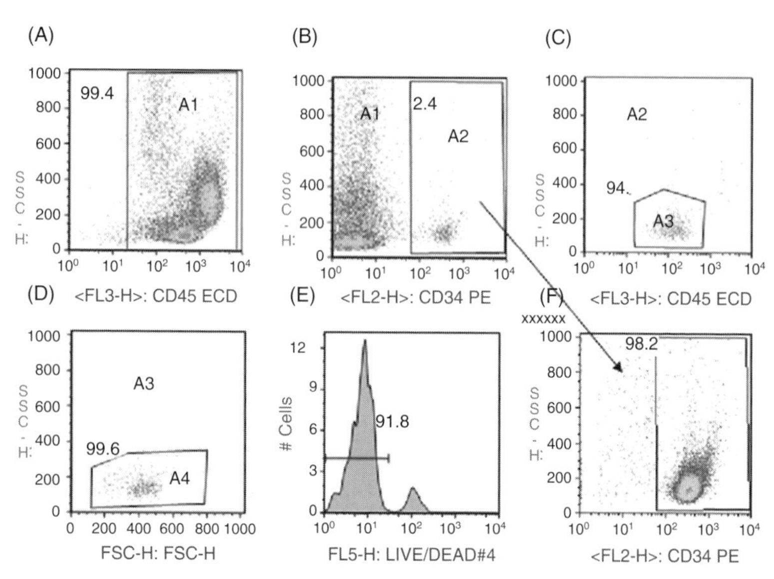

Fig. 11.7. Flow cytometry analysis of clinical leukapheresis sample (after compensation). Numbers in panels are percentages of gated cell subpopulations. Panels A–E show sample before sorting. Panel F shows sorted fraction. (A) Dot plot of side scatter versus CD45-ECD. A1: CD45+ leukocytes in the total population; (B) Dot plot of side scatter versus CD34-PE. A2: the CD34+ cells in the subpopulation A1; (C) Dot plot of side scatter versus CD45-ECD; A3: CD45 low to medium and SSC low in the subpopulation A2; (D) Dot plot of side scatter versus forward scatter to gate A4 in the subpopulation A3; (E) Dot plot of side scatter versus LIVE/DEAD Reduced Biohazard Cell Viability Kit #4 to check the viability of the cells in A4. Here the purity of CD34+ is defined as: (# events gated in A4)/(# events gated in A1); (F) Dot plot of side scatter versus CD34-PE. A2: the CD34+ cells in the subpopulation A1 (cell fraction after QMS separation) [7].

11.6.2. Progenitor cell yield (recovery) and nonprogenitor cells retention frequency analysis by cell counter

The yield (also called recovery) of the progenitor cells (Y) is calculated from

$$Y = \frac{N_p \times P}{N_o \times O} \tag{11.2}$$

where N_o and N_p are the total cell numbers in the original sample and the sorted cell product, respectively, and O and P is the progenitor cell purity in the original and sorted product samples, determined by the FCM analysis as described above. Coulter Z1 counter (Beckman-Coulter Corp., Hialeah, FL) with a 70-μm aperture was used to count the cell numbers of cell suspension with a set cell diameter range between 5 and 20 μm. The nonprogenitor cell retention frequency (d) is calculated by

$$d = \frac{N_p \times (1 - P)}{N_o \times (1 - O)} \tag{11.3}$$

(as suggested by Dr. Diether Recktenwald, BD Biosciences, San Jose, CA). Through combination of Eqs. (11.2) and (11.3) one obtains the following expression for d:

$$d = \frac{Y \times O \times (1 - P)}{P \times (1 - O)} \tag{11.4}$$

Results of Coulter Counter analysis and stained smears of progenitor cells, before and after enrichment, are illustrated in Fig. 11.8. The impact of initial purity on the final purity when progenitor cells are isolated by negative cell depletion is shown in Fig. 11.9.

11.6.3. Progenitor cell morphology analysis by cytospin and microscope

Cytospin analysis was accomplished following a typical protocol. Specifically, the cell sample suspension was centrifuged for 5 min at 1300 rpm. Then, the cell pellet was resuspended in 100-μl labeling buffer, and added to the cytospin funnel for 5 min

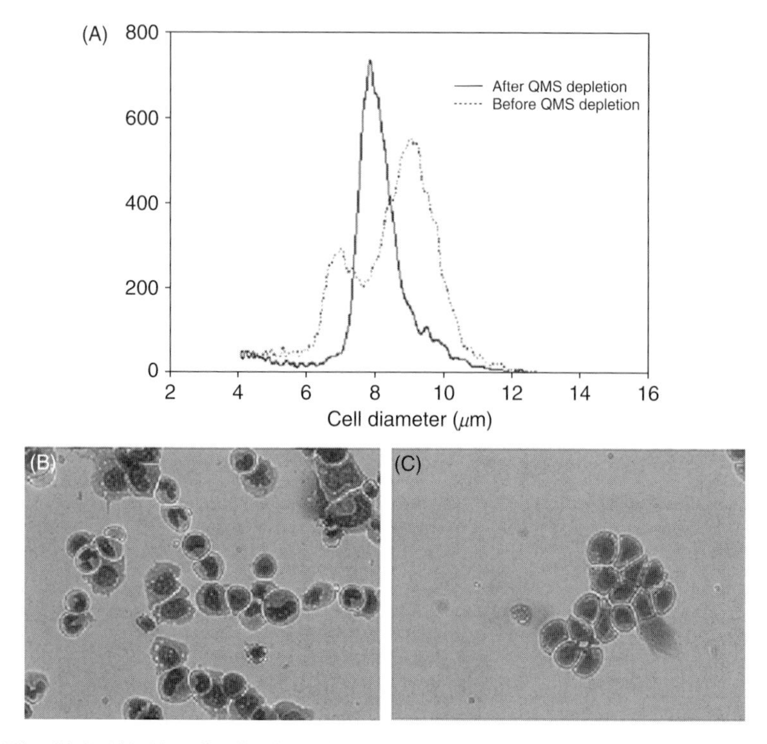

Fig. 11.8. (A) Size distribution of the leukapheresis sample before (dash line) and after (solid line) non-progenitor cell depletion using QMS. The morphology of the leukapheresis sample before (B) and after (C) nonprogenitor cell depletion. Note more uniform size distribution and more homogeneous morphology in the nonprogenitor cell-depleted sample, consistent with the enrichment of the progenitor cells in the sample.

centrifugation at 1500 rpm. The cells were layered evenly onto the glass slide and stained with Fisher Hema 3 Manual Staining System (Fisher Diagnostics, Fisher Scientific Company, LLC, Middletown, VA). The stained slides are washed and then air dried for the subsequent microscopic evaluation.

11.6.4. Progenitor cell function analysis by cell colony forming unit (CFU) assay

CFU analysis is used to determine if there is a deterioration of the progenitor cell proliferation and differentiation capacity following

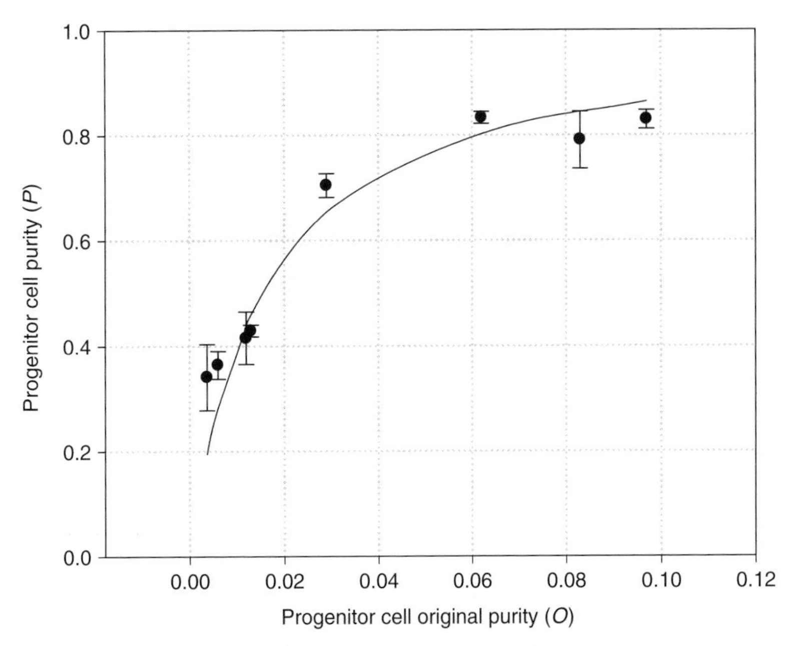

Fig. 11.9. The purity of the blood progenitor cells in the fraction depleted of nonprogenitor cells by the QMS depends on the original purity in the patient sample ($n = 8$, each sample is in duplicate or triplicate). The solid line was calculated based on the yield and the frequency of contaminating non-progenitor cells in the sample (discussed in text). The close agreement between measured and calculated progenitor cell purities shows internal consistency and high precision of the analytical methods used to determine progenitor cell content [6] (reprinted with permission).

the QMS separation process. Aliquots of $(0.2–1) \times 10^4$ leukapheresis cells without separation and $(2–8) \times 10^2$ enriched progenitor cell product were respectively plated on 35-mm Petri dishes (StemCell Technologies, Vancouver, Canada) with a commercial semi-solid clonogenic culture assay medium based on methyl cellulose (StemCell Technologies, Vancouver, Canada). Each sample has two repeats. The dishes were incubated at 37 °C and 5% CO_2 in humidified atmosphere. After 14 days of culture, colonies were scored based on the average of the two dishes using an inverted microscope. Colonies containing more than 200 erythroblasts in a

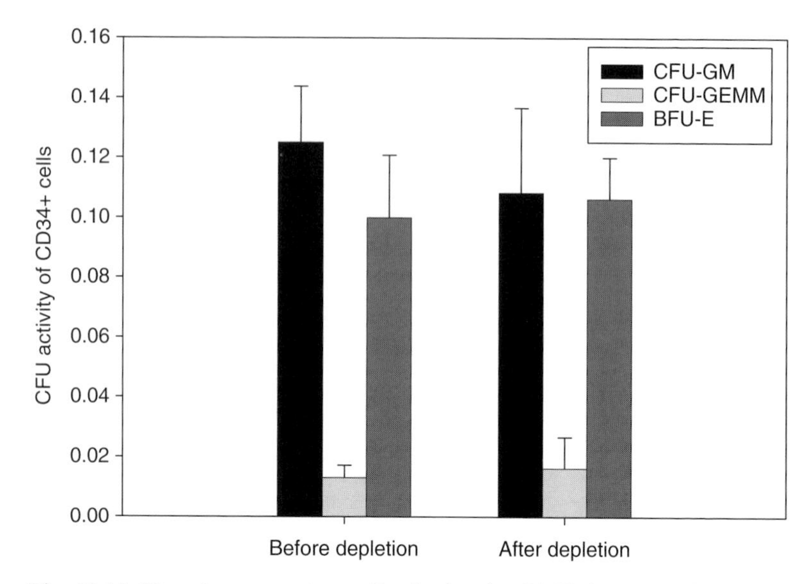

Fig. 11.10. Negative progenitor cell selection by QMS does not affect CFU activity of the progenitor cells (after depletion) as compared to the original sample (before depletion), as determined by 14 days cell culture. CFU activity of CD34+ cell = (# colonies after 14 days culture)/(# CD34+ cells before culture) [6] (reprinted with permission).

single or multiple clusters are subclassified as burst-forming unit-erythroid (BFU); Colonies including at least 20 granulocyte cells, macrophages, or cells of both lineages are counted as colony-forming unit-granulocyte, macrophage (CFU-GM); Colonies composed of erythroblasts and cells of at least two other recognizable lineages are defined as colony-forming unit-granulocyte, erythroid, macrophage, megakaryocyte (CFU-GEMM), which are usually derived from multipotential progenitor cells. Figure 11.10 shows a typical result.

11.7. Companies and brand names mentioned in Chapters 11 and 12

Axis-Shield PoC AS, P.O. Box 6863 Rodeløkka, N-0504, Oslo, Norway, tel. +47-22-04-2000, Norway, website: http://www.axis-shield-poc.com

Beckman-Coulter, Inc., 4300 N. Harbor Boulevard, Po.O. Box 3100, Fullerton, CA 92834, tel. 1-800-742-2345, website: http://www.beckmancoulter.com

Becton, Dickinson and Company (BD), 1 Becton Drive, Franklin Lakes, NJ 07417, tel. 201-847-6800, website: http://www.bd.com

Cytospin centrifuge or Shandon Cytospin® is available from Thermo Fisher Scientific.

Dynal Biotech and Dynal brand products are available from Invitrogen Corp.

Ficoll-Paque™ and Ficoll-Paque™ Plus are available from StemCell Technologies, Inc.

Fisher Scientific brand products are available from Thermo Fisher Scientific, Inc.

Gibco, Gibco-BRL, Gibco-BRL Life Technologies brand products are available from Invitrogen Corp.

Gibco® cell culture systems and media are available from Invitrogen Corp.

Immunicon Corporation, 3401 Masons Mill Road, Suite 100, Huntingdon Valley, PA 19006, tel. 877-822-0777, website: http://www.immunicon.com

Invitrogen Corporation, 1600 Faraday Avenue, P.O. Box 6482, Carlsbad, CA 92008, tel. 760-603-7200, website: http://www.invitrogen.com

Johnson & Johnson, One Johnson & Johnson Plaza, New Brunswick, NJ 08933, website: http://www.jnj.com

Lymphoprep™ is available from Axis Shield Point-of-Care (PoC)

Miltenyi Biotec Bergisch Gladbach, Frierich-Ebert-Strasse 68, 51429 Bergisch Gladbach, Germany, tel. +49-2204-8306-0, website: http://www.miltenyibiotec.com

Pfizer Corporation, 235 East 42nd Street, New York, NY 10017, tel. 212-733-2323, website: http://www.pfizer.com

Pharmacia brand products are available from Pfizer Corp.

Sigma, Aldrich, Fluka and Supelco brand products are available from Sigma-Aldrich Corporation.

Sigma-Aldrich Corp., 3050 Spruce Street, St. Louis, MO 63103, tel. 1-800-521-8956, website: http://www.sigmaaldrich.com

StemCell Technologies, Inc., Vancouver, BC, Canada, tel. 1-800-667-0322, website : http://www.stemcell.com
Thermo Fisher Scientific, Inc., 81 Wyman Street, Waltham, MA 02454, website: http://www.thermo.com
Veridex, LLC, a Johnson & Johnson Company, 33 Technology Drive, P.O. Box 4920, Warren, NJ 07059, tel. 1-877-837-4339, website: http://www.veridex.com

References

[1] Dynal, T. (2007). Cell Isolation Kit: https://www.invitrogen.com/content/sfs/manuals/57_113.11_Dynal_T_Cell_Negative_Isolation_Kit.pdf.
[2] CD34 Progenitor Cell Selection System: https://www.invitrogen.com/content/sfs/manuals/CD34%20Progenitor%20Cell%20Selection%20System.pdf
[3] Sutherland, DR., Keating, A., Nayar, R., Anania, S. and Stewart, AK. (1994). Sensitive detection and enumeration of CD34+ cells in peripheral blood and cord blood by flow cytometry. Exp Hematol *22*, 1003–10.
[4] Miltenyi Biotec at: http://www.miltenyibiotec.com/pg.326.9,f3ccf60dfebe8e2f44119070975db868,index.html
[5] Melnik, K., Nakamura, M., Comella, K., Lasky, LC., Zborowski, M. and Chalmers, JJ (2001). Evaluation of eluents from separations of CD34+ cells from human cord blood using a commercial, immunomagnetic cell separation system. Biotechnol. Prog. *17*, 907–16.
[6] Jing, Y., Moore, LR., Schneider, T., Williams, PS., Chalmers, JJ., Farag, SS., Bolwell, B. and Zborowski, M. (2007). Negative selection of hematopoietic progenitor cells by continuous magnetophoresis. Exp. Hematol. *35*, 662–72.
[7] Jing, Y., Moore, LR., Williams, PS., Chalmers, JJ., Farag, SS., Bolwell, B. and Zborowski, M. (2006). Blood progenitor cell separation from clinical leukapheresis product by magnetic nanoparticle binding and magnetophoresis. Biotechnol. Bioeng. *96*, 1139–54.
[8] Sutherland, DR., Anderson, L., Keeney, M., Nayar, R. and Chin-Yee, I. (1996). The ISHAGE guidelines for CD34+ cell determination by flow cytometry. International Society of Hematotherapy and Graft Engineering. J. Hematother *5*, 213–26.

Laboratory Techniques in Biochemistry and Molecular Biology, Volume 32
Magnetic Cell Separation
M. Zborowski and J. J. Chalmers (Editors)

CHAPTER 12

New challenges and opportunities

Maciej Zborowski,[1] P. Stephen Williams,[1] Lee R. Moore,[1] Jeffrey J. Chalmers[2] and Peter A. Zimmerman[3]

[1]*Department of Biomedical Engineering, Lerner Research Institute, Cleveland Clinic, Cleveland, OH 44195, USA*
[2]*Department of Chemical and Biomolecular Engineering, University Cell Analysis and Sorting Core, The Ohio State University, OH 43210, USA*
[3]*The Center for Global Health & Diseases, Case Western Reserve University, Cleveland, OH 44106, USA*

12.1. Introduction

Magnetic cell separation instrumentation and applications continue to be an active area of research and development [1, 2]. The progress is being fueled by the need for faster and better separation methods on a large scale, in anticipation of the new cellular therapies that reach the stage of preclinical trials, and by the need of better diagnostic tools based on rare cancer cell detection. Over the course of the past 20 years, magnetic cell separation has become an accepted laboratory technique of immunoseparation with wide-ranging applications, such as detection of circulating fetal cells in maternal blood [3, 4], circulating tumor cells as markers of disease progression and survival in the course of cancer therapy [5], bone marrow purging of residual cancer cells for autologous transplantation, or purging of immunocompetent cells (T cells) for allogeneic transplantation as a measure to eliminate graft-versus-host-disease

DOI: 10.1016/S0075-7535(06)32012-8

(GvHD) [6, 7]. Further evolution of the magnetic cell separation methods is made possible by the availability of stronger magnets at reasonable cost, and better magnetic micro- and nanoparticles as reagents for cell labeling.

The selection of the magnetic separation methods under development, described in this chapter, is by no means complete. Rather, it is meant as a snapshot of active research of the authors of this volume. In particular, there is a considerable ongoing effort in miniaturizing the magnetic cell separators that is not described here. The interested readers are referred to the published reports [8–12].

12.2. Magnetophoresis and magnetic capture of malaria-infected erythrocytes

The current magnetic cell separation strategies rely on attaching synthetic, magnetic beads to cells in order to pull cells from suspension [13]. The specificity of the separation is based on cell immunophenotype and use of antibodies that distinguish between characteristic cell surface antigen markers. In this respect, the current practice of the magnetic cell separation relies on immunocytochemistry. Here we describe cell separation by the magnetic field that does not require attachment of the synthetic beads to the cell but relies solely on the intrinsic magnetic properties of the cell. We selected a model of a malaria parasite-infected erythrocyte because of a number of important factors: (1) the model is pertinent to research on malaria and finding a cure; (2) the physical properties of the erythrocyte are well known, including magnetic susceptibility of hemoglobins from the pioneering work of Pauling in 1936 and later studies, such as functional nuclear magnetic resonance imaging [14], which provided a firm basis for the physical explanation of the observed effect; (3) the crystal structure of the malaria pigment, hemozoin, has been recently elucidated and its relation to the magnetic susceptibility of the hemozoin is now well established; (4) the accumulation of the hemozoin crystals over time of the parasite blood development has been clearly correlated to distinct parasite forms, which facilitates quantitative correlation between the separation of the infected erythrocyte and the hemozoin load.

The effect of the parasite biology and its host, the erythrocyte, on cell motion in the magnetic field does not require cell manipulation or chemical treatment to achieve cell magnetization. There are practical applications of the strategies presented here that may lead to magnetic fractionation tools of live erythrocytes that are sensitive to cell biology and improve our understanding of parasite heme management in relation to antimalarial drug susceptibility and resistance. It has been reported that within infected erythrocytes, *Plasmodium falciparum* consumes from 50% [15] to 80% [16] of the cytosolic hemoglobin (molar concentration of 5 mM in the erythrocyte's volume of 88.4 μm^3) during a 48-hour time period releasing an equivalent of 10- to 16-mM free heme, a potent biological toxin [17]. For survival, the parasite compartmentalizes free heme in the digestive vacuole (volume of 4 μm^3, resulting in an estimated 22-fold increase in concentration reaching 350–400 mM [16, 18]), and polymerizes this toxin into insoluble hemozoin [19–24] (Fig. 12.1).

The erythrocyte magnetic susceptibility is related to the hemoglobin susceptibilities as described by Pauling and Coryell [25–27], Savicki et al. [28], Cerdonio [29, 30], Spees et al. [31], and other groups [14, 32, 33], and to the hemozoin susceptibility [34]. The hemozoin heme electron configuration corresponds to that of a ferriheme with five unpaired electrons, $S = 5/2$ [34]. The relative contributions of the hemoglobin and hemozoin to the erythrocyte volume magnetic susceptibility are calculated as weighed fractions of their respective susceptibilities, as discussed in Chapter 1 [35]. The increase in the infected erythrocyte hemozoin content is illustrated in Fig. 12.1. The magnitude of iron spin in free heme or amorphous heme aggregates is less certain. The iron ion of hemin, a ferric protoporphyrin IX chloride, exists in two states, $S = 5/2$ and 1/2 (hematin is the ferric protoporphyrin IX hydroxide) [36]. Thus, the erythrocyte magnetic susceptibility may be a function of not

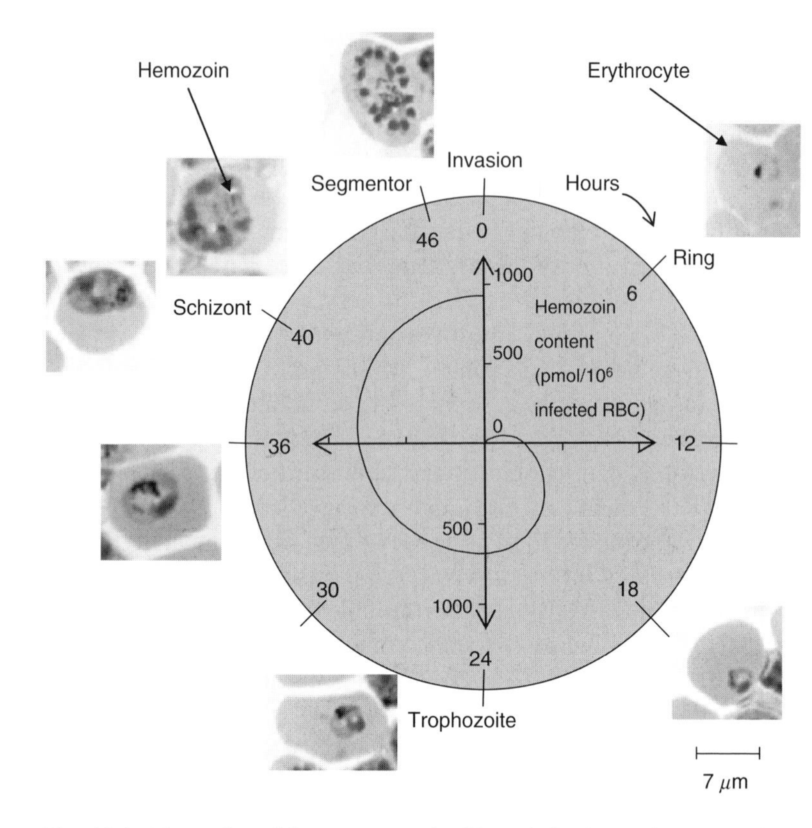

Fig. 12.1. *Plasmodium falciparum* parasite life cycle in the erythrocyte and the associated increase in the intraerythrocytic hemozoin content. Note the hemozoin particle appearance as brown inclusions in the erythrocyte microscopic images [41].

only the distribution of heme molecules between oxyhemoglobin ($S = 0$) and hemozoin ($S = 5/2$) but also the presence of free heme and amorphous heme aggregates (for which S may vary from $S = 0$ to $S = 5/2$). However, that fraction is relatively small and does not exceed 5% [37].

The magnetophoretic mobility measurements were performed on parasite cultures in sorbitol-synchronized, infected erythrocytes. Heparinized blood was washed with RPMI-1640 (Invitrogen/Gibco, Carlsbad, CA) and stored at 50% hematocrit at 4 °C prior to use. *P. falciparum* HB3 clone in AA erythrocytes were cultivated at 5% hematocrit in albumax II complete medium (RPMI-1640 supplemented with 25 mg/ml HEPES, 2 mg/ml sodium bicarbonate, 5% albumax II (Invitrogen/Gibco) [38]. All cultures were maintained at 37 °C in an atmosphere of 5% CO_2, 5% O_2, and 90% N_2, with once or twice daily medium changes. Parasitized erythrocytes with ring forms were treated with 5% D-sorbitol. This procedure was repeated 4 hour after the first treatment to eliminate other forms and thus synchronize the culture [39]. After treatment, the cells were washed twice with complete medium, resuspended and cultured for 24, 36, 38, and 45 hour to allow continued development to early trophozoite, late trophozoite, early schizont, and late schizont stages, respectively. A maximum expected level of parasitemia for these conditions is ~10%. Parasitized erythrocytes were counted per 10^4 erythrocytes. Five hundred parasitized erythrocytes were observed and evaluated to estimate the developmental stage proportion.

The magnetophoretic mobility of uninfected erythrocytes was distributed normally (Fig. 12.2A). A small, negative value of the mean peak net susceptibility indicated that the erythrocytes were oxygenated and in equilibrium with the ambient air [40]. A highly symmetrical distribution around the mean indicated contribution of random errors of measurement not related to the magnetic field, such as the effects of gravitational sedimentation. The cutoff mobility, set at the 95th percentile of cumulative mobility frequency distribution, was $m_0 = 0.75 \times 10^{-6}$ mm^3 sec/kg (Fig. 12.2A). Erythrocyte magnetophoretic mobility and the corresponding net magnetic susceptibility distributions differed markedly between predominantly ring and predominantly schizont cultures, Fig. 12.2B and C. In the predominantly rings and early trophozoites sample, the majority of the cells, 89%, had mobility below the cutoff mobility m_0. The "magnetic" cell fraction, 11% in Fig. 12.2B, was comparable to the infected cell fraction in the sample, 12%, as determined by the differential cell counting (not shown) [41]. The unexpected presence of a few highly mobile cells was explained as an artifact of the experimental procedure and considered outliers. The 25 to 75

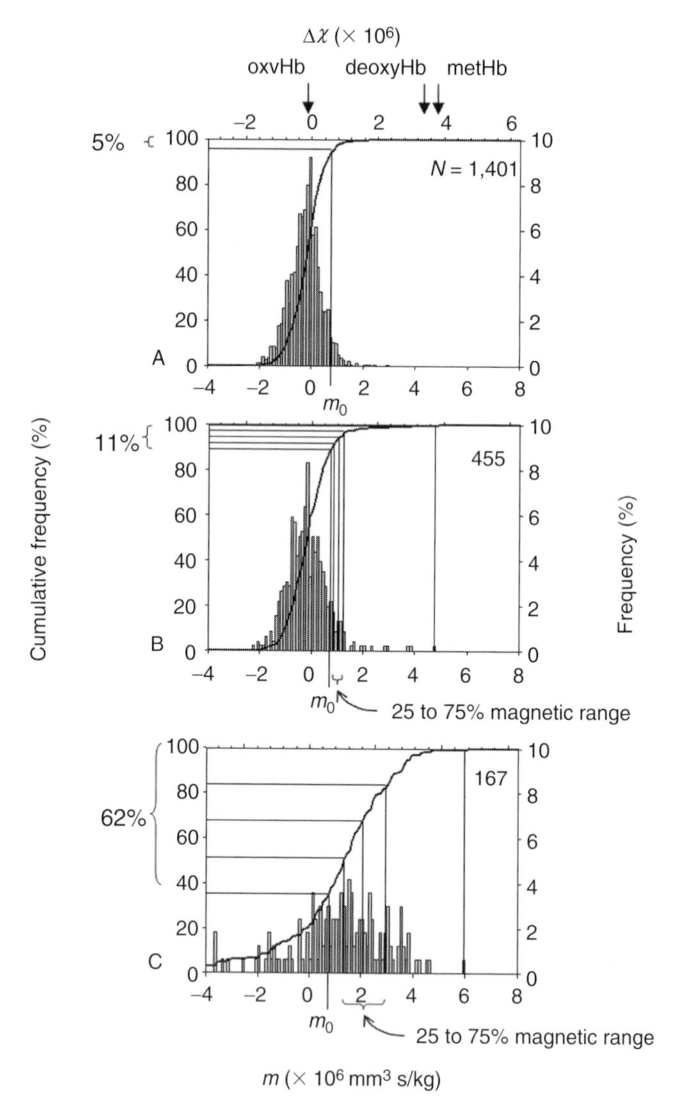

Fig. 12.2. Magnetophoretic mobility, m, and the corresponding net volume magnetic susceptibility, $\Delta\chi$, histogram of the control (uninfected) erythrocyte suspension (A) and the infected erythrocytes (B and C). The cutoff mobility, $m_0 = 0.75 \times 10^{-6}$ mm^3 sec/kg, was set at 95 percentile cumulative mobility

percentile magnetic mobilities were grouped in a narrow range from 0.85×10^{-6} to 1.25×10^{-6} mm^3 sec/kg, Fig. 12.2B.

In comparison, the culture with predominantly late schizonts (Fig. 12.2C) contained 62% of cells that increased magnetophoretic mobility higher than the cutoff mobility, m_0. The 25 to 75 percentile mobility range was from 1.35×10^{-6} to 2.95×10^{-6} mm^3 sec/kg, Fig. 12.2C, and was significantly higher than that in the sample devoid of schizonts shown in Fig. 12.2B. The magnetophoretic mobility of the erythrocyte enriched in the schizont form did not exceed that of the normal, deoxygenated erythrocytes and that of the erythrocytes with the hemoglobin converted to high-spin methemoglobin, as indicated in Fig. 12.2A, determined during our previous studies [40]. This observation agrees with the published magnetic susceptibility data for the constituents of the infected erythrocyte [41].

An important impediment to effective malaria treatment and control is the absence of low-cost diagnostic tools and strategies capable of evaluating infection status rapidly in rural settings where the majority of malaria cases are encountered. As a result, generalized treatment of malarial and bacterial infections follows symptoms-based diagnosis [42]. This approach is certain to contribute to selection favoring drug resistant parasites and bacteria. Although the conventional blood smear serves as the "gold standard" tool for malaria diagnosis (individual diagnosis $\approx$ US \$0.12–0.40) [42], it is widely acknowledged that molecular tools are faster antigen-based rapid diagnostic tools (RDTs) [43, 44] or provide significantly greater sensitivity and specificity (polymerase chain reaction, PCR) [45–47]. However, molecular techniques are unlikely to become "gold standard" malaria diagnostic methods.

(thick line and the left ordinate axis). Reference data for oxygenated, deoxygenated, and methemoglobin-converted erythrocytes are also shown (arrows). N is the number of tracked cells. (B) The erythrocyte suspension with predominantly early ring and trophozoite forms, and (C) predominantly late schizont fraction. Note shift toward higher magnetic cell fraction and higher peak magnetophoretic mobility with the parasite infection stage [41].

Current antigen-based RDTs are expensive (individual diagnosis ≈ US$0.60–2.50) [42], do not assess *P. vivax*, *P. malariae*, or *P. ovale* with specificity, and have been observed to be less sensitive than the blood smear [42, 48]. PCR-based diagnosis (individual diagnosis ≈ US$0.50–1.00) [42] requires a laboratory with electricity and expensive equipment, and is most expedient when analysis is performed on large numbers of samples in a 96-well plate format [49].

In an attempt to overcome some problems inherent to blood smear microscopy, a magnet-based approach to concentrate malaria parasites and augment detection of malaria-infected erythrocytes by microscopy has been developed [35]. This system, malaria magnetic deposition microscopy (MDM), exploits the magnetic properties of hemozoin. Unlike previous systems requiring elution of cells from steel mesh [50, 51], MDM captures parasitized erythrocytes in a narrow magnetic field and deposits them directly onto a small region of a polyester slide, which is then immediately ready for fixation and staining. By concentrating parasites, MDM increases the sensitivity of diagnosis and decreases the time it takes to read the slide. Here we demonstrate the ability of MDM to concentrate parasites of all four human malaria parasite species, including efficient capture of *P. falciparum* gametocytes.

Malaria MDM is based on an open-gradient magnetic field separator and a thin-film magnetophoresis process developed for cell analysis during past studies [52–56]. The magnetic field was designed to maximize local Maxwell stress gradients that drive cell separation from the flowing suspension (Fig. 12.3), optimizing erythrocyte capture from the cell suspension [57]. A fringing field of an interpolar gap (described in Chapter 4), combined with a thin flow channel pressed against the interpolar gap was used as a means of cell capture from suspension. The magnetic field was generated by a permanent magnet assembly comprising ferrite magnets (Dexter Magnetic Technologies, Inc., Elk Grove Village, IL) and a pair of 1016 low-carbon steel pole pieces (in-house model C designation) [52]. The interpolar gap width was 1.27 mm, the magnetic field strength, H, measured at the midline of the interpolar gap at the magnet surface was 1.135×10^6 amperes/

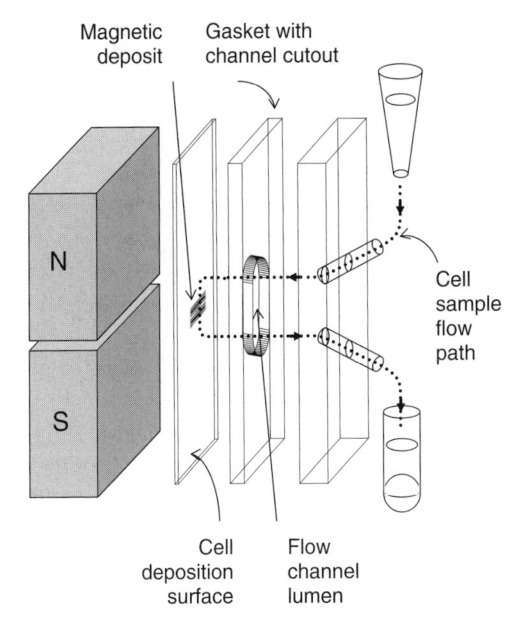

Fig. 12.3. Components of the malaria MDM device and the sample flow path. Note the position of the erythrocyte deposition band next to the magnet pole piece tips, represented diagrammatically, resulting from action of the magnetic field. Five such modules were integrated into a single MDM apparatus. Not to scale [35].

meter (A/m), the magnetic field intensity, *B*, was 1.426 tesla (T), the magnetic field gradient was 804 T/m, and the magnitude of the Maxwell stress gradient was 1.824×10^9 A T/m^2. The direction of the resulting magnetic force acting on erythrocytes was essentially perpendicular to the deposition surface along the width of the interpolar gap, and its magnitude rapidly decreased with distance from the interpolar gap (as shown in Chapter 4). The flow channel lumen (6.4×0.25 mm^2) including sample inlet and outlet tubing was formed to meet a cutout in a silicone rubber gasket separating a thin, inner polyester sheet (75-μm thick, or one half the thickness of #1 glass coverslips, Clear Polyester, McMaster-Carr, Aurora, OH) from a thick, outer acrylic cover (4 mm). The interpolar gap formed a magnetic barrier to malaria-infected cells and was sufficiently long to accommodate five flow channels. Cell suspensions (500 μl each) were delivered in a continuous manner into flow channels by syringes connected to inlet tubing, and evacuated from flow channels by outlet tubing leading to waste containers (Fig. 12.3). All five samples were held in 1-ml sterile disposable syringes (Becton-Dickenson, Franklin Lakes, NJ) mounted on the syringe pump (Sage Instruments, Cambridge, MA). Each syringe was connected to its matching flow chamber through a 1500-μl pipette tip, 10-mm-long Tygon tubing [inner diameter (ID) 1.59 mm; outer diameter (OD) 3.18 mm (Norton Performance Plastic Co., Akron, OH)], and 60-mm-long Teflon tubing [ID 0.79 mm, OD 1.59 mm (Zeus, Inc., Boise, ID)]. The same type Teflon tubing carried eluate fractions to sample collection tubes. The syringe pump was modified by fitting it with a five-syringe receptacle and by extending a pusher plate (not pictured) to accommodate all five-syringe plungers. The position of each plunger relative to the pusher plate was individually adjusted by thumb screws to allow simultaneous delivery of all five samples. Before each experiment, the flow chambers and connecting tubings were primed with phosphate-buffered saline (PBS, pH 7.2) so that cells entered the magnetic deposition zone in a fully developed flow. The flow channel dimensions and the volumetric flow rate of the cell sample were selected to maximize the cell interaction with the magnetic field, and consequently increase the likelihood of depositing mobile cells magnetophoretically on the thin plastic sheet surface. The flow channel crosssection was 6.4×0.25 mm^2, the volumetric flow rate was 0.7 ml/hour, the resulting average linear velocity of fluid across the interpolar gap region was 1.2 mm/sec, and the average fluid volume element residence time in the interpolar gap region (taken as twice the interpolar gap width, or 2.54 mm) was ~2 sec.

After the entire cell suspension volume was pumped across the fringing field of the interpolar gap, the flow channel was disassembled and the plastic sheet was evaluated for the presence of cells in the area exposed to the fringing field. The resulting, expected deposition pattern of infected erythrocytes formed a well-defined band on the plastic sheet visible by an unaided eye approximating the breadth of the flow channel and the width of the interpolar gap (compare Figs. 12.3 and 12.4). Cells captured on the polyester slides were fixed for 30 sec in 100% methanol and subsequently stained with 4% Giemsa. Once dried, the slides were mounted between a standard glass slide and coverslip with Permount™ (Fisher Scientific, Pittsburgh, PA). The slides were visualized and photographed under oil immersion at 100× power.

Blood for this study was obtained from primates at the Centers for Disease Control and Prevention (CDC), Division of Parasitic Diseases, Atlanta, GA and Yerkes National Primate Research Center, Atlanta, GA. *P. falciparum*-infected blood was drawn from an *Aotus nancymai* monkey with a peripheral blood parasitemia of 2.7%. *P. vivax* and *P. malariae*-infected blood samples were drawn from *A. vociferans* monkeys with peripheral blood parasitemias of 0.1% and 0.4%, respectively. *P. ovale*-infected blood was taken from a chimpanzee (*Pan troglodytes*) with a parasitemia of 0.2%. All infected primate blood donors in the study had been splenectomized. Blood was collected in heparin tubes and diluted 1:6 with PBS, pH 7.2, before loading 500 μl of the diluted sample into the syringe pump. To produce a *P. falciparum*/*P. vivax* mixed infection, equal volumes of whole blood from each infection were mixed and

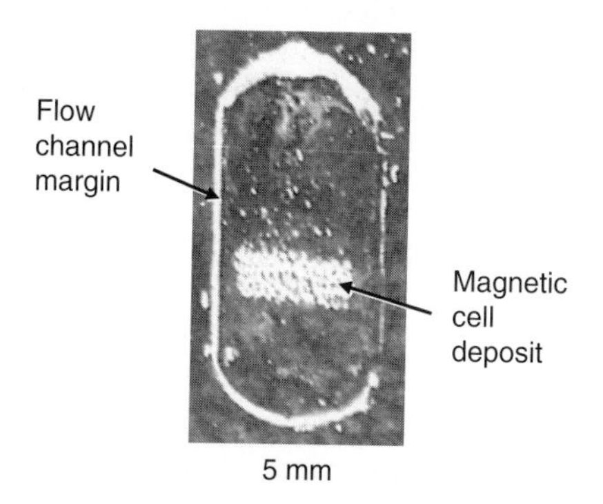

Fig. 12.4. An unaided eye appearance of the magnetic deposition, collected in the interpolar gap area (Fig. 12.3), from a *P. falciparum*-parasitized blood sample [35].

then immediately diluted 1:6 with PBS. All samples were processed fresh, within 6 hour of the time that blood samples were drawn. Protocols for infecting monkeys with malaria parasites were approved by the CDC Institutional Animal Care and Use Committee according to Public Health Service Policy.

P. falciparum-infected blood samples were enriched 40-fold from a parasitemia of 2.7% (panel A) to nearly 100% (Fig. 12.5, Panel B). *P. vivax*-infected blood samples were enriched up to 250-fold, from an initial parasitemia of 0.1% (Panel C) to clusters with 25% (Panel D) infected erythrocytes. *P. malariae*-infected blood samples were enriched from 0.4% to 100% infected erythrocytes, at least a 250-fold concentration (Panel E to F). *P. ovale*-infected blood samples were enriched up to 375-fold from an initial parasitemia of 0.2% to clusters containing 75% infected erythrocytes (Panel G to H). Additionally, we observed that MDM successfully concentrated *P. simium*-infected erythrocytes (data not shown). This observation along with earlier reports, where magnetic columns were used to enrich murine malaria parasite (*P. berghei*) ookinetes [58], suggests that magnetic capture methods are generally applicable to *Plasmodium* species.

In most malarious regions of the world, multiple *Plasmodium* species are present in the population [59] and mixed-species infections within individuals are common [49]. Therefore, to determine how a mixed *Plasmodium* species infection would be evaluated by malaria MDM analysis, we performed a mixing experiment with *P. falciparum*- and *P. vivax*-infected blood samples (equal volumes of each sample), prepared slides, and evaluated the mixture as previously described for the unmixed samples. In our results shown in Fig. 12.6, we have provided evidence to show that malaria MDM concentrated infected cells of both *Plasmodium* species on one slide. Additionally, *P. falciparum*- and *P. vivax*-infected erythrocytes appear to have been captured in proportions similar to the initial parasitemia (~20:1).

Consistent with hemoglobin digestion, liberation, and crystallization of free heme into hemozoin (Fig. 12.1), we observed that

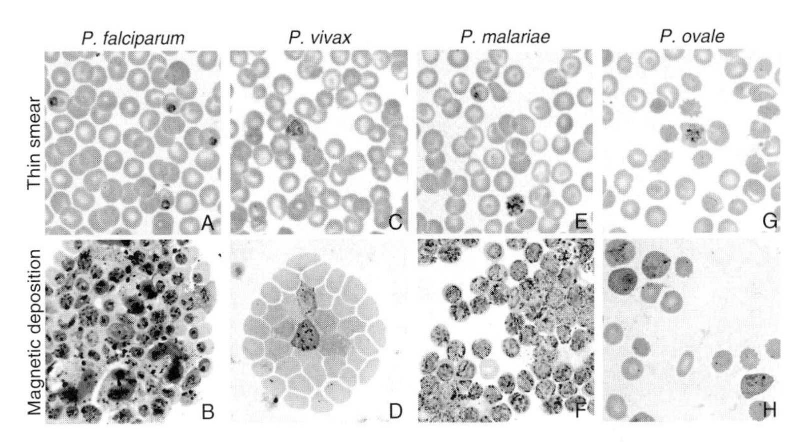

Fig. 12.5. Malaria MDM concentrates *Plasmodium*-infected erythrocytes. *P. falciparum* (A, B), *P. vivax* (C, D), *P. malariae* (E, F), and *P. ovale* (G, H) infections comparing conventional thin blood smear (top row) and malaria MDM (bottom row) for each *Plasmodium* species were prepared from infected nonhuman primate blood samples. Individual parasitemias determined by the Earle and Perez method were 2.7% for *P. falciparum*, 0.1% for *P. vivax*, 0.4% for *P. malariae*, and 0.2% for *P. ovale*. All slides were stained using standard Giemsa staining procedures and examined using a 100× oil immersion objective. In part B, "M" = macrophage ([35], with permission from *American Journal of Tropical Medicine and Hygiene*). (See Color Insert.)

trophozoites, schizonts, and gametocytes of all species were captured by MDM, but there was a noticeable underrepresentation of ring-stage parasites. This observation is consistent with earlier reports where infected erythrocytes containing mature trophozoites and schizonts were captured on a magnetized steel mesh, while ring-stage parasites were collected in a flow-through fraction [50, 51]. This is an important limitation of the current method. Diagnosis of *P. falciparum* infection relies on observation of rings owing to the sequestration of trophozoites and schizonts in postcapillary venules away from peripheral blood flow ordinarily monitored by vena-puncture or finger-prick blood collection techniques. Moreover, as parasitemia is known to fluctuate at regular intervals, and the early developmental stages can comprise the majority of infected erythrocytes,

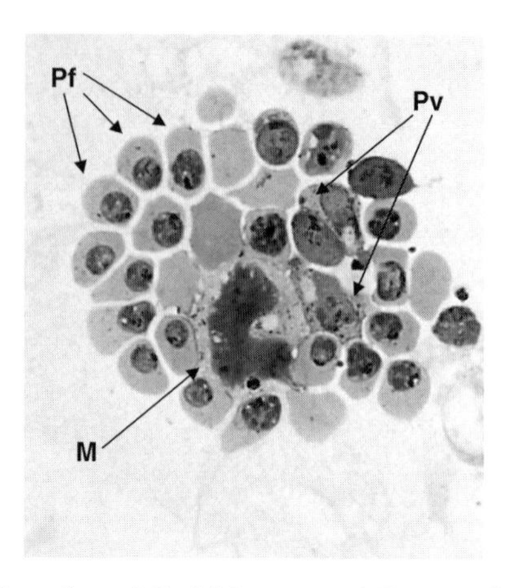

Fig. 12.6. MDM detection of *P. falciparum* and *P. vivax* from a mixed blood sample. Equal volumes of blood from *P. falciparum* (initial parasitemia of 2.7%) and *P. vivax* (initial parasitemia of 0.1%) infected monkeys were mixed and then subjected to MDM analysis. Giemsa stained slides show MDM concentration of *P. falciparum* (Pf), *P. vivax* (Pv), and macrophages (M) containing hemozoin [35]. (See Color Insert.)

capture of ring and early trophozoites would greatly improve the ability to estimate parasitemia. With the enrichment of gametocytes it may be possible to estimate better gametocytemia and evaluate malaria transmission potential within endemic populations. Finally, we observed that hemozoin-laden macrophages were also captured by MDM from *P. falciparum*-infected blood (Figs. 12.5B and 12.6).

As can be observed in Figs. 12.5 and 12.6, clustering of cells can compress erythrocyte membranes; however, this did not distort parasite morphology, staining, or infected versus uninfected erythrocyte size characteristics familiar to malaria microscopists. Slides from samples with lower parasitemia tend to have smaller and fewer cell clusters than high-level infections. Unlike the typical blood smear pattern of evenly spaced cells, MDM deposits infected erythrocytes in cell clusters (uninfected erythrocytes can be found within these clusters at low frequencies) in proximity to the interpolar gap region between the edges of the magnetic bars (Fig. 12.3). This greatly assists microscopists in locating infected cells and in restricting the region of the slide to be evaluated. Additionally, because MDM enriches capture of parasitized erythrocytes, it should facilitate differentiation of species- and stage-specific morphological features by providing opportunity for comparison among a greater number of infected cells. These characteristics of malaria MDM slide preparations should contribute to more rapid evaluation of blood slides. Similarly, we anticipate that malaria MDM will also contribute to more thorough and efficient teaching of microscopy-based malaria diagnosis.

12.3. Magnetic flow cell sorting

The current, commercially available magnetic cell separators operate in batch mode, which in principle is difficult to scale up and stage. Moreover, batch mode separations do not allow a quantitative magnetic cell fractionation by the amount of the magnetic reagent attached to the cell. Quantitative cell fractionation may become important for cell function studies, and for early progenitor cell isolation. Continuous systems based on an open gradient, dipole and quadrupole magnetic field-geometries have been studied [6, 60–62]. The expected advantages of such a system include flexibility in scale-up and staging of the separation process, as well as the potential for a quantitative cell fractionation based on the cell surface marker expression [63, 64]. The properties of the quadrupole magnetic flow sorter (QMS) system are described briefly in this section. In particular, we have investigated the relationship between the cell magnetophoretic mobility, and the purity, recovery, and throughput of the sorted cell fractions [6, 65, 66]. The cell magnetophoretic mobility has been defined by analogy to the cell electrophoretic mobility (used in free-flow cell electrophoresis) as a ratio of the field-induced cell velocity to the strength of the magnetic force, S_m (reviewed in Chapter 5). Cell separation results from differences in cell magnetophoretic mobilities in the continuous QMS system. The concept of the cell magnetophoretic mobility allows for separation of variables describing properties of the cell from those of the external magnetic field (Chapter 5). In brief, assuming that the magnetically labeled cell behaves as an ideal, induced magnetic dipole, the magnetophoretic mobility of the cell, m, depends on the binding of the antibody to the cell and the magnetic susceptibility of the magnetic label, but not on the external magnetic field. On the other hand, the magnetic force strength, S_{m}, depends on the field strength of the source and the field geometry but is independent of the cell (and the magnetic label) properties. Such separation of variables greatly simplifies the theoretical and experimental treatment of the continuous cell sorting process, and is the basis of cell sorting process analysis described below.

Magnetophoretic mobilities of the cell populations are provided by cell-tracking velocimetry (CTV) measurements (reviewed in Chapter 8). The theory of QMS operation was developed on the basis of the theory of the split-flow thin (SPLITT) fractionation, developed by Giddings and coworkers [67, 68]. Experimental verification was performed with reference magnetic bead standards and white blood cells [6, 66, 69]. The review provided in this section is intended to illustrate a conceptual framework necessary for the design and engineering of a continuous cell sorting device based on magnetophoresis. The experiments were performed on white blood cell fraction samples (leukapheresis product) used for hematopoietic stem cell enrichment.

12.3.1. Continuous cell sorting in the quadrupole field

The separation element of the QMS separator consists of two concentric cylinders (the inside cylinder was solid) surrounded by four pole pieces generating the magnetic quadrupole field, as illustrated in Fig. 12.7. QMS inlets and outlets are connected to the

QMS separation area by suitable flow manifolds, equipped with inlet and outlet flow splitters, respectively. The sources of the magnetomotive force were four pieces of neodymium–iron–boron (Nd–Fe–B) permanent magnets. The quadrupole field geometry was achieved by using four specially shaped soft iron pole pieces connected to the permanent magnets. The force field strength inside the quadrupole field has a centrifugal character, that is, it is a linear function of the distance from the field's axis of symmetry. There are

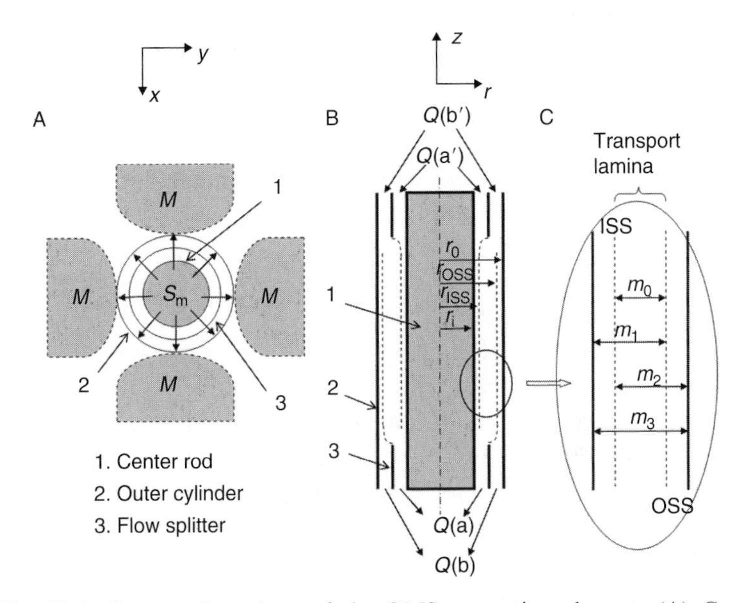

Fig. 12.7. Cross-section views of the QMS separation element. (A) Cross section showing disposition of magnet pole tips (M), flow channel lumen bounded by the surfaces of the solid center rod (1), the outer cylinder (2), the flow splitter (3), and the radial direction of the magnetic force field (S_m). (B) Longitudinal section showing radial coordinates of the flow channel surfaces, r_o and r_i, radial coordinates of ISS, r_{ISS}, originating at the tip of the inlet flow splitter and the OSS, r_{OSS}, ending at the tip of the outlet flow splitter. Note designations for the inlet and outlet volumetric flow rates, Q. (C) Fragment of channel longitudinal section showing transport lamina and the magnetophoretic transport distances associated with characteristic cell mobilities, m, as discussed in the text. Typically, $m_0 < m_1 < m_2 < m_3$ [62].

no tangential or axial components of the force strength inside an ideal quadrupole field, Fig. 12.7 (also reviewed in Chapter 4). The magnetic field in the QMS separation element is the source of the radial movement of the magnetically labeled cell, whose regular pattern is amenable to mathematical analysis, as described below. Because of the distributed character of the cell magnetophoretic mobility, m, the radial cell velocity, u_{m}, was also distributed for any given ρ (defined below) in the magnetic field accessible to cells. Therefore, in calculating the retrieval factors at outlets **a** and **b** (Fig. 12.8), we used sets of trajectories corresponding to different mobilities, with weighting factors equal to the mobility frequency as measured by the CTV. An additional source of the distribution of the cells between the outlets **a** and **b** was the random distribution of cell initial radial position when entering the magnetic field region (Fig. 12.9). The combination of the radial (magnetic) and axial (convective) velocity components determined the final position of the cell in the fluid stream at the outlet flow splitter, and therefore the cell transport to outlet either *a* or *b*.

The characteristics of the magnetic system include a mean magnetic field in the channel annulus of 1.028 T and a mean force field strength, S_{m}, of 1.45×10^2 T A/mm^2. A prototype separation channel (item no 1059, SHOT, Inc., Greenville, IN) was used. The prototype test channels were manufactured for reproducible operation, but not for high throughput. The axial separation zone of the channel was 15.24 cm; its inner cylinder radius was 0.69 cm and its core rod radius was 0.47 cm. The inlet splitter has ID = 10.7 mm and OD = 12.1 mm, and outlet splitter has ID = 11.4 mm and OD = 12.1 mm. Two double-syringe pumps (type "33" syringe pump, Harvard Instrument, Inc., Cambridge, MA) were used: one pump controlled the inlet streams **a**′ and **b**′, while the other pump controlled the outlet aspiration flow stream **a** and another inlet dispensing stream **a**″ (Fig. 12.8). The stream **b** was not connected to a pump to allow for the pressure equilibration within the system. Both dispensing streams **a**′ and **a**″ were connected to a diagonal switching valve (Upchurch Scientific, Inc., Oak Harbor, WA) to

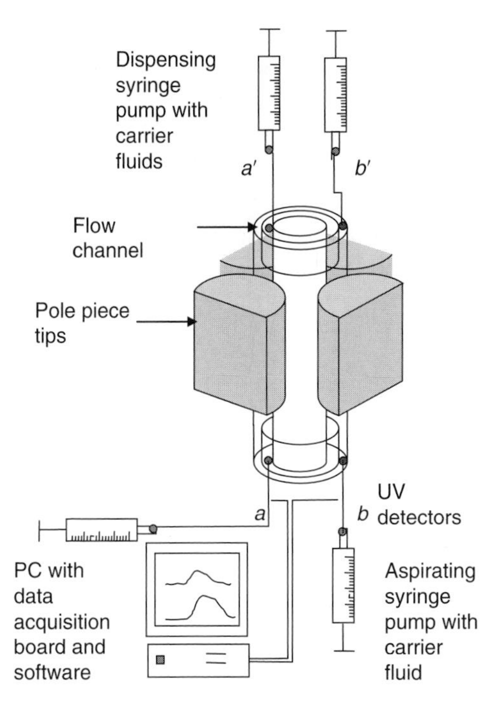

Fig. 12.8. The schematic diagram of the basic QMS cell sorter system. The actual implementation may vary between experiments as discussed in text.

allow the switch of inlet fluid between these two streams without inducing a flow disturbance. Before starting, the QMS sorter was flushed bottom up with degassed PBS to remove all the bubbles in the system. Then the inlet syringe **a′** filled with labeled leukapheresis suspension and syringes **a″** and **b′** filled with PBS were connected to the pumps. The experiment was initialized by starting the inlet pumps, allowing the inlet streams **a″** and **b′** to enter the system. At the same time, the outlet streams **a** and **b** were open until the system pressure was balanced. The sample syringe pump for **a′** was then started. After evidence of stable operation, the diagonal valve was switched so that the flow of inlet stream **a′** was directed to the channel. Just before the sample in syringe **a′** was depleted, the diagonal valve was switched back to **a″** in order to flush the system

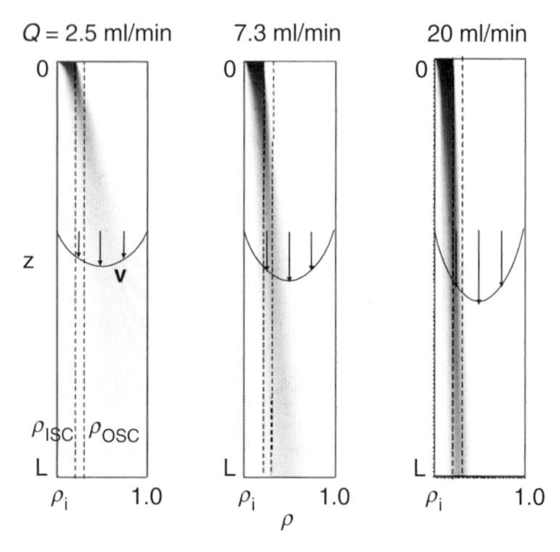

Fig. 12.9. The cell concentration distribution inside the QMS separation element. Note decreasing transport of cells across the transport lamina (between ρ_{ISC} and ρ_{OSC}, where $\rho = r/r_o$, compare Fig. 12.7) with the increasing total flow rate, Q (the direction of flow is from top to bottom; the flow velocity profile, v, is not to scale). The inlet flow rate ratio was $Q(a')/Q = 0.1$; in outlet flow rate ratio was $Q(a)/Q = 0.2$.

with carrier. Finally, the pumps were stopped simultaneously. The volumes and cell concentrations of collected fluids were determined by weight and Z-1 Coulter Counter, respectively, for the calculation of progenitor cell recovery. The progenitor cell purity was determined with the aid of flow cytometry (FCM).

The inlet flow rate ratio, $Q(a')/Q$ and the outlet flow rate ratio, $Q(a)/Q$, determined the positions of the flow stream surfaces originating at the edges of the inlet and outlet flow splitters inside the QMS channel, respectively. These characteristic stream surfaces were referred to as "inlet splitting cylinder" (ISC) and "outlet splitting cylinder" (OSC) in analogy to the SPLITT fractionation theory developed by Giddings [67, 70]. The volume of flow enveloped by the ISC and OSC was referred to as "the transport lamina," and it represented a resistive element to transport during the separation process in the QMS channel, Figs. 12.7 and 12.9. The effect of flow imperfections in the QMS system has been observed as a residual cell transport across the transport lamina even in the absence of the magnetic field (dubbed "nonspecific cell crossover").

12.3.2. Predicted sorter output based on cell magnetophoresis

The trajectory of a cell with magnetophoretic mobility, m, is given for the longitudinal position in the channel, z, as a function of the dimensionless radial variable ρ, defined as $\rho = r/r_o$, where r_o is the outer radius of the channel. The trajectory is calculated from $z = 0$ to $z = L$, the region of the magnetic field, which is assumed to coincide with the distance between the channel flow splitter. The equation, previously reported [68], is

$$z(\rho) = \frac{QI_1[\rho_1, \rho]}{2\pi A_1 m S_{mo} r_o (1 - \rho_i^2)} \tag{12.1}$$

where Q is the total flow rate and S_{mo} is the magnetic force field strength evaluated at the inner surface of the outer wall. $A_1 = 1 + \rho_i^2 - A_2$ and $A_2 = (1 - \rho_i^2)/\ln(1/\rho_i)$ are the parameters related to the annular channel geometry; $\rho_i = r_i/r_o$, where r_i is inner radius of the channel annulus; $\rho_1 = r_1/r_o$ and $\rho = r/r_o$ correspond to the initial radial positions at $z = 0$ and at some arbitrary distance z. $I_1[\rho_1, \rho]$ is an integral solution as given by

$$I_1[\rho_1, \rho] = [4 \ln \rho - 2\rho^2 + 2A_2(\ln \rho)^2]_{\rho_1}^{\rho} \tag{12.2}$$

For an ideal quadrupole field,

$$S_{mo} = \frac{B_o^2}{\mu_0 r_o} \tag{12.3}$$

where B_o is the magnetic field intensity at r_o and μ_0 is the magnetic permeability of free space. In the quadrupole field, S_m varies linearly with radial position: it is 0 at the aperture center and reaches a maximum at the cylinder wall,

$$S_m = S_{mo}\rho \tag{12.4}$$

By applying Eq. (12.1), the trajectory of a cell of known magnetophoretic mobility is determined for its initial position, selected Q, and fixed magnet and channel specifications.

12.3.3. Calculation of cell recovery and purity of QMS separation

Inlet and outlet splitting surfaces, ISS and OSS, describe the virtual flow streamlines originating and ending at the edges of the inlet and outlet flow splitters, respectively (Figs. 12.7 and 12.9). The ISS divides the inlet flows **a′** (feed) and **b′** (carrier), and the OSS divides the outlet flows **a** (depleted fraction) and **b** (enriched fraction). Once the inlet flow rate ratio $Q(\mathrm{a}')/Q$ and outlet flow rate ratio $Q\,(\mathrm{a})/Q$ are selected, the position of the ISS and OSS are obtained by solving two equations for the upper limit of integration ρ_{ISS} or ρ_{OSS}:

$$\frac{Q(\mathrm{a}')}{Q} = \frac{I_2[\rho_{\mathrm{i}}, \rho_{\mathrm{ISS}}]}{A_1(1-\rho_{\mathrm{i}}^2)} \tag{12.5}$$

$$\frac{Q(\mathrm{a})}{Q} = \frac{I_2[\rho_{\mathrm{i}}, \rho_{\mathrm{OSS}}]}{A_1(1-\rho_{\mathrm{i}}^2)} \tag{12.6}$$

where the integral solution is

$$I_2[\rho_{\mathrm{i}}, \rho_{\mathrm{ISS}}] = [2\,\rho^2 - \rho^4 + 2A_2\rho^2 \ln\rho - A_2\rho^2]_{\rho_{\mathrm{i}}}^{\rho_{\mathrm{ISS}}} \tag{12.7}$$

In order for a cell to be recovered in the enriched **b** fraction, it must migrate across the virtual fluid shell bounded by ρ_{ISS} to ρ_{OSS}; this shell is called the transport lamina, and its thickness δ is given by

$$\delta = (\rho_{\mathrm{OSS}} - \rho_{\mathrm{ISS}})r_{\mathrm{o}} \tag{12.8}$$

The cell assignment to a specific outlet is determined by evaluating its radial position just before the splitter and applying the conditional statements

$$\rho(z = L) < \rho_{\mathrm{OSS}} \quad \text{outlet } \mathbf{a} \tag{12.9}$$

$$\rho_{\text{OSS}} \leq \rho(z = L) < 1 \quad \text{outlet} \quad \mathbf{b} \tag{12.10}$$

$$1 = \rho(z \leq L) \quad \text{wall} \tag{12.11}$$

In cases where the mobility is high or the flow rate is low, a cell might have sufficient time to migrate radially to the cylinder wall, at $\rho = 1.0$, before it exits the channel. In this case, we assume the cell remains on the wall and is not recovered in either outlet. This condition is described by Eq. (12.11).

The above equations are applied to a computer program written in Maple (Maplesoft, Waterloo, Ontario). It requires an input of mobility distribution, obtained from measurements by CTV. The distribution comprises mobility-frequency data, which are individually processed to obtain outlet recovery, defined as the number of cells in an outlet divided by the number fed in at **a**′. The recovery subunits are weighted by their frequency in the sample and summed to obtain overall recovery in **a**, **b**, and **w**: *F*(a), *F*(b), and *F*(w).

The sample mobility distribution may be gated into magnetically positive and negative cells, where the location of the gate is determined by comparison with an unmanipulated negative control population, a technique widely used in FCM. The purity of the gated magnetically positive cells (or target cells) in the leukapheresis sample feed is referred to as *P*(f). The positive cell fraction is processed separately through the program to obtain the positive cell recovery in outlet **a**, *F*(a+), and outlet **b**, *F*(b+). The negative cell fraction is similarly processed to obtain *F*(a−) and *F*(b−). The recovery of negative cells in outlet **b**, *F*(b−), is also referred to as nonspecific crossover, *X*. These contaminating cells determine the purity of the target cells recovered in the enriched fraction. Purity, *P*, is the ratio of positive cells to total cells in a given outlet; therefore, the purity of target cells in outlet **a**, *P*(a), and outlet **b**, *P*(b), are calculated by the following equations:

$$P(\mathrm{a}) = \frac{F(\mathrm{a}+)P(\mathrm{f})}{F(\mathrm{a})} \tag{12.12}$$

$$P(\mathrm{b}) = \frac{F(\mathrm{b+})P(\mathrm{f})}{F(\mathrm{b})} \tag{12.13}$$

The impact of nonspecific crossover on P(b) is shown by [71]

$$P(\mathrm{b}) = 1 - \frac{X(1 - P(\mathrm{f}))}{F(\mathrm{b})} \tag{12.14}$$

As high purity in **b** is almost always of paramount importance, it is instructive to see what causes a reduction in P(b): high nonspecific crossover, X, low purity in the feed P(f), and low recovery at outlet **b**, F(b).

12.3.4. Determination of the flow rate parameters for high resolving power and throughput separation

The m distribution is used to predict the cell "fate" during the QMS separation under defined flow rate parameters (Fig. 12.9). Comparing the sample mobility distribution to the four critical mobilities, m_0 through m_3, gives rise to the following: (1) if $0 \leq m < m_0$, then the cells exit at outlet **a**; (2) if $m_1 \leq m < m_2$, then the cells exit at outlet **b**; (3) if $m_3 \leq m$, then the cells will be lost to the outer wall; (4) if $m_0 \leq m < m_1$, then the cells will distribute between outlets **a** and **b**; and (5) if $m_2 \leq m < m_3$ then the cells will distribute between outlet **b** and the wall. By selecting the critical mobilities for the CTV m distribution, the total flow rate Q and the position of ISS and OSS are solved by combination of following equations:

$$m_0 = \frac{Q}{2\pi r_\mathrm{o} L S_\mathrm{mo}} \frac{I_1[\rho_\mathrm{ISS}, \rho_\mathrm{OSS}]}{A_1(1 - \rho_\mathrm{i}^2)} \tag{12.15}$$

$$m_1 = \frac{Q}{2\pi r_\mathrm{o} L S_\mathrm{mo}} \frac{I_1[\rho_\mathrm{i}, \rho_\mathrm{OSS}]}{A_1(1 - \rho_\mathrm{i}^2)} \tag{12.16}$$

$$m_2 = \frac{Q}{2\pi r_\mathrm{o} L S_\mathrm{mo}} \frac{I_1[\rho_\mathrm{ISS}, 1]}{A_1(1 - \rho_\mathrm{i}^2)} \tag{12.17}$$

$$m_3 = \frac{Q}{2\pi r_o L S_{mo} A_2} \tag{12.18}$$

Subsequently, $Q(a')/Q$ and $Q(a)/Q$ are calculated using Eqs. (12.5) and (12.6). Once the flow rate conditions are fixed, the separation resolving power and throughput are determined theoretically. Resolving power (RP) for the QMS system was defined as:

$$RP = \frac{m_1}{\Delta m} = \frac{m_1}{m_1 - m_0} = \frac{I_1[\rho_i, \rho_{OSS}]}{I_1[\rho_i, \rho_{ISS}]} \tag{12.19}$$

The system throughput (TP) is defined as the number of cells introduced to the system and sorted per unit time:

$$\mathrm{TP} = CQ(a') = 2\pi r_0 L S_{mo} m_1 C \frac{I_2[\rho_i, \rho_{ISS}]}{I_1[\rho_i, \rho_{OSS}]} \tag{12.20}$$

where C is the number concentration of the cells in the feed stream (i.e., the number of cells per unit volume) and $Q(a')$ is the sample feed stream flow rate. Combination of Eqs. (12.3), (12.19), and (12.20) gives

$$\mathrm{TP} = 2\pi L \frac{B_o^2}{\mu_0} m_1 C \frac{I_2[\rho_i, \rho_{ISS}]}{I_1[\rho_i, \rho_{ISS}]} \frac{1}{RP} \tag{12.21}$$

12.3.5. The cell model system

The magnetophoretic cell sorting in the QMS system was tested on a small amount of clinical leukapheresis products that were generously donated for research by patients or normal donors of the Cleveland Clinic Foundation (CCF) Taussig Cancer Center who signed the informed consent form, approved by the CCF IRB. The leukapheresis cells were collected, washed with Ca^{2+}-, Mg^{2+}-free Dulbecco's PBS (Media Core Facility, CCF), containing 0.5% bovine serum albumin (BSA, Sigma, St. Louis, MO) and 2-mM ethylene diamine tetraacetic acid (EDTA, Sigma, St. Louis, MO) by centrifugation at 400 × g for 10 min. The cells were then resuspended in ACK-lysing buffer (Quality Biologicals, Inc., Gaithersburg, MD) and incubated at room temperature for 7 min. After lysis of the red blood cells, the sample was washed twice by adding PBS and centrifuging at 400 × *g* for 10 min. Washed leukapheresis sample was labeled with CD34 Progenitor Cell Isolation Kit™, Miltenyi Biotec, according to manufacturer's recommendations (see Chapter 11 for a sample protocol).

12.3.6. Analysis of cell fluorescence intensity distribution by FCM

Samples were stained using three fluorescence dyes, including CD34 PE (8G12) (Miltenyi Biotec), CD45 ECD (J33) (IOTest Beckman Coulter, Inc., Fullerton, CA), and LIVE/DEAD Reduced Biohazard Cell Viability Kit No. 4 (Molecular Probes, Invitrogen Corporation, Carlsbad, CA). For each test, 2 million cells (or less, depending on the cell number in the sorted fraction) were stained with 20 μl of CD34 PE and 10 μl of CD45 ECD, followed by incubation at 4 °C for 30 min. The cells were washed and resuspended in PBS. One microliter of LIVE/DEAD reagent was then added to the cell pellet followed by incubation at room temperature for 15 min. The cells were washed again and resuspended in 0.5% formaldehyde (Polysciences, Inc., Warrington, PA) for fixation. The fixed cells were measured by FCM (BD Bioscience, San Jose, CA), where a total of 10,000–50,000 cells were counted for each sample. Sequential gating was used to identify CD34+ cells by their expression of CD34, characteristic low to intermediate CD45 antigen expression, and characteristic light scatter properties. A typical leukapheresis sample CD34 analysis result by FCM is shown in Fig. 12.10.

12.3.7. Analysis of the cell magnetophoretic mobility distribution by CTV

Magnetophoretic mobility measurements were made on an instrument referred to as a CTV (reviewed in Chapter 8) [72, 73]. The schematics of CTV are shown in Fig. 8.1. Briefly, CTV measures

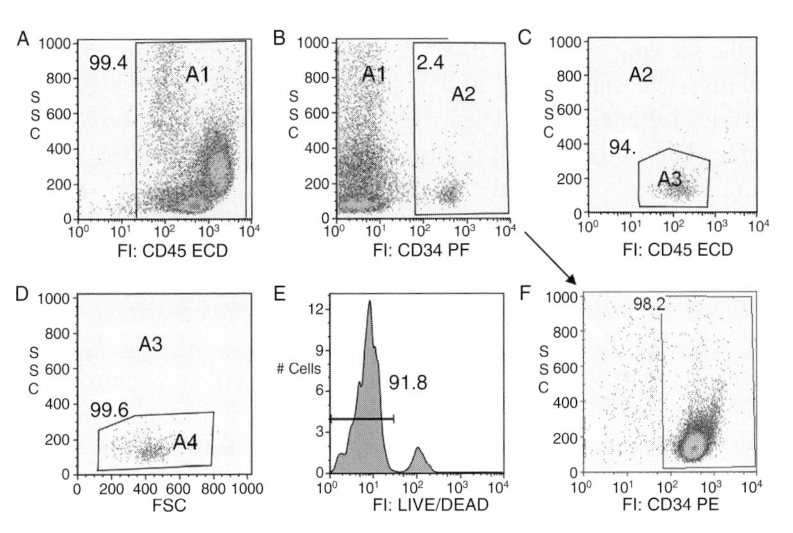

Fig. 12.10. FCM analysis of clinical leukapheresis sample. Numbers in panels are percentages of gated cell subpopulations. Panels A–E show sample before QMS sorting. Panel F shows QMS sorted fraction. (A) Dot plot of side scatter versus CD45-ECD fluorescence intensity (FI). A1: CD45+ leukocytes in the total population. (B) Dot plot of side scatter versus CD34-PE FI. A2: the CD34+ cells in the subpopulation A1. (C) Dot plot of side scatter versus CD45-ECD FI; A3: CD45 low to medium and SSC low in the subpopulation A2. (D) Dot plot of side scatter versus forward scatter to gate A4 in the subpopulation A3. (E) Dot plot of side scatter versus FI of LIVE/DEAD Reduced Biohazard Cell Viability Kit No. 4 to check the viability of the cells in A4. Here the purity of CD34+ is defined as: (number of events gated in A4)/(number of events gated in A1) × 100%. (F) Dot plot of side scatter versus CD34-PE FI. A2: the CD34+ cells in the subpopulation A1 (after QMS separation) [6]. (See Color Insert.)

the magnetically induced motion of cells, on a cell-by-cell basis, in a well-defined magnetic field. The cell suspension to be analyzed was pumped into a rectangular borosilicate glass channel and placed in a specially designed magnet with a magnetic field intensity B of 1.41 T and a mean magnetic force field strength S_m of 146 T A/mm^2 in the viewing area (1.71 × 1.28 mm^2, width × height). The horizontal cell velocity (perpendicular to the sedimentation velocity induced by gravity) induced by the magnetic field is nearly constant in the viewing area. The displacements of cells are recorded using an inverted microscope and a 30-Hz Cohu CCD 4915 camera (Cohu Electronics, San Diego, CA), and then processed by a computer. The output is cell m population statistics (including mean, standard deviation, and 95% confidence interval) and a cell m distribution histogram.

12.3.8. Comparison of magnetic and fluorescent cell fractions

The specificity of the cell magnetizing reagent to the progenitor cells was examined by a comparison of magnetically induced cell motion (by CTV) with cell fluorescence (by FCM). For the secondary antibody to be specific to the primary antibody, it is necessary that the percentage of the "magnetically positive" cells (measured by CTV) is equal to the percentage of the "fluorescence-positive" cells (measured by FCM). The positive cell fractions were determined based on comparison to the respective negative control populations. As shown in Fig. 12.11, the fluorescent cell fraction coincides with the magnetic cell fraction: 4.0% versus 4.5%, respectively (confirmed in repeated tests, not shown).

12.3.9. Flow rate optimization

The QMS-sorting process was controlled by adjusting three flow parameters: total flow rate Q, inlet flow rate ratio $[Q(a')/Q]$, and outlet flow rate ratio $[Q(a)/Q]$. The selection of the flow rates was guided by the magnetophoretic mobility and the original purity of the target progenitor cells. The relatively low m of the CD34+ limits the choice of m_1 [Eq. (12.16)]. Moreover, the original purity of the CD34+ cells is relatively low, 0.5–5%. Consequently, the use of high resolving power [or small Δm, Eq. (12.19)] is necessary to achieve the high purity separation because positive and negative peaks lie close together (Fig. 12.11). An additional confounding factor is the variability of the m distribution of CD34+ cells

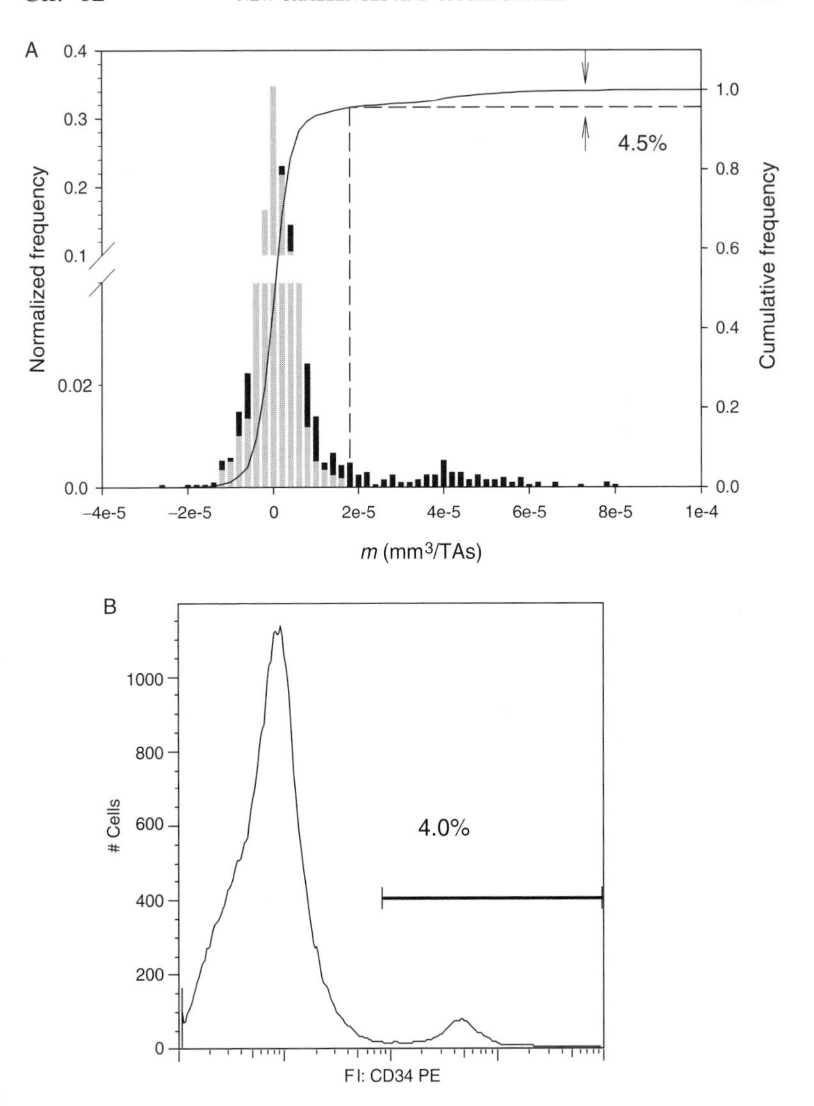

Fig. 12.11. (A) Superposition of control (gray) and test (black) magnetophoretic mobility (m) histograms for leukapheresis sample labeled with CD34 Progenitor Cell Isolation Kit™, Miltenyi Biotec, according to manufacturer's recommendations; the magnetically positive cell fraction (putative CD34+ cells) is identified by cell mobilities that are higher than those found in the

between donors (data not shown). To achieve high purity isolation for a majority of the samples, robust conditions or high m_1 were required. Thus, only higher m cells were deflected to outlet **b** to maintain high CD34+ cell purity at the cost of a reduction in recovery, as many target cells of lower m were left to outlet **a**. All things considered, the following critical mobility values were selected: $m_1 = 5 \times 10^{-5}$ mm^3/(T A sec) and $m_2 = 1.4 \times 10^{-4}$ mm^3/(T A sec) [Eqs. (12.16), (12.17), and Fig. 12.11]. We set the resolving power RP $= m_1/\Delta m$ to 2.5 [Eq. (12.19)]. It follows that $m_0 = 3 \times 10^{-5}$ mm^3/(T A sec) [Eq. (12.19)] and $m_3 = 1.6 \times 10^{-4}$ mm^3/(T A sec) ($m_3 = m_2 + m_1 - m_0$, Fig. 12.7). The required total flow rate $Q = 7.6$ ml/min was calculated using Eq. (12.20). From Eqs. (12.5) and (12.6), we obtain the required inlet and outlet flow rate ratios $Q(\text{a}')/Q = 0.09$ and $Q(\text{a})/Q = 0.25$, respectively. With these calculated flow rates, exploratory experiments were done for the purpose of further optimization of the flow parameters (data not shown), leading to an increased total flow rate of 10 ml/min, a decreased inlet flow rate ratio of 0.06 and an increased outlet flow ratio of 0.4 relative to the calculated values.

12.3.10. Sorted cell assay by both FCM and CTV

Seven milliliters of leukapheresis sample at a concentration of 3×10^6 cells/ml were injected into the system using the flow parameters selected above. The feed and cells eluted from outlets **a** and **b** were measured by FCM (Fig. 12.12A–C) and CTV (Fig. 12.12D–F). The FCM analysis showed that the purity of the "fluorescence positive" fraction that defines the CD34+ cells, increased from 1.8% to 97% in the enriched fraction **b** and decreased from 1.8% to 0.44% in the depleted fraction **a**. This was corroborated by the CTV measurements. Based on a mobility gate of 2×10^{-5} mm^3/T A sec, the purity

---

negative control sample. (B) FCM histograms of the same sample labeled with antiCD34-PE antibody. Note agreement between percentages of the positive cells in A and B, suggesting that these are the same cells [6].

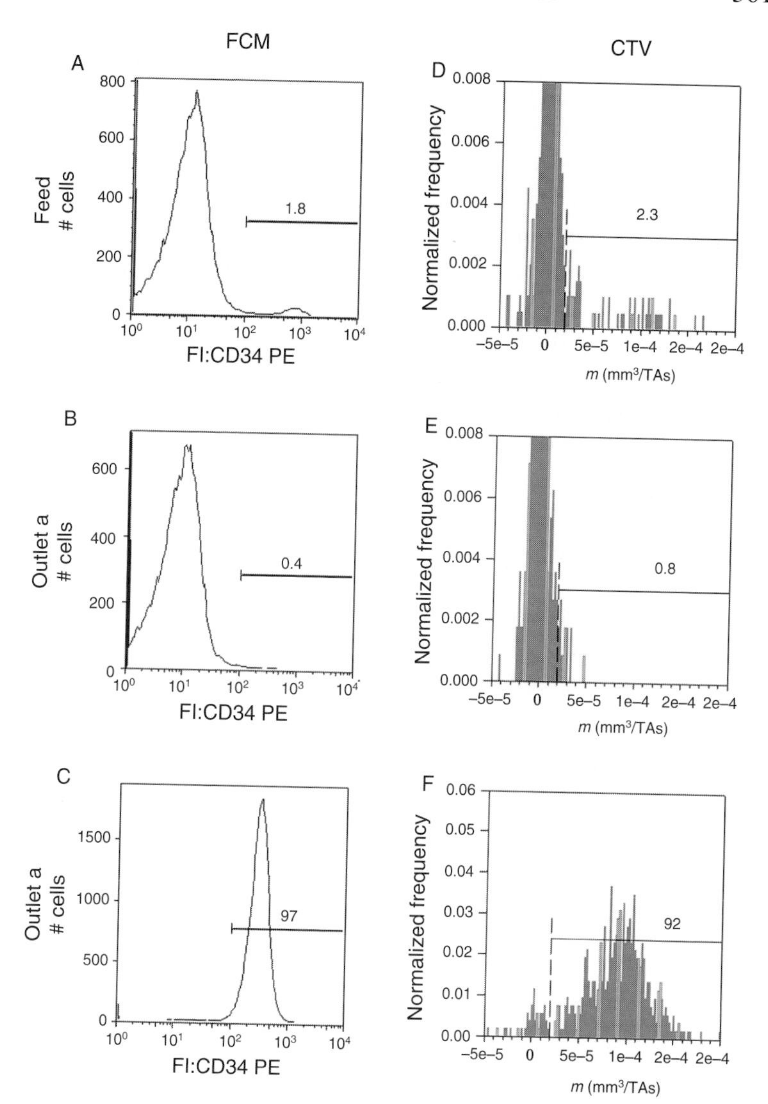

Fig. 12.12. (A–C) Cell FI histograms of the leukapheresis sample before QMS enrichment and after enrichment collected from outlets a and b, respectively. (D–F) Magnetophoretic mobility (m) histograms of the same samples. Note comparable content of positive cells (CD34+) measured by the two different methods (FCM and CTV), and CD34+ cell enrichment in outlet b [6].

of the magnetically positive fraction (putative CD34+ cells) increased from 2.3% to 90% in the enriched fraction **b** and decreased from 2.3% to 0.8% in the depleted fraction **a**.

12.3.11. Experimental verification of the predicted decrease of CD34+ cell recovery with the increasing Q(*a*)/Q

The dependence of the magnetic cell fraction recovery on $Q(\mathrm{a})/Q$, measured experimentally (Fig. 12.13A) agreed with the recovery calculated on the basis of cell magnetophoresis [Eqs. (12.9–11)]. Increasing $Q(\mathrm{a})/Q$ increases transport lamina thickness, δ (Fig. 12.7), though not linearly. As the transport lamina acts as a resistive barrier to transport, its increasing thickness impedes the migration of magnetic cells into **b**; this is followed by decreasing $F(\mathrm{b}+)$ with the $Q(\mathrm{a})/Q$ increase. As fewer cells are able to cross the transport lamina, they are left behind in **a**; therefore, $F(\mathrm{a}+)$ increases. The agreement between data points and the theoretical curves in Fig. 12.13A is very good considering complex magnetic and convective transport phenomena involved in the separation process.

12.3.12. CD34– cell nonspecific crossover decreases with the increasing Q(*a*)/Q

The nonspecific crossover [parameter X in Eq. (12.14)] was dependent on flow rate ratio (Fig. 12.13B). As mentioned above, nonspecific crossover, X, is synonymous with the recovery of negative (contaminating) cells into outlet **b**, $F(\mathrm{b}-)$. Except for some crossover at $Q(\mathrm{a})/Q = Q(\mathrm{a}')/Q = 0.1$, ideally, there should be no crossover at the conditions of the experiment because the negative cells have insufficient mobility to cross the transport lamina. The measured crossover deviates from theory, especially at low flow rate ratios. The theoretical simulation is based on the assumption of noninteracting cells, perfect channel design, and perfect laminar flow [74],

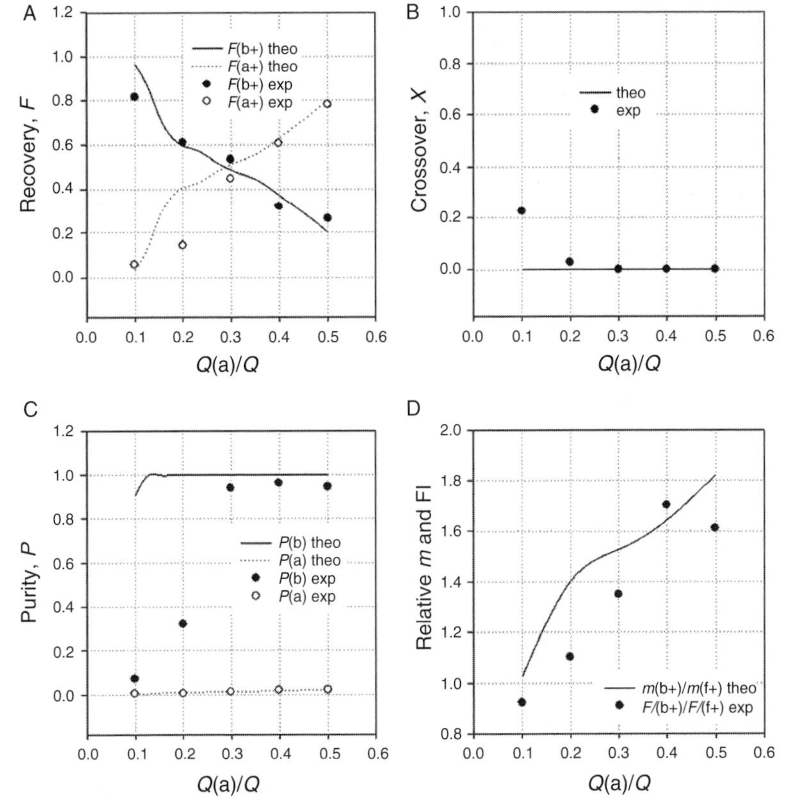

Fig. 12.13. (A) CD34+ cell recovery in outlets a and b after QMS sorting. Theoretical lines were obtained based on cell sample magnetophoretic mobility distribution (shown in Fig. 12.12D). Note close agreement with the experiment. (B) CD34– cell crossover. Note deviation from experiment when transport lamina becomes thin at $Q(a)/Q \leq 0.2$. (C) CD34+ cell purity in outlet a and b fractions by QMS sorting. Note deviation of theory from experiment when the nonspecific crossover, X, becomes significant [compare with panels A, B, and Eq. (12.14)]. (D) CD34+ cells mean FI ratio and the mean magnetophoretic mobility ratio (outlet b to feed). Note that the trend in the experimental data follows closely that predicted from the theory of the cell magnetophoresis (solid line) [6].

consequently, no crossover. In the real separation, the small transient regions where the flow streams merge and then divide, the presence of the hydrodynamic lift force, high cell concentration and geometric imperfections are all factors that may lead to the degradation of separation [75, 76]. These effects decrease with increasing thickness of the transport lamina, δ, as shown by good agreement with $Q(\mathrm{a})/Q = 0.3$ and higher (Fig. 12.13B). To avoid crossover, experiments must be conducted with larger values of δ, and therefore $Q(\mathrm{a})/Q$, than prescribed by theory for ideal behavior. This is why the selected experimental flow conditions differed somewhat from theoretical prediction, as mentioned earlier.

12.3.13. CD34+ cell purity increases with the increasing $Q(a)/Q$

Equation (12.14) illustrates that crossover reduces $P(\mathrm{b})$. The theory predicts no significant crossover, with $P(\mathrm{b})$ equal to unity for all $Q(\mathrm{a})/Q > Q(\mathrm{a}')/Q$ values. However, it should not be surprising that where significant crossover was measured, as for $Q(\mathrm{a})/Q = 0.1$ and 0.2 in Fig. 12.13B, a commensurate reduction in $P(\mathrm{b})$ was also seen, Fig. 12.13C. At higher $Q(\mathrm{a})/Q$, experimental crossover is low (Fig. 12.13B), so the purity data well matches the theory. The purity rises in outlet **a** with increasing $Q(\mathrm{a})/Q$, because the thickening transport lamina prevents the negative cells from crossing the OSS; thus, they are diverted to outlet **a**. This is also consistent with the effect of increasing separation between m distributions in outlets **a** and **b** with the increasing $Q(\mathrm{a})/Q$, discussed above.

12.3.14. Mean CD34+ cell fluorescence intensity increases with the increasing $Q(a)/Q$

Figure 12.13D shows the dependence on $Q(\mathrm{a})/Q$ of experimental mean, relative cell fluorescence intensity (FI) in outlet **b** (the ratio of the mean FI of CD34 fluorescence positive progenitor cells in outlet **b** to the mean FI of CD34 fluorescence positive progenitor cells in the feed) and the theoretical mean, relative magnetophoretic mobility (the ratio of the mean m of magnetically positive progenitor cells in the outlet **b** to the mean m of magnetically positive progenitor cells in feed). Cell antibody-binding capacity (ABC) relates directly to the number of antibodies binding to the surface molecules on individual cells [61, 77, 78]. Thus, ABC is proportional to cell FI with immunofluorescent labeling and cell magnetophoretic mobility with immunomagnetic labeling. The close agreement between the theory (solid line) and the experimental data in Fig. 12.13D further demonstrates that the continuous magnetophoretic cell sorting is capable of fractionation based on antigen expression (the *ABC* value) of the magnetically tagged cell in a well-controlled and quantitative manner.

12.4. Magnetic field-flow fractionation

Magnetic field-flow fractionation (MgFFF) is an analytical and characterization technique for particulate magnetic materials. It is a separation by elution technique similar to chromatography in operation but not in mechanism. It separates particles according to the strength of their interaction with an applied magnetic field. Particles may be nanosized or up to tens of microns in diameter. However, the mechanism of separation is different for these two extremes, and a single analysis cannot deal with this full range of applicability. Analyses are generally confined to either submicron or supramicron ranges. The following discussion will be mainly concerned with the application of magnetic FFF to submicron particles. In the submicron particle size range, the technique has the potential for quantitative magnetic characterization. Separation and characterization of submicron particles depends only on the strength of their interaction with a magnetic field. The technique is therefore insensitive to nonmagnetic coatings such as dextran and antibodies to cell surface markers. Given the magnetic properties of the magnetic component of the particles, it is capable of determining the mass distribution of magnetic material in a particulate sample. This type of analysis could be invaluable for quality control purposes in the production of immunospecific magnetic nanoparticles.

Other commonly available methods of determining magnetic susceptibility or magnetization of particulate materials are capable of providing only a bulk measurement for the sample. Either the force on the total sample induced by an inhomogeneous magnetic field is measured, or the current induced in a circuit close to the sample moving in a magnetic field is measured. If the number of particles in the sample is known, then the mean strength of interaction of a single particle with the field may be obtained. Then, with knowledge of the magnetic properties of the magnetic component of the particles, the mean magnetic material content of the particles may be determined. This provides no information on the distribution of the strength of the individual particle interactions with the field. A notable exception to these bulk measurement techniques is the CTV [79], which can provide the distribution of the field-induced velocity of micron-sized particles moving in an isodynamic magnetic field. If the particles have a very narrow-size distribution and both particle size and fluid viscosity are known then the distribution of the magnetic susceptibility or magnetization may be obtained. Again, if the magnetic properties of the magnetic component of the particles are known, then the mass distribution of the magnetic material in the particles may be determined.

The direct measurement of field-induced velocity of magnetic nanoparticles is not possible using CTV because the particles are too small to be observed. Brownian motion might also interfere with velocity measurements. An indirect method has been used where a known number of nanoparticles are bound to the surface of larger beads using specific binding chemistry. The field-induced velocity of the labeled beads then provides the information of the strength of interaction of the nanoparticles with the field [80]. However, the relatively large viscous drag of the micron-sized beads requires a large number of nanoparticles per bead (of the order of 10^3–10^5) to induce measurable motion. Again, due to sampling theory, only a mean value for the magnetic moment may be obtained. The distribution in velocity of the labeled beads would be due to the variation in the number of sites on the beads rather than the variation in the magnetic properties of the nanoparticles.

12.4.1. Field-flow fractionation

The two closely related separation techniques, FFF [81–84] and SPLITT fractionation [85–88], were invented by J. Calvin Giddings at the University of Utah. The initial concept for each of the systems was based on separation within a fluid that is driven along a thin, planar, parallel-walled channel with a field applied across the thin dimension of the channel, perpendicular to the flow. The majority of systems still utilize channels having such geometry. The materials separated range from polymeric materials in solution, through colloids, up to particles of tens of microns in diameter.

FFF is a zonal separation technique that is similar to chromatography in operation. A small sample is introduced into a flow of carrier fluid entering the FFF channel. As mentioned above, the channel is a thin, parallel-walled closed duct, across the thickness of which a field of some nature is applied. However, the channel need not be planar, as we shall see later, and the optimum channel geometry depends on the nature of the field employed. The sample is separated into its components as they are carried along the length of the channel at differing velocities [81, 83, 89]. As in chromatography, a detector may be placed at the channel outlet to register the elution of the sample components from the channel at different times.

There are two principal modes of operation for FFF: the so-called normal mode and the steric mode. The normal mode is referred to as such because it corresponds to the initial concept for the FFF separation mechanism [81]. This mode was well established by the time steric operation was demonstrated [90–92]. In both modes of operation the particles or macromolecules are driven by the field across the channel thickness to accumulate next one of the walls, termed the accumulation wall. In the case of the normal mode, the particles or macromolecules are small enough that there is a significant diffusion away from this region of increased concentration next to the wall. A dynamic equilibrium is established where transport toward the wall due to interaction with the applied field is balanced by this back-diffusion. Such dynamic equilibrium distributions are shown in Fig. 12.14 for two monodisperse sample components A and B. There is a constant exchange of particles within each zone thickness, and all particles sample fluid streamlines in the parabolic fluid velocity profile according to a weighting function that varies in parallel with the concentration profile. A monodisperse set of particles consequently moves along the channel as a coherent zone at this averaged velocity. Particles that interact strongly with the field form thin zones adjacent to the accumulation wall, while those that interact less strongly form

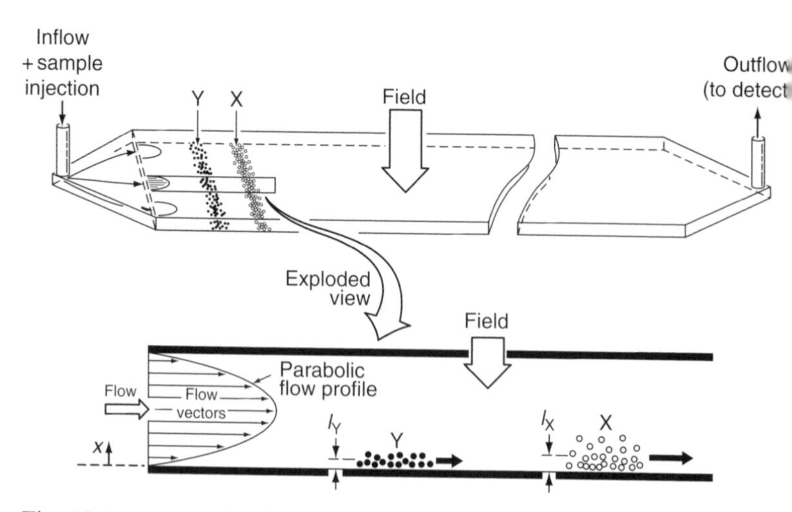

Fig. 12.14. Schematic of a field-flow fractionation channel. The expanded edge-on view illustrates the mechanism of differential elution velocity. The more diffuse distribution for the particles X can sample faster flow vectors as well as the slow vectors close to the wall, and the zone has a higher mean elution velocity than zone Y.

thicker zones. Particles occupying thicker zones sample faster streamlines further from the wall as well as those adjacent to the wall, and their averaged velocities are therefore higher.

Figure 12.14 illustrates a separation of two monodisperse sample components. Component B, having the thicker zone, migrates more quickly than component A. The result is that particles elute in the order of increasing elution time with increasing strength of interaction with the field. This is equally true for polydisperse samples where the shape of the elution curve (particle concentration as a function of elution time) is a reflection of the polydispersity.

For the analysis of a sample it is common practice to stop the fluid flow immediately after sample introduction to the channel to allow the sample components to approach their steady state distributions next to the accumulation wall. The flow is then resumed to carry out the separation. With the imposition of the sheared flow on a particle distribution next to the accumulation wall, there is inevitably a small departure from the equilibrium distribution under stopped flow conditions. This is manifested as the so-called nonequilibrium contribution to zone spreading [93, 94], which will be discussed later. In a well-designed FFF system this tends to be the dominant contribution to zone spreading.

Particles larger than about 1 μm in diameter are separated in the steric mode of FFF, so-called because it is the physical size of the particles that primarily determines how far they protrude into the fluid stream adjacent to the accumulation wall. Particles of such a size exhibit negligible diffusional transport. Consequently, larger particles protrude into faster streamlines and elute before smaller ones. In steric elution, particles of a given size are in fact held firmly at some fixed distance from the accumulation wall by a balance of the force due to interaction with the applied field and an opposing hydrodynamic lift force [95–99]. The range of different stream velocities sampled during elution is therefore negligible, and there is no significant nonequilibrium contribution to zone spreading. Consequently, steric separations tend to be very efficient and can be obtained very quickly [96, 100–102].

12.4.2. Separation in parallel plate channels

Most implementations of both FFF and SPLITT fractionation have employed plane, parallel-plate geometry. The first FFF channels were constructed using two flat-faced blocks, clamped around a thin spacer (usually Teflon, Mylar, or polyimide) out of which the channel outline had been cut. Tubes through the blocks conveyed the carrier solution to and from the ends of the channel. Planar channels are suited to those fields that can be applied uniformly across the thickness of the channel. Such channels are still in use for gravitational, thermal, electrical, and flow FFF. The design was modified for sedimentation FFF where the channel is wrapped around a centrifuge basket. However, the centrifugal field in this case still acts uniformly across the channel thickness.

The first implementation of SPLITT fractionation used a channel formed by spacers and stream splitters placed between flat glass plates. A schematic of the system is shown in Fig. 12.15, where the thickness has been expanded in order to better illustrate the mechanism of separation. In the ideal implementation the field acts to drive the particles at constant velocities across the planar channel thickness. In parallel plate SPLITT fractionation, the virtual surfaces between fluid elements entering inlets **a**′ and **b**′ and between fluid elements conveyed to different outlets are planar. They are referred to as the inlet and outlet splitting planes (ISP and OSP, respectively). It is important that the splitting planes are parallel to each other and to the channel walls. The transport distance between the ISP and the OSP is then constant throughout the system, and this is necessary for optimum resolution of separations. Fractionation of glass microspheres according to size was achieved using such a channel and the Earth's gravitational field [86, 87]. Many other successful applications of gravitational SPLITT followed [103–106]. A parallel-plate electrical SPLITT channel has also been used for separation of charged species [107, 108]. The same principles apply to centrifugal SPLITT fractionation [109–111]. As in the case of sedimentation FFF, the channel remains

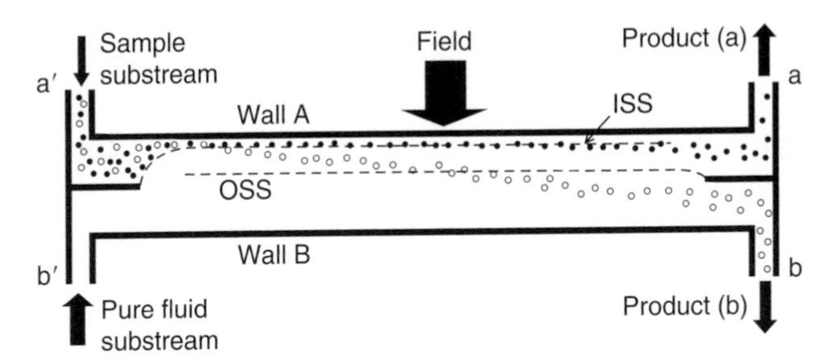

Fig. 12.15. Schematic of SPLITT channel.

essentially parallel walled, but is wrapped around a custom-designed centrifuge basket. The carrier input and output tubes are attached to the channel via rotating seals at the centrifuge axis. Sample throughput increases in proportion to transverse migration velocity and therefore to field strength. The increased field strength obtainable with a centrifuge is therefore beneficial for all separations that exploit differences in sedimentation velocities. Centrifugal SPLITT is also applicable to particulate samples that are either too small or do not differ sufficiently in density from the suspending solution for gravitational SPLITT to be suitable.

A parallel plate magnetic SPLITT system has also been successfully demonstrated [112–115]. The channel was placed parallel to the gap of a dipole magnet. The length and breadth of the gap for most of the experiments was equal to the length and breadth of the channel (10 cm and 5 mm, respectively). The parallel-plate geometry does allow for adjustment of the magnetic field strength and field gradient by varying the distance of the channel from the dipole gap. This flexibility was utilized to optimize the separations, but the variation in field strength and field gradient across the channel breadth must have been considerable when the magnet was close to the channel.

A parallel-plate magnetic SPLITT fractionation system has also been proposed that uses a high-gradient magnetic field approach to draw paramagnetic nanoparticles across the channel thickness

[116]. It is referred to as high-gradient magnetic field (HGMF)–SPLITT fractionation. Finely milled stainless steel wool powder was applied to one wall of the SPLITT channel and the channel placed in the warm bore of a superconducting magnet that could generate uniform fields of up to 8 T. The variation of magnetic field across the channel thickness is far from the ideal isodynamic field or a constant gradient field. As in high-gradient magnetic separator (HGMS) batch separators, the field varies very rapidly with distance from the steel wool powder substrate, and it is not surprising that the system tended to trap particles rather than separating them in a flow through mode of operation. If the external field is reduced so as to lessen the tendency to trap particles, then those that enter the channel further from the high-gradient wall are too weakly influenced by the field to migrate across the transport region between the ISP and the OSP.

12.4.3. Separation in cylindrical and annular channels

The principles of SPLITT fractionation have been carried over to cylindrical or annular geometry for axisymmetric, radial fields. The channel must be cylindrical or, more commonly, annular and the stream splitters cylindrical. Channel walls and splitters must be concentric to ensure uniform initial sample lamina thickness and constant migration distances between inlet and outlet splitting surfaces. Ideally, these splitting surfaces are cylindrical and are referred to as the inlet and outlet splitting cylinders (ISC and OSC, respectively). The radial field must drive sample components across the channel thickness either toward the central axis or toward the outer wall. In the case of outward migration, the channel must be of annular geometry since the field would necessarily decay to zero at the axis, and sample components must be exposed to finite field strength at their initial placement. To date, the only successful implementation of this type of SPLITT geometry is the QMS [68, 117–123]. The theory developed for optimization of flow rate conditions for the transport mode of planar SPLITT fractionation was adapted to optimization of cell separation in the QMS [68, 118, 122].

To implement FFF, the field must act radially outward in the case of cylindrical geometry, and radially in either direction for annular systems. The approach to zero field strength at the axis of a cylindrical channel presents an impediment to complete sample relaxation to the accumulation wall. Nevertheless, almost all of the experimental work to date using radial fields has involved cylindrical channels. This is due to the ready availability of very uniform porous hollow fibers. They have been used for an early implementation of electrical FFF [124], but otherwise exclusively for a type of flow FFF [125–133]. Annular systems have been the subject of several theoretical studies [134–137] but, previous to the development of quadrupole magnetic FFF, just one experimental work involving electrical FFF [138].

The development of SPLITT fractionation took place almost two decades after the initial work in FFF. The considerable experience gained in parallel-plate FFF over this period allowed the rapid development of parallel-plate SPLITT systems. In many cases the FFF channels could be adapted for SPLITT operation simply by including the stream splitters and an additional spacer in the assembly, and the addition of a second inlet and outlet. The inclusion of two additional pumps and a second detector completed the adaptation of an FFF system for SPLITT operation. Similarly, the experience gained in the development of the QMS contributed to the relatively rapid development of quadrupole magnetic FFF. The experience in design of other forms of FFF directed the modification of the QMS channel to obtain an FFF channel. The splitters had to be removed along with one each of the inlets and outlets. The extra-channel dead volumes of the inlet and outlet manifolds had to be reduced. While this dead volume is irrelevant to SPLITT operation, it can contribute significantly to zone broadening in FFF. As will be explained later, it was decided that the use of a helical channel held many advantages over a simple annular channel. Finally, as will be evident from the following discussion, the ability to program a magnetic field decay during sample elution required the design of a quadrupole electromagnet.

12.4.4. *Earlier implementations of magnetic FFF*

There have been some previous attempts at developing MgFFF but, for various reasons, they have not resulted in much success. Vickrey and Garcia-Ramirez [139] used a tubular channel with a transverse field gradient that is far from an ideal combination [84]. A 304-cm Teflon tube of 0.15 cm internal diameter was coiled against the windings of a 400-G (0.04 T) solenoid, close to the plane of one end of the iron core. Field strength and gradient therefore varied throughout the system, and were too weak in any case to influence the retention of their Ni–BSA complex sample.

Schunk et al. [140, 141] used a parallel-plate channel with a magnetic field provided by an electromagnet. The channel was made of 0.5-in.-thick (1.27 cm) and 0.125-in.-thick (0.32 cm) glass plates separated by a 250-μm spacer. The thinner wall was placed in contact with one end of the iron core of the magnet to maximize field strength and field gradient in the channel. A maximum field strength of about 275 G (0.0275 T) and field gradient of about 21 G/mm (2.1 T/m) were obtainable, with less than 1% variation in field strength along the length of the channel. They were able to separate singlet from doublet 0.8-μm rod-shaped iron oxide particles used in the recording industry (i.e., fairly large, high susceptibility particles). The relatively weak field strength and field gradient would be insufficient to retain particles that are much smaller than those separated.

In 1986, Semenov and Kuznetzov [142] suggested a very different approach to the design of a MgFFF system. They proposed that a ferromagnetic wire be placed at the axis of a tubular channel that was mounted perpendicular to a uniform magnetic field. The field would then magnetize both the wire and the particles to be separated. They presented calculations that suggested that retention of both paramagnetic and diamagnetic particles would be possible in such a system. The coaxial channel geometry is also far from ideal for FFF. The force on retained particles would rapidly increase as the wire is approached, which would tend to induce particle capture. Also, a fraction of the small surface of the wire would have to serve as the accumulation wall, which would make the system highly susceptible to overloading. In the same year, Semenov [143] proposed a parallel-plate MgFFF channel in which the field would be provided by the induced magnetization of parallel ferromagnetic wires arrayed uniformly in the surface of one of the walls. The wires were to lie in the direction of flow and the field was to be applied across the channel breadth, perpendicular to the wires and the flow direction. He presented a theoretical study of the proposed system and concluded that it would be more efficient than the coaxial system. Such an arrangement would help to alleviate overloading, although the strong spatial variation in force on the particles would remain. It will be appreciated that the use of ferromagnetic wires in a magnetic field results in very short range HGMF that are more suited to particle capture as in the HGMS and HGMF-SPLITT systems. The rapid variation of field gradient adjacent to the ferromagnetic wires is not conducive to the sustaining of steady state diffusive zones of magnetic particles.

Also in 1986, Mori [144] attempted to demonstrate MgFFF using a system very similar to that of Vickrey and Garcia-Ramirez [139]. He used a Teflon tube of 0.5 mm internal diameter and length 270 cm, 210 cm of which was coiled against a pole of a 9000-G (0.9 T) electromagnet. Very slight retardation of Ni^{2+} complexes with bovine serum albumin (BSA), egg albumin, and ethylene diamine tetraacetic acid (EDTA) was observed. Although Mori used a stronger field strength than in the earlier system [139], and the tube was placed against the pole piece rather than the electrical coil, the field gradient would not have been very high and the tubular channel geometry with transverse field gradient is not suited to FFF.

From 1994, Ohara and coworkers [145–153] have pursued an approach to MgFFF similar in concept to that proposed by Semenov [143]. The majority [145–152] of their publications describe theoretical studies and simulations. Tsukamoto et al. [146] considered a set of stacked channels with ferromagnetic wires embedded in the dividing walls. The stacking of channels was to maximize throughput of the system. The wires were assumed to be parallel to one another and in the direction of fluid flow. An external magnetic field was to be applied across the channel thicknesses, perpendicular to the wires and to the fluid flow. Note that the direction of applied field differs from that proposed by Semenov [143]. The application of field across channel thickness would generate very strong, localized field gradients at the channel wall, directly above the wires. Application of the field across the channel breadth would result in a repulsive force on paramagnetic particles directly above the wires, and an attraction to the broader regions between the wires. This would mitigate the tendency for overloading to some extent. It was assumed that wires of 10-μm diameter would be embedded in 10-μm-thick walls with 10-μm-thick channels between them, and that a 10-T magnetic field would be applied. The feasibility of separating 20-nm particles of uranium from plutonium was examined. It is apparent from this, and from following publications, that the system would suffer from the very localized distribution in concentration close to the wires due to the rapidly increasing field gradient as the wires are approached. Subsequent papers by Ohara and coworkers [147–153] consider single channel systems, and refer to the technique as magnetic chromatography, even though no partition between phases is involved. Regularly spaced wires were assumed to be embedded within the two walls of the channel, each wire being matched by another directly across the channel thickness. In 2002, Mitsuhashi et al. [153] reported an experimental implementation. The channel was 45- cm long, 2.7-cm wide, and 190-μm thick. Wires were embedded in both channel walls and lay in the direction of fluid flow. They were of rectangular cross section, being 150-μm broad and 30-μm thick. They were embedded into the walls to a depth of 30 μm, so that they were flush with the wall surfaces, and spaced 150 μm apart. Rather than the wires of one wall being arranged exactly opposite the wires of the other, they were offset by 150 μm so that wires in each wall lay opposite spaces between wires of the other. An external field of 3 T was applied. They were able to show slight retention of some transition metal salts, although band spreading was extremely high.

A theoretical study of a system referred to as MgFFF was presented in 2001 [154]. However, although the system involved a flow of fluid in a parallel-plate channel with a transverse magnetic field, the model did not describe FFF in its accepted sense. The modeling was more relevant to SPLITT fractionation in transport mode that included particle diffusion. The convection and dispersion of submicron paramagnetic particles were calculated as they were carried across the channel thickness. No modeling of the magnetic field was included, and the magnetic field was assumed to be simply isodynamic.

Another attempt at implementing MgFFF using capillary tubular channel with transverse magnetic field has recently been presented [155]. A short, cylindrical NdFeB permanent magnet (3.8-cm long and 7.6 cm in diameter) was used. A fused-silica capillary tube of 250 μm internal diameter was wrapped three times around the beveled edge of one circular pole of the magnet, giving a total channel length of 71.8 cm under the influence of the field. The field strength was measured to be 0.36 T at the beveled edge of the pole, and the field gradient may be estimated to be around 230 T/m (based on the reported fall of field strength by about 0.080 T across the outer tube diameter of 350 μm). The magnetic field was therefore much higher than that used by Vickrey and Garcia-Ramirez [139] and the field gradient much greater than in the systems of either Vickrey and Garcia-Ramirez [139] or Mori [144]. Samples of maghemite (mean diameters of 3.5, 6.2, and 7.8 nm) and $CoFe_2O_4$ (8.6 and 12.5 nm) in hexane suspension were eluted through the system. Only the 12.5-nm $CoFe_2O_4$ particles were significantly retained relative to the suspending fluid, corresponding to a retention ratio of around 0.6 (see Section 12.4.7.3). Smaller particles, being little influenced by the magnetic field, eluted close to the void time. The authors were therefore able to demonstrate a separation of retained 12.5-nm $CoFe_2O_4$ from non-retained 6.2-nm maghemite particles.

12.4.5. Interaction of particles with fields

In general, particles may be characterized in terms of the strength of their interaction with a field; for example, a particle's mass determines the strength of its interaction with a gravitational or centrifugal field; the electric charge on a particle determines the strength of its interaction with an electric field. Following the terminology of FFF [156], this relationship may be written in the following general form:

$$\mathbf{F} = \phi \mathbf{S} \tag{12.22}$$

where **F** is the force exerted on the particle, **S** is a quantitative measure of the field, and ϕ is a so-called particle–field interaction parameter. [Note that **F** and **S** are vector quantities depicted in bold face, and that in the Eq. (12.22) the force **F** is assumed to act in the same direction as **S**.] In the case of a centrifugal field, ϕ represents the particle buoyant mass and **S** the acceleration.

As explained in Chapter 3, a similar approach may be employed to describe particle interaction with magnetic fields, although the interaction may be a little more complicated than that of buoyant mass with acceleration fields (in the Newtonian domain). This is because the magnetic dipole moment of the particle, which is the particle property that interacts with the magnetic field, may be induced by the field itself. Just as buoyant mass may be reduced to a per unit volume quantity (density difference between the particle and the suspending medium), magnetic dipole moment may be reduced to per unit volume quantity (the particle magnetization **M**).

12.4.6. Force on particles in a magnetic field

The force $\mathbf{F}_m$ acting on a particle of volume V_m and induced magnetization $\mathbf{M}$ due to a magnetic induction $\mathbf{B}$ is given by

$$\mathbf{F}_m = V_m(\mathbf{M} \cdot \nabla)\mathbf{B} \tag{12.23}$$

Note that $\mathbf{F}_m$, $\mathbf{M}$, and $\mathbf{B}$ are vector quantities and are therefore depicted in bold face. In the case of a paramagnetic material, magnetization $\mathbf{M}$ varies linearly with magnetic field $\mathbf{H}$, as described by the equation $\mathbf{M} = \chi_m \mathbf{H}$, in which χ_m is the volumetric magnetic susceptibility of the material and is a constant (typically between 10^{-3} and 10^{-5}). In the case of a ferromagnetic or superparamagnetic material, magnetization approaches an upper limit called the saturation magnetization $\mathbf{M}_s$, so that χ_m is not constant. The value of χ_m is much greater than 1 for ferromagnetic materials, and typically can have values between 50 and 100,000. Particles may be suspended in a weakly diamagnetic medium such as an aqueous electrolyte solution which has a negative susceptibility of order 10^{-5}. In this case, $\mathbf{H} = \mathbf{B}/\mu_0(1 + \chi_f) \approx \mathbf{B}/\mu_0$, where χ_f is the volumetric susceptibility of the medium and μ_0 is the magnetic permeability of free space, and we obtain

$$\mathbf{F}_m = \frac{V_m \chi_m}{\mu_0}(\mathbf{B} \cdot \nabla)\mathbf{B} = \frac{V_m \chi_m}{2\mu_0}\nabla(\mathbf{B} \cdot \mathbf{B}) = V_m \chi_m \frac{\nabla B^2}{2\mu_0} \tag{12.24}$$

Here we assumed that the magnetic particles are free to rotate in the magnetic field. The force is therefore dependent on the product $V_m \chi_m$ of the particle, and $\nabla B^2/2\mu_0$ describing the magnetic field. In the terminology of FFF, force is given by the product of a particle–field interaction parameter ϕ and a measure of the field strength $\mathbf{S}$. A direct or indirect measure of the force on a particle situated in a magnetic field of known $\nabla B^2/2\mu_0$, representing $\mathbf{S}_m$, therefore yields a value for $V_m \chi_m$, which is simply the particle–magnetic field interaction parameter ϕ_m.

It was mentioned earlier that very small particles of ferromagnetic material that are sub-single domain in size (i.e., all the magnetic moments tend to be aligned with one another) behave superparamagnetically. They attain a saturation magnetization $\mathbf{M}_s$ that aligns with $\mathbf{B}$ at relatively modest field strengths. In this case, the magnetization is a nonlinear function of field strength, and the force is given by

$$\mathbf{F}_m = V_m(\mathbf{M} \cdot \nabla)\mathbf{B} = V_m M \nabla B = \frac{V_m M}{H} \frac{\nabla B^2}{2\mu_0} \tag{12.25}$$

At high field strengths, where magnetization approaches a constant saturation value, the force varies with ∇B rather than with ∇B^2. If we assume our measure of field strength $\mathbf{S}_m$ still corresponds to $\nabla B^2/2\mu_0$, then $\phi_m = V_m M_s/H$, and ϕ_m is not constant.

In all of the systems considered here, the particulate materials are in a fluid suspension, so that any measure of a force exerted should strictly take into account the magnetic property of the suspending fluid. Magnetic labels are composites of different materials. There must be magnetic and antibody components to the label, but there can be other components too. The magnetic component may be dispersed within another material, or may be distributed on the surface of a polymer bead. There may be a stabilizing coating on the magnetic material to prevent aggregation of the suspended beads or to prevent their chemical breakdown. If the label is to have any utility, the strongly magnetic component will dominate the overall value of ϕ, but the magnetic properties of the other components should also be considered. (All materials have some degree of magnetic susceptibility, and we shall use the term "weakly magnetic" to refer to these other materials.) For a composite particle having a (strongly) paramagnetic component, it may be shown that

$$\mathbf{F}_m = V_p(\varphi_m \chi_m + \varphi_w \chi_w - \chi_f)\frac{\nabla B^2}{2\mu_0} = V_p(\chi_p - \chi_f)\frac{\nabla B^2}{2\mu_0} = V_p \Delta\chi \frac{\nabla B^2}{2\mu_0} \tag{12.26}$$

where V_p is the volume of the particle, φ_m and φ_w are the volume fractions of the strongly magnetic and weakly magnetic components, χ_m, χ_w, and χ_f are the susceptibilities of the strongly and weakly magnetic components and the fluid, respectively, χ_p is the overall susceptibility of the particle, and $\Delta\chi$ is the difference in susceptibility of the particle and the suspending fluid. Note that we gather all weakly magnetic components together in the interests of simplicity, and their net behavior may be paramagnetic or diamagnetic. If both $\varphi_w\chi_w$ and χ_f are negligible compared to $\varphi_m\chi_m$ then the force equation reduces to

$$\mathbf{F}_m = V_p\varphi_m\chi_m \frac{\nabla B^2}{2\mu_0} = V_m\chi_m \frac{\nabla B^2}{2\mu_0} \tag{12.27}$$

where V_m is the volume of the magnetic component in the particle. If the magnetic component is superparamagnetic, and exhibits non-linear dependence of magnetization on field strength, the force is given by

$$\begin{aligned} \mathbf{F}_m &= V_p(\varphi_m M + \varphi_w\chi_w H - \chi_f H)\nabla B \\ &= V_m M\nabla B + V_p(\varphi_w\chi_w - \chi_f)\frac{\nabla B^2}{2\mu_0} \end{aligned} \tag{12.28}$$

Again, if both $\varphi_w\chi_w$ and χ_f are negligible then this equation reduces to

$$\mathbf{F}_m = V_m M\nabla B \tag{12.29}$$

where again V_m is the volume of the magnetic component in the particle. In both cases, retention time will tend to increase with the volume of magnetic component, or the cube of its effective spherical diameter.

12.4.7. Theory for retention in quadrupole MgFFF

A general theory for retention in longitudinal flow along an annular channel has already been presented by Davis and Giddings [157]. They were able to derive approximate analytical solutions for a field strength that varies with radial distance from the axis raised to an arbitrary power. The solutions correspond to limiting behavior at high retention.

12.4.7.1. Velocity profile in an annulus

The fluid velocity v_z along the length of an annular channel as a function of dimensionless radial distance ρ from the axis is described by the equation [158]

$$v_z(\rho) = \frac{2\langle v_z \rangle}{A_1}[\ln \rho + A_2(1 - \rho^2)] \qquad (12.30)$$

where ρ is the ratio of radial distance r from the axis to the radius r_o of the outer wall of the annulus, A_1 and A_2 are functions of ρ_i, the ratio of the radius r_i of the inner wall to r_o. These are given by

$$A_1 = \frac{\rho_i^2 + 1}{\rho_i^2 - 1}\ln \rho_i - 1; \quad A_2 = \frac{\ln \rho_i}{\rho_i^2 - 1} \qquad (12.31)$$

Note that the form of Eq. (12.30) differs from that given in some other publications [68, 118, 158] and the definitions of A_1 and A_2 consequently differ from those given in these other references. The form of Eqs. (12.30) and (12.31) were rearranged for consistency with the following cases. Angular fluid velocity ω is given by the equation

$$\omega = \frac{2\langle \omega \rangle}{B_1}\left(\ln \rho + B_2\left(1 - \frac{1}{\rho^2}\right)\right) \qquad (12.32)$$

with B_1 and B_2 given by

$$B_1 = \frac{4\rho_i^2(\ln \rho_i)^2}{(1 - \rho_i^2)^2} - 1; \quad B_2 = \frac{\rho_i^2 \ln \rho_i}{1 - \rho_i^2} \qquad (12.33)$$

The azimuthal fluid velocity v_θ is given by

$$v_\theta = \frac{2\langle v_a \rangle}{C_1}\rho\left(\ln \rho + C_2\left(1 - \frac{1}{\rho^2}\right)\right) \qquad (12.34)$$

in which the constants C_1 and C_2 are given by

$$C_1 = \frac{4}{(\rho_i^2 - 1)}\left(\frac{2}{3}\frac{\rho_i^2 \ln \rho_i}{1+\rho_i} + \frac{1-\rho_i^3}{9}\right); \quad C_2 = B_2 = \frac{\rho_i^2 \ln \rho_i}{(1-\rho_i^2)} \tag{12.35}$$

The channel used for quadrupole MgFFF occupies an annular space that is axisymmetric with the quadrupole field. It can occupy the full annulus in which the fluid flows along the length of the channel and the velocity profile is described by Eq. (12.30). For reasons to be explained later, the channel may alternatively follow a helical path within the annular space. In the latter case, the velocity profile will correspond more closely to Eqs. (12.32) and (12.34). Each of the velocity profiles can be approximated by a cubic equation in the radial distance from the accumulation wall. For example, for the longitudinal velocity:

$$v_z = 6\langle v_z\rangle[(1+\upsilon_z)\xi - (1+\mu\upsilon_z)\xi^2 + (\mu - 1)\upsilon_z\xi^3] \tag{12.36}$$

in which

$$\mu = \frac{7+8\rho_i}{2+3\rho_i} \tag{12.37}$$

and υ_z is a parameter that is included to match the first derivatives of the velocity profiles at the accumulation wall. Equivalent equations to Eq. (12.36) may be written for both v_ω and v_θ:

$$v_\omega = 6\langle v_\omega\rangle[(1+\upsilon_\omega)\xi - (1+\mu\upsilon_\omega)\xi^2 + (\mu - 1)\upsilon_\omega\xi^3] \tag{12.38}$$

$$v_\theta = 6\langle v_\theta\rangle[(1+\upsilon_\theta)\xi - (1+\mu\upsilon_\theta)\xi^2 + (\mu - 1)\upsilon_\theta\xi^3] \tag{12.39}$$

The value for μ remains unchanged in these equations, and is given by Eq. (12.37). Analytical solutions may be found for the three fitting parameters:

$$\upsilon_z = \frac{(\rho_i - 1)(1-2A_2)}{3A_1} \tag{12.40}$$

$$v_{\omega} = \frac{(\rho_i - 1)(1 + 2B_2)}{3B_1} \qquad (12.41)$$

$$v_{\theta} = \frac{(\rho_i - 1)(1 + 2C_2)}{3C_1} \qquad (12.42)$$

12.4.7.2. Concentration profile in an annulus with quadrupole field
The concentration profile depends on the paramagnetic or superparamagnetic properties of the particles and the variation of B within the annulus. In a quadrupole magnetic system magnetic field intensity B is axisymmetric and it increases linearly with distance from the axis. This may be expressed by the equation $B = B_o r/r_o = B_o \rho$, where B_o is the field intensity at the outer wall. The field gradient ∇B is constant within the aperture of the quadrupole. It is directed radially outward from the axis and has a value of B_o/r_o. This is quite different to the rapid variation of field gradient found in HGMS devices which are designed to trap magnetic particles. The constant field gradient in the quadrupole allows for the maintenance of a steady-state distribution of diffusing magnetic nanoparticles adjacent to a channel wall which is a requirement for elution in the normal mode of FFF.

A steady-state particle concentration profile is obtained when the field driven flux of particles toward the outer wall is balanced by diffusion away from the resultant region of higher concentration. The radial flux of particles at radial distance r in an axisymmetric field is given by

$$J_r = uc - D\frac{dc}{dr} \qquad (12.43)$$

where u and c are the local field-induced particle velocity and particle concentration, respectively, D is the particle diffusion coefficient (assumed to be constant for the range of concentrations expected), and dc/dr is the local concentration gradient. At steady state, the radial flux must be zero at all r, so that

$$\frac{\mathrm{d}c}{\mathrm{d}r} = \frac{uc}{D} = \frac{F_{\mathrm{m}}c}{kT} \tag{12.44}$$

The last form of the expression for $\mathrm{d}c/\mathrm{d}r$ is obtained by replacing D with the Stokes–Einstein expression kT/f, and u with F_{m}/f, where k is the Boltzmann constant, T is the absolute temperature, F_{m} is the force on a single particle due to its interaction with the field, and f is the particle friction coefficient. The steady-state concentration profile is obtained by rearranging this equation and integrating across the annular thickness

$$\int_{c_0}^{c} \frac{\mathrm{d}c}{c} = \frac{1}{kT} \int_{r_{\mathrm{i}}}^{r} F_{\mathrm{m}} \mathrm{d}r = \frac{r_{\mathrm{o}}}{kT} \int_{\rho_{\mathrm{i}}}^{\rho} F_{\mathrm{m}} \mathrm{d}\rho \tag{12.45}$$

where $\rho_{\mathrm{i}} = r_{\mathrm{i}}/r_{\mathrm{o}}$.

For paramagnetic particles the radial force is proportional to their magnetization which increases linearly with distance from the axis, in parallel with the field strength. The local force F_{m} is related to the force at the accumulation wall by

$$F_{\mathrm{m}} = F_{\mathrm{mo}}\rho \tag{12.46}$$

Substituting Eq. (12.46) into Eq. (12.45) and integrating yields the result

$$c = c_{\mathrm{o}} \exp\left(-\frac{1}{\lambda}\frac{1-\rho^2}{1-\rho_{\mathrm{i}}^2}\right) \tag{12.47}$$

where λ is the so-called FFF retention parameter and is the ratio of thermal energy kT and the work W required to drive a particle across the channel thickness against the influence of the field, that is,

$$\lambda = \frac{kT}{W} = \frac{2kT}{F_{\mathrm{mo}} r_{\mathrm{o}} (1-\rho_{\mathrm{i}}^2)} \tag{12.48}$$

In the case of magnetically saturated particles, the force is constant across the channel thickness, and the integration of Eq. (12.45) results in the following concentration profile

$$c = c_o \exp\left(-\frac{1}{\lambda}\frac{(1-\rho)}{(1-\rho_i)}\right) = c_o \exp\left(-\frac{\xi}{\lambda}\right) \quad (12.49)$$

in which c_o is the concentration at the outer wall, and ξ is the ratio of distance x from the accumulation wall and the channel thickness w. In this case, λ is given by

$$\lambda = \frac{kT}{W} = \frac{kT}{F_m r_o (1-\rho_i)} = \frac{kT}{F_m w} \quad (12.50)$$

In practice, particles will not generally be either paramagnetic or magnetically saturated. As will be explained later, the field strength is often gradually reduced during the elution of a particulate sample, and particles which may be close to magnetically saturated at the initial field strength may be far from saturation when they elute from the channel. Also to be taken into account is the fact that particles are confined to a relatively small fraction of the thin annular channel thickness, and the difference between the concentration profiles will be relatively subtle. Therefore, the concentration profile may be approximated by the following equation

$$c = c_0 \exp\left(-\frac{1}{\lambda_0}\frac{1-\rho}{1-\rho_i}\right) = c_0 \exp\left(-\frac{\xi}{\lambda_0}\right) \quad (12.51)$$

where c_0 is the particle concentration at the accumulation wall (where $x = 0$), and λ_0 is the value of λ under conditions at the accumulation wall, that is,

$$\lambda_0 = \frac{kT}{F_{mo} w} = \frac{kT}{V_m M(B_o) \nabla B} \quad (12.52)$$

in which $M(B_o)$ is the particle magnetization at the field strength adjacent to the accumulation wall. The symbols c_0 and λ_0 in Eqs. (12.51) and (12.52) have been given the subscript "0" for consistency with the FFF literature where the symbols refer to conditions at the accumulation wall where $x = 0$. For the annular channel, the position corresponding to $x = 0$ is of course the same as that where $r = r_o$. In the remaining part of this chapter, the retention parameter λ will correspond to the value of λ_0 as defined by Eq. (12.52).

12.4.7.3. Retention ratio

The retention ratio R for a sample of monodisperse particles is defined as the ratio of the time for the carrier fluid to pass along the length of the channel (the void time) to the time for the sample zone to migrate along the channel (the elution time). A monodisperse collection of particles migrates as a coherent zone because of the relatively fast exchange of particles throughout the zone thickness. Each particle samples the range of stream velocities many times as the zone moves along the channel. Zone velocity therefore corresponds to a weighted average of the stream velocities over the channel cross section, the weighting being proportional to local particle concentration. For a parallel-plate channel the retention ratio is given by the equation:

$$R = \frac{\langle cv \rangle}{\langle c \rangle \langle v \rangle} \tag{12.53}$$

where the angle brackets represent an averaging over the channel cross section. Therefore,

$$< c >= \frac{\int_0^w c \mathrm{d}x}{\int_0^w \mathrm{d}x} = \frac{1}{w} \int_0^w c \mathrm{d}x \tag{12.54}$$

Equivalent expressions yield values for $\langle v \rangle$ and $\langle cv \rangle$. For an annular channel, the averaging must be carried out over an element of the channel cross-sectional volume, taking into account the increasing contribution to volume with radial distance from the axis [157]. The averaged concentration, for example, is given by

$$\langle c \rangle = \frac{2\pi \int_{r_o}^{r_i} rc\mathrm{d}r}{2\pi \int_{r_o}^{r_i} r\mathrm{d}r} = \frac{2\int_1^{\rho_i} \rho c \mathrm{d}\rho}{(1-\rho_i^2)} \tag{12.55}$$

It was mentioned earlier that depending on the design of the channel, the flow along the channel may be longitudinal or approximately azimuthal. The resulting expression for retention ratio will depend on the flow direction and the two extreme cases are therefore considered below.

Longitudinal flow. The retention ratio for migration along the length of an annulus may be written as

$$R_z = \frac{\langle c v_z \rangle}{\langle c \rangle \langle v_z \rangle} \tag{12.56}$$

Therefore we have

$$R_z = \frac{\int_1^{\rho_i} \rho c(\rho) v_z(\rho) \mathrm{d}\rho}{\langle v_z \rangle \int_1^{\rho_i} \rho c(\rho) \mathrm{d}\rho} = \frac{\int_0^1 [1-\xi(1-\rho_i)] c(\xi) v_z(\xi) \mathrm{d}\xi}{\langle v_z \rangle \int_0^1 [1-\xi(1-\rho_i)] c(\xi) \mathrm{d}\xi} \tag{12.57}$$

Substituting Eq. (12.51) for concentration profile and the cubic approximation Eq. (12.36) for longitudinal velocity profile, and integrating yields the following analytical solution:

$$\begin{aligned} R_z = {} & 6\lambda[(3\mu-8)(1+\upsilon_z)+\lambda(46-16\mu+6(5+\mu-\mu^2)\upsilon_z) \\ & -6\lambda^2(15-5\mu-2(4-13\mu+4\mu^2)\upsilon_z)-120\lambda^3(3-4\mu+\mu^2)\upsilon_z \\ & +\exp\left(\frac{-1}{\lambda}\right)\{(7-2\mu)(1+(2-\mu)\upsilon_z)+2\lambda(22-7\mu+(51-45\mu+9\mu^2)\upsilon_z) \\ & +6\lambda^2(15-5\mu+(52-54\mu+12\mu^2)\upsilon_z)+120\lambda^3(3-4\mu+\mu^2)\upsilon_z\}]/ \\ & [3\mu-8-5\lambda(\mu-3)-\exp(-1/\lambda)(7-2\mu-5\lambda(\mu-3))] \end{aligned} \tag{12.58}$$

For acceptably well retained materials where λ is less than about 0.15 or 0.20, the terms in $\exp(-1/\lambda)$ become insignificant and the expression reduces to the more manageable form

$$R_z = 6\lambda[(3\mu - 8)(1 + \upsilon_z) + \lambda(46 - 16\mu + 6(5 + \mu - \mu^2)\upsilon_z) \\ -6\lambda^2(15 - 5\mu - 2(4 - 13\mu + 4\mu^2)\upsilon_z) - 120\lambda^3(3 - 4\mu + \mu^2)\upsilon_z]/ \\ [3\mu - 8 - 5\lambda(\mu - 3)] \tag{12.59}$$

Azimuthal flow. In the case of azimuthal flow, the retention ratio is defined in terms of the angular fluid velocity [159]:

$$R_\omega = \frac{\langle c\omega\rangle}{\langle c\rangle\langle\omega\rangle} = \frac{\int_1^{\rho_i} c\omega\rho\,\mathrm{d}\rho}{\langle\omega\rangle\int_1^{\rho_i} c\rho\,\mathrm{d}\rho} \tag{12.60}$$

At this point we make the observation that the azimuthal velocity $v_\theta = \omega r = r_o\omega\rho$. It may also be shown that

$$\frac{1}{\langle\omega\rangle} = \frac{C_1}{B_1}\frac{r_o}{\langle v_\theta\rangle} \tag{12.61}$$

Making the appropriate substitutions and changing the variable of integration to ξ, we obtain

$$R_\omega = \frac{C_1}{B_1\langle v_\theta\rangle}\frac{\int_0^1 c(\xi)v_\theta(\xi)\mathrm{d}\xi}{\int_0^1 [1 - \xi(1 - \rho_i)]c(\xi)\mathrm{d}\xi} \tag{12.62}$$

This may be solved analytically on substituting the appropriate cubic approximation for v_θ:

$$R_\omega = \frac{6C_1\lambda(3\mu - 8)}{B_1}[1 + \upsilon_\theta - 2\lambda(1 + \mu\upsilon_\theta) + 6\lambda^2(\mu - 1)\upsilon_\theta \\ +\exp(-1/\lambda)\{1 + 2\upsilon_\theta - \mu\upsilon_\theta + 2\lambda(1 + 3\upsilon_\theta - 2\mu\upsilon_\theta) - 6\lambda^2(\mu - 1)\upsilon_\theta\}]/ \\ [3\mu - 8 - 5\lambda(\mu - 3) - \exp(-1/\lambda)\{7 - 2\mu - 5\lambda(\mu - 3)\}] \tag{12.63}$$

For sufficiently retained material, this reduces to the relatively simple form

$$R_{\omega} = \frac{6C_1\lambda(3\mu - 8)}{B_1} \frac{[1 + \upsilon_\theta - 2\lambda(1 + \mu\upsilon_\theta) + 6\lambda^2(\mu - 1)\upsilon_\theta]}{[3\mu - 8 - 5\lambda(\mu - 3)]} \tag{12.64}$$

Figure 12.16 shows the dependence of both R_z and R_ω on λ for (a) $\rho_i = 0.5$ and (b) $\rho_i = 0.9$. Also shown for comparison in each figure is the dependence of R on λ for a parallel-plate channel. The region of relevance to FFF analysis lies in the range of R from around 0.02 to around 0.5. The values of ρ_i were chosen to illustrate the differences between the retention ratios. For typical annular magnetic FFF channels ρ_i tends to be greater than 0.9, and the retention ratios more closely approach the parallel-plate solution.

12.4.7.4. Nonequilibrium band spreading

This contribution to band spreading arises out of the effect of the sheared fluid flow on the concentration distribution of a sample component. Those particles that are furthest from the accumulation

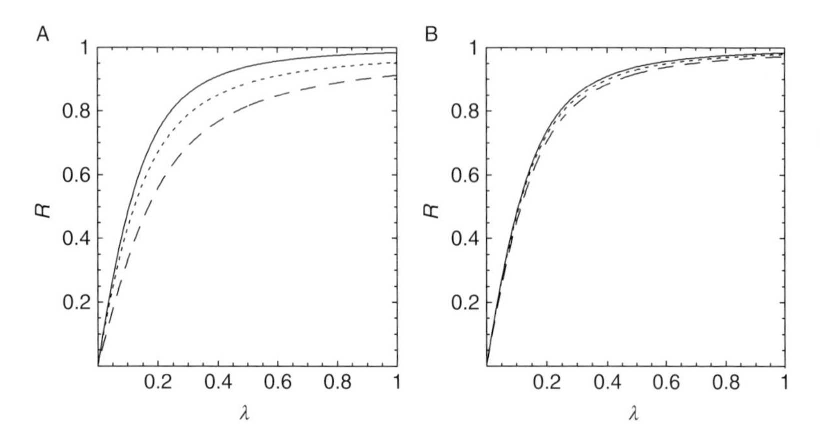

Fig. 12.16. The dependence of retention ratio R on the FFF retention parameter λ for inner to outer annulus radius ratios of (A) 0.5 and (B) 0.9. In each figure, the full line shows the dependence for a parallel-plate channel, the dotted line shows the dependence for longitudinal flow in an annular channel, and the dashed line shows the dependence for azimuthal flow.

wall are driven ahead of particles that are closer to the wall. The equilibrium concentration profile corresponding to stop-flow conditions is thereby disturbed throughout the zone. The effects of field and diffusion act to reestablish equilibrium, but this is not achieved instantaneously. Therefore under flowing conditions the concentration distribution is always disturbed to a greater or lesser degree.

For the ideal parallel-plate system, the nonequilibrium contribution to theoretical plate height is given by [93, 94]

$$H = \frac{Xw^2\langle v\rangle}{D} \tag{12.65}$$

where D is the diffusion coefficient of the sample particles, and X (we use the upper case X here, rather than the conventional FFF parameter X, to avoid confusion with magnetic susceptibility) is a function of λ. In the limiting case of high retention, $X \approx 24\lambda^3$. This contribution to plate height therefore increases rapidly with λ. In FFF, other contributions to plate height tend to be small, or can be made small for a well-designed system. High overall separation efficiency therefore requires that λ be small. Conversely, poor retention results in poor separation efficiency. Note that the particle diffusion coefficient depends on the overall size of the particle, which may or may not be related to the magnetic content of the particle.

In annular FFF, X is a function of λ, ρ_i, and the position dependence of the radial force. Davis [160] derived approximate expressions for X for longitudinal flow along an annular channel with dependence of field-induced migration velocity on radial distance from the axis raised to an arbitrary power. The solution for X will also dependent on the direction of flow in the annulus, with linear and azimuthal flow being the two extremes. For a given magnetic quadrupole FFF system, X therefore remains a strong function of λ, and poor retention will again result in poor separation efficiency. The detailed solutions for nonequilibrium contributions to band spreading in MgFFF are beyond the scope of the present discussion.

12.4.7.5. Data reduction

For both constant field and programmed operation, the elution time t_r is obtained by solving the integral equation

$$V^0 = \int_0^{t_r} R\dot{V}\mathrm{d}t \tag{12.66}$$

where V^0 is the channel volume, R is the retention ratio and $\dot{V}$ is the volumetric flow rate (either or even both of these may vary with time), and the integration is carried out over time from the start of elution to elution time t_r. The variance in elution time due to nonequilibrium effects is given by [161]

$$\sigma_t^2 = \frac{w^2 \int_0^{t_r} XR\dot{V}^2 \mathrm{d}t}{DR_r^2 \dot{V}_r^2} \tag{12.67}$$

where R_r and $\dot{V}_r$ are the values of R and $\dot{V}$ at the elution time. As mentioned earlier, the diffusion coefficient may not be related to the magnetic content of the particles.

A data reduction program has been written to transform a fractogram of detector response versus elution time into a mass or number distribution in either V_m or d_m, given an assumed or measured dependence of magnetization of the core material on field strength. It is applicable to constant field and flow operation and to conditions of programmed field decay. It takes a numerical approach that accounts for actual monitored variation of field strength and flow rate [161]. It can therefore account for deviations of field strength from a preset program that follows some mathematical functional decay.

12.4.7.6. Equipment

The necessary equipment has been described in a number of publications from our laboratory [162–164]. A brief description of the major components is given below.

Channel design. It is possible to build a channel that utilizes the full annular space between an outer tube and an inner core, as in the QMS system. However, the use of the full annulus for the channel demands a high degree of uniformity in annular thickness around the circumference. Any variation in thickness would contribute significantly to band spreading, as would any variation in flow through the radial tubes (due to construction flaws or partial, or even full, plugging). In addition, any nonuniformity in the magnetic field strength around the annular circumference would also contribute to band spreading. A solution would be to employ just one radial tube for inlet and outlet, and to restrict the channel to a fraction of the complete annulus. A better solution is to use a single radial tube at the inlet and outlet, and to spiral the channel several times around the core rod, as shown in Fig. 12.17. This would result in a much longer channel that makes use of the quadrupole magnetic field. Any circumferential variation in magnetic field would be sampled equally by all particles as they migrate along the channel. The increased length will also offer higher separation efficiency.

This channel is machined into the surface of a cylindrical Delrin™ rod (du Pont), that is inserted within a tight-fitting internally polished stainless steel (316 grade) tube. This is accomplished by cooling the components in liquid nitrogen, assembling, and then allowing their return to room temperature. The band spreading due to multipath effects (nonuniformity in flow velocity across the breadth of the channel) is expected to be lower than for the full annular channel, for the reasons explained above.

Quadrupole electromagnet. It has been found that magnetic nanoparticles generally vary significantly in the magnetic content of

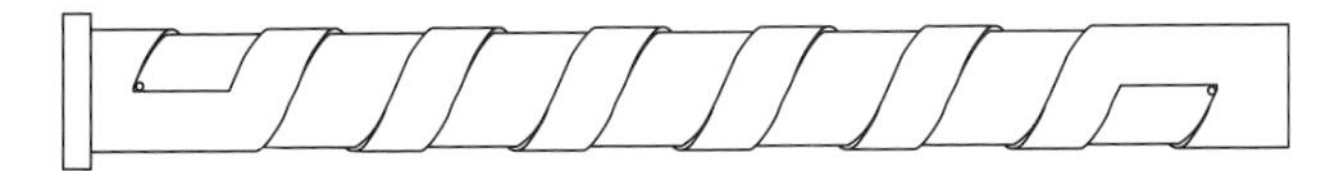

Fig. 12.17. The helical channel insert. It is machined to a depth of 250 μm into the surface of a Delrin™ rod. Inlet and outlet tubes are machined into the ends and are threaded to accept standard HPLC fittings.

the particles, or in other words, they are found to be polydisperse in magnetic content. Some particles would be very strongly retained even to the point of not eluting, while some would be only slightly retained. Field programming in FFF achieves the same objectives as temperature programming in gas chromatography or solvent programming in liquid chromatography. Conditions are gradually changed during the analysis to hasten the elution of the more strongly retained sample components. In SdFFF, this is easily accomplished by slowing the centrifuge rotation rate during the analysis. In the case of MgFFF, programmed reduction of field strength will require a programmed reduction of current to the quadrupole electromagnet. With calibration of B_o as a function of current, it is possible to program B_o according to an optimum function for uniform fractionating power (a measure of relative resolving power) [156, 165–169].

An electromagnet allows for the selection of optimum field strengths for given samples and the programmed decay of field strength during the elution of broadly polydisperse samples. It consists of four copper coils, each nominaly 1900 m of AWG 18 coated copper wire ($\phi_{unisolated} = 1.024$ mm), four pole pieces made of 1018 low carbon, cold-rolled steel. The copper coils had measured electrical resistances between 35.6 and 35.8 Ω. The pole tips were machined to the required hyperbolic profile to generate a radially symmetric magnetic field. The field return paths were facilitated with a square yoke of 1018 low carbon, cold-rolled steel. The pole pieces were designed and assembled to form a 16-mm aperture into which the FFF channel could be inserted. The power for the electromagnet is provided by a Xantrex HPD60-5 (Xantrex Technology, Inc., British Columbia, Canada) regulated DC power supply with a nominal maximum current of 5 A at 60 V. The four coils generate a quadrupole field with a maximum field strength of 0.71 T at the pole tips at a total magnetization current of 5 A corresponding to 1.25 A to each of the coils wired in parallel.

Ancillary Equipment. Ancillary equipment, such as fluid pumps, sample injectors, and detectors are typically of the types used in liquid chromatography. A constant current power supply is required for the electromagnet, as mentioned above. This is controllable using a computer that can vary the current with time to obtain some desired field decay function. A gauss/tesla meter is useful for calibrating the electromagnet field as a function of current, and also for monitoring the field strength close to the channel during an analysis. The monitored field strength is recorded by a computer along with the detector output using an analog-to-digital converter. A correction is applied to the monitored field strength to account for the small displacement of the probe from the actual position of the channel accumulation wall.

12.4.7.7. Example of a Magnetic Nanoparticle Analysis

Figure 12.18A shows an elution curve of detector response as a function of time (known as a fractogram) obtained for a sample of magnetite nanoparticles coated with dextran. These were kindly provided by BD Biosciences Pharmingen (San Jose, CA). They were obtained from Skold Technology and did not have antibodies conjugated to them. They had a nominal size of 230 $\pm$ 150 nm diameter (information provided by the company). This information was in good agreement with quasielastic light scattering (QELS) measurements. A 20-μl volume of sample was introduced to the channel at a low sample loading flow rate of 0.10 ml/min and the initial magnetic field strength of 57 mT. The flow was then interrupted for 30 min to allow the sample particles to approach their steady-state distribution across the channel thickness. The channel flow was then reinstated at the higher rate of 0.5 ml/min and the field decay program initiated for the analysis to take place. Also shown in the figure is the curve showing the measured field decay during the analysis (referring to the right axis). The field decay follows the commonly used power programmed decay function [165]. This is given by the equation:

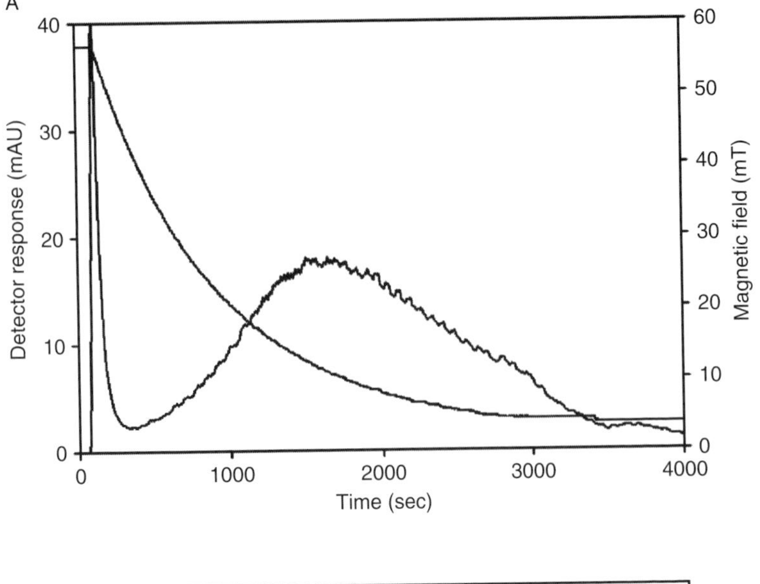

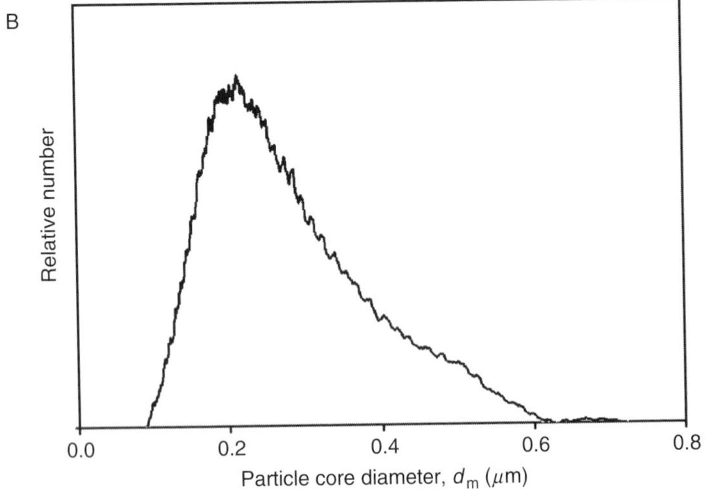

Fig. 12.18. (A) Fractogram for Skold nanoparticles (lot 80) together with th field decay curve as monitored during elution and referring to right axi (B) The particle core size distribution calculated from the fractogram Reprinted with permission from Carpino et al. [164].

$$B(t) = B(0)\left(\frac{t_1 - t_a}{t - t_a}\right)^p \tag{12.68}$$

in which $B(t)$ is the field strength at the accumulation wall as a function of time t, $B(0)$ is the initial field strength, t_1 is an initial period during which the field is held constant, t_a is a second time constant, and p is a positive real number. For the example shown in Fig. 12.18A, the initial field strength was set to 57 mT, and the field decay follows the power program with $t_1 = 12$ min, $t_a = -96$ min, and $p = 8$. The initial hold period was overridden and set to 2 min. The fractogram shows a narrow peak corresponding to the void (nonretained) time followed by a broad peak for the sample.

The data reduction procedure requires a magnetization curve for the magnetic nanoparticle core material. Because such a curve was unavailable for the particular sample in hand, a magnetization curve for uncoated magnetite nanoparticles was taken from the literature [170]. A curve fitting to these data resulted in the following relationship between M and B:

$$M = \frac{9.152 \times 10^6 B}{1 + 27.30B - 0.9229B^2} \tag{12.69}$$

where M has units of A/m and B units of T. The result of transformation of the fractogram to a number distribution of equivalent spherical core diameters of magnetite is shown in Fig. 12.18B. This shows a broad distribution from about 0.1 μm to 0.6 μm, with a peak maximum at about 0.2 μm. The calculated core diameters are larger than expected, given that 95% of the particles were determined by QELS to have an overall size between 0.075 and 0.39 μm. This may indicate some particle aggregation in the channel due to dipole–dipole interaction or some other physical attraction. The fact that a magnetization curve was assumed for the particles would not account for the extent of the apparent overestimation of the core sizes. On the other hand, the magnetic FFF elution curves are found to be highly reproducible, and the technique has been shown to be capable of distinguishing between different lots of magnetic nanoparticles from the same source [162].

12.4.7.8. Sample overloading and particle–particle interaction

In MgFFF, sample overloading may have dramatic effects. Ove loading is generally manifested as a variation of elution time wi sample concentration. It could also result in skewed peaks as we as reduced sample recovery. A sample is overloaded when particl particle interaction becomes significant. During stop-flow relax tion, the sample concentration adjacent to the accumulation wa rises to $\sim c_s/\lambda$, where c_s is the concentration of the introduc sample. This concentration effect can influence the local viscosi and perturb the fluid velocity profile in the region of the samp zone [171, 172]. In the case of permanently magnetized particles in magnetic field, there is also a resistance to rotation of the particl which effectively increases the viscosity of the suspension. Th effect can even be apparent for suspensions of superparamagnet subdomain-sized particles, depending on the relative magnitude time constants for Brownian rotation and magnetization rot tion compared to a characteristic time for the sheared flow (s Rosensweig [173]). This is known as the magnetoviscous effe [174]. At high concentrations there may be an increased tenden for chaining of particles that would affect their field-induced velo ity as well as the local fluid viscosity. All these effects can reduced to insignificance by sufficient dilution of the sample.

References

[1] Karumanchi, R. S. M. S., Doddamane, S. N., Sampangi, C. a Todd, P. W. (2002). Field-assisted extraction of cells, particles, a macromolecules. Trends Biotechnol. *20*, 72–8.

[2] Pankhurst, Q. A., Connolly, J., Jones, S. K. and Dobson, J. (200 Applications of magnetic nanoparticles in biomedicine. J. Phys. Appl. Phys. *36*, R167–81.

[3] Guetta, E., Gutstein-Abo, L. and Barkai, G. (2005). Trophobla isolated from the maternal circulation: In vitro expansion and potent application in non-invasive prenatal diagnosis. J. Histochem. Cytoche *53*, 337–9.

[4] Guetta, E., Simchen, M. J., Mammon-Daviko, K., Gordon, D., Aviram-Goldring, A., Rauchbach, N. and Barkai, G. (2004). Analysis of fetal blood cells in the maternal circulation: Challenges, ongoing efforts, and potential solutions. Stem Cells Dev. *13*, 93–9.

[5] Budd, G. T., Cristofanilli, M., Ellis, M. J., Stopeck, A., Borden, E., Miller, M. C., Matera, J., Repollet, M., Doyle, G. V., Terstappen, L. W. and Hayes, D. F. (2006). Circulating tumor cells versus imaging–predicting overall survival in metastatic breast cancer. Clin. Cancer Res. *12*, 6403–9.

[6] Jing, Y., Moore, L. R., Williams, P. S., Chalmers, J. J., Farag, S. S., Bolwell, B. and Zborowski, M. (2007). Blood progenitor cell separation from clinical leukapheresis product by magnetic nanoparticle binding and magnetophoresis. Biotechnol. Bioeng. *96*, 1139–54.

[7] Kantor, A. B., Gibbons, I., Miltenyi, S. and Schmitz, J. (1998). Magnetic cell sorting with colloidal superparamagnetic particles. In: Cell Separation Methods and Applications (Recktenwald, D. and Radbruch, A., eds.). Marcel Dekker, Inc., New York, pp. 153–73.

[8] Smistrup, K., Hansen, O., Bruus, H. and Hansen, M. F. (2005). Magnetic separation in microfluidic systems using microfabricated electromagnets – experiments and simulations. J. Magn. Magn. Mater. *293*, 597–604.

[9] Ramadan, Q., Samper, C., Poenar, D. and Yu, C. (2006). Microcoils for transport of magnetic beads. Appl. Phys. Lett. *88*, 032501.

[10] Pekas, N., Granger, M., Tondra, M., Popple, A. and Porter, M. D. (2005). Magnetic particle diverter in an integrated microfluidic format. J. Magn. Magn. Mater. *293*, 584–8.

[11] Blankenstein, G. (1997). Microfabricated flow system for magnetic cell and particle separation. In: Scientific and Clinical Applications of magnetic Carriers (Hafeli, U., Schuett, W., Teller, J. and Zborowski, M., eds.). Plenum Press, New York, pp. 233–45.

[12] Nath, P., Moore, L. R., Zborowski, M., Roy, S. and Fleischman, A. J. (2006). A method to obtain uniform magnetic-field energy density gradient distribution using discrete pole pieces for a microelectromechanical-system-based magnetic cell separator. J. Appl. Phys. *99*, 08R905.

[13] McCloskey, K. E., Moore, L. R., Hoyos, M., Rodriguez, A., Chalmers, J. J. and Zborowski, M. (2003). Magnetophoretic cell sorting is a function of antibody binding capacity. Biotechnol. Prog. *19*, 899–907.

[14] Fabry, M. E. and San George, R. C. (1983). Effect of magnetic susceptibility on nuclear magnetic resonance signals arising from red cells: a warning. Biochemistry *22*, 4119–25.

[15] Orjih, A. U. and Fitch, C. D. (1993). Hemozoin production by *Plasmodium falciparum*: Variation with strain and exposure to chloroquine. Biochim. Biophys. Acta *1157*, 270–4.

[16] Becker, K., Tilley, L., Vennerstrom, J. L., Roberts, D., Rogerson, S. and Ginsburg, H. (2004). Oxidative stress in malaria parasite-infected erythrocytes: Host-parasite interactions. Int. J. Parasitol. *34*, 163–89.
[17] Kumar, S. and Bandyopadhyay, U. (2005). Free heme toxicity and its detoxification systems in human. Toxicol. Lett. *157*, 175–88.
[18] Francis, S. E., Sullivan, D. J., Jr. and Goldberg, D. E. (1997). Hemoglobin metabolism in the malaria parasite *Plasmodium falciparum*. Annu. Rev. Microbiol. *51*, 97–123.
[19] Goldberg, D. E., Slater, A. F. G., Cerami, A. and Henderson, G. B. (1990). Hemoglobin degradation in the malaria parasite *Plasmodium falciparum:* An ordered process in a unique organelle. Proc. Natl. Acad. Sci. USA *87*, 2931–5.
[20] Bendrat, K., Berger, B. J. and Cerami, A. (1995). Haem polymerization in malaria. Nature *378*, 138–9.
[21] Pagola, S., Stephens, P. W., Bohle, D. S., Kosar, A. D. and Madsen, S. K. (2000). The structure of malaria pigment beta-haematin. Nature *404*, 307–10.
[22] Noland, G. S., Briones, N. and , Sullivan, D. J., Jr. (2003). The shape and size of hemozoin crystals distinguishes diverse Plasmodium species. Mol. Biochem. Parasitol. *130*, 91–9.
[23] Egan, T. J. (2002). Physico-chemical aspects of hemozoin (malaria pigment) structure and formation. J. Inorg. Biochem. *91*, 19–26.
[24] Dzekunov, S. M., Ursos, L. M. and Roepe, P. D. (2000). Digestive vacuolar pH of intact intraerythrocytic *P. falciparum* either sensitive or resistant to chloroquine. Mol. Biochem. Parasitol. *110*, 107–24.
[25] Pauling, L. and Coryell, C. D. (1936). The magnetic properties and structure of hemoglobin, oxyhemoglobin and carbonmonoxyhemoglobin. Proc. Natl. Acad. Sci. USA *22*, 210–6.
[26] Coryell, C., Stitt, F. and Pauling, L. (1937). The magnetic properties and structure of ferrihemoglobin (methemoglobin) and some of its compounds. J. Am. Chem. Soc. *59*, 633–42.
[27] Pauling, L. (1977). Magnetic properties and structure of oxyhemoglobin. Proc. Natl. Acad. Sci. USA *74*, 2612–3.
[28] Savicki, J. P., Lang, G. and Ikeda-Saito, M. (1984). Magnetic susceptibility of oxy- and carbonmonoxyhemoglobins. Proc. Natl. Acad. Sci. USA *81*, 5417–9.
[29] Cerdonio, M., Congiu-Castellano, A., Calabrese, L., Morante, S., Pispisa, B. and Vitale, S. (1978). Room-temperature magnetic properties of oxy- and corbonmonoxyhemoglobin. Proc. Natl. Acad. Sci. USA *75*, 4916–9.
[30] Cerdonio, M., Morante, S., Torresani, D., Vitale, S., DeYoung, A. and Noble, R. W. (1985). Reexamination of the evidence for paramagnetism

in oxy- and carbonmonoxyhemoglobins. Proc. Natl. Acad. Sci. USA *82*, 102–3.

[31] Spees, W. M., Yablonskiy, D. A., Oswood, M. C. and Ackerman, J. J. (2001). Water proton MR properties of human blood at 1.5 Tesla: Magnetic susceptibility, T(1), T(2), T*(2), and non-Lorentzian signal behavior. Magn. Reson. Med. *45*, 533–42.

[32] Plyavin, Y. A. and Blum, E. Y. (1983). Magnetic parameters of blood cells and high gradient paramagnetic and diamagnetic phoresis. Magnetohydrodynamics *19*, 349–59.

[33] Weisskoff, R. M. and Kiihne, S. (1992). MRI susceptometry: image-based measurement of absolute susceptibility of MR contrast agents and human blood. Magn. Reson. Med. *24*, 375–83.

[34] Bohle, D. S., Debrunner, P., Jordan, P. A., Madsen, S. K. and Schulz, C. E. (1998). Aggregated heme detoxification byproducts in malarial trophozoites: beta-hematin and malaria pigment have a single S = 5/2 iron environment in the bulk phase as determined by EPR and magnetic Moessbauer spectroscopy. J. Am. Chem. Soc. *120*, 8255–6.

[35] Zimmerman, P. A., Thomson, J. M., Fujioka, H., Collins, W. E. and Zborowski, M. (2006). Diagnosis of malaria by magnetic deposition microscopy. Am. J. Trop. Med. Hyg. *74*, 568–72.

[36] Bartoszek, M., Balanda, M., Skrzypek, D. and Drzazga, Z. (2001). Magnetic field effect on hemin. Physica. B *307*, 217–23.

[37] Egan, T. J., Combrinck, J. M., Egan, J., Hearne, G. R., Marques, H. M., Ntenteni, S., Sewell, B. T., Smith, P. J., Taylor, D., van Schalkwyk, D. A. and Walden, J. C. (2002). Fate of haem iron in the malaria parasite *Plasmodium falciparum*. Biochem. J. *365*, 343–7.

[38] Cranmer, S. L., Magowan, C., Liang, J., Coppel, R. L. and Cooke, B. M. (1997). An alternative to serum for cultivation of *Plasmodium falciparum* in vitro. Trans. R. Soc. Trop. Med. Hyg. *91*, 363–5.

[39] Lambros, C. and Vanderberg, J. P. (1979). Synchronization of Plasmodium falciparum erythrocytic stages in culture. J. Parasitol. *65*, 418–20.

[40] Zborowski, M., Ostera, G. R., Moore, L. R., Milliron, S., Chalmers, J. J. and Schechter, A. N. (2003). Red blood cell magnetophoresis. Biophys. J. *84*, 2638–45.

[41] Moore, L. R., Fujioka, H., Williams, P. S., Chalmers, J. J., Grimberg, B., Zimmerman, P. A. and Zborowski, M. (2006). Hemoglobin degradation in malaria-infected erythrocytes determined from live cell magnetophoresis. FASEB J. *20*, 747–9.

[42] Anonymous (2000). WHO/MAL/2000.1091 New Perspectives in Malaria Diagnosis. World Health Organization, Geneva.

[43] Shiff, C. J., Premji, Z. and Minjas, J. N. (1993). The rapid manual Para-Sight-F test. A new diagnostic tool for *Plasmodium falciparum* infection. Trans. R. Soc. Trop. Med. Hyg. *87*, 646–8.

[44] Palmer, C. J., Lindo, J. F., Klaskala, W. I., Quesada, J. A., Kaminsky, R., Baum, M. K. and Ager, A. L. (1998). Evaluation of the OptiMAL test for rapid diagnosis of *Plasmodium vivax* and *Plasmodium falciparum* malaria. J. Clin. Microbiol. *36*, 203–6.

[45] Snounou, G., Viriyakosol, S., Zhu, X. P., Jarra, W., Pinheiro, L., do Rosario, V. E., Thaithong, S. and Brown, K. N. (1993). High sensitivity of detection of human malaria parasites by the use of nested polymerase chain reaction. Mol. Biochem. Parasitol. *61*, 315–20.

[46] Mehlotra, R. K., Lorry, K., Kastens, W., Miller, S. M., Alpers, M. P., Bockarie, M., Kazura, J. W. and Zimmerman, P. A. (2000). Random distribution of mixed species malaria infections in Papua New Guinea. Am. J. Trop. Med. Hyg. *62*, 225–31.

[47] McNamara, D. T., Thomson, J. M., Kasehagen, L. J. and Zimmerman, P. A. (2004). Development of a multiplex PCR-ligase detection reaction assay for diagnosis of infection by the four parasite species causing malaria in humans. J. Clin. Microbiol. *42*, 2403–10.

[48] Moody, A. (2002). Rapid diagnostic tests for malaria parasites. Clin. Microbiol. Rev. *15*, 66–78.

[49] Zimmerman, P. A., Mehlotra, R. K., Kasehagen, L. J. and Kazura, J. W. (2004). Why do we need to know more about mixed Plasmodium species infections in humans? Trends. Parasitol. *20*, 440–7.

[50] Paul, F., Roath, S., Melville, D., Warhurst, D. C. and Osisanya, J. O. S. (1981). Separation of malaria-infected erythrocytes from whole blood: Use of a selective high gradient magnetic separation technique. Lancet *318* 70–1.

[51] Nalbandian, R. M., Sammons, D. W., Manley, M., Xie, L., Sterling, C. R. Egen, N. B. and Gingras, B. A. (1995). A molecular-based magnet test for malaria. Am. J. Clin. Pathol. *103*, 57–64.

[52] Fang, B., Zborowski, M. and Moore, L. R. (1999). Detection of rare MCF-7 breast carcinoma cells from mixtures of human peripheral leukocytes by magnetic deposition analysis. Cytometry *36*, 294–302.

[53] Zborowski, M., Fuh, C. B., Green, R., Baldwin, N. J., Reddy, S. Douglas, T., Mann, S. and Chalmers, J. J. (1996). Immunomagnetic isolation of magnetoferritin-labeled cells in a modified ferrograph. Cytometry *24*, 251–9.

[54] Zborowski, M., Fuh, C. B., Green, R., Sun, L. and Chalmers, J. J. (1995). Analytical magnetapheresis of ferritin-labeled lymphocytes. Anal. Chem. *67*, 3702–12.

[55] Zborowski, M., Malcheski, P. S., Savon, S. R., Green, R., Holl, G. S. and Nose, Y. (1991). Modification of ferrography method for analysis of lymphocytes and bacteria. Wear *142*, 135–49.

[56] Zborowski, M., Malchesky, P. S., Jan, T. F. and Hall, G. S. (1992). Quantitative separation of bacteria in saline solution using lanthanide Er(III) and a magnetic field. J. Gen. Microbiol. *138*, 63–8.

[57] Zborowski, M. (1997). Physics of the magnetic cell sorting. In: Scientific and Clinical Aplications of Magnetic Microcarriers: An Overview (Hafeli, U., Schutt, W., Teller, J. and Zborowski, M., eds.). Plenum Press, New York, pp. 205–31.

[58] Carter, V., Cable, H. C., Underhill, B. A., Williams, J. and Hurd, H. (2003). Isolation of *Plasmodium berghei* ookinetes in culture using Nycodenz density gradient columns and magnetic isolation. Malar. J. *2*, 35.

[59] Snounou, G. and White, N. J. (2004). The co-existence of Plasmodium: Sidelights from falciparum and vivax malaria in Thailand. Trends Parasitol. *20*, 333–9.

[60] Ghebremeskel, A. N. and Bose, A. (2000). A flow-through, hybrid magnetic-field-gradient, rotating wall device for magnetic colloidal separations. Sep. Sci. Technol. *35*, 1813–28.

[61] Schneider, T., Moore, L. R., Jing, Y., Haam, S., Williams, P. S., Fleischman, A. J., Roy, S., Chalmers, J. J. and Zborowski, M. (2006). Continuous flow magnetic cell fractionation based on antigen expression level. J. Biochem. Biophys. Methods *68*, 1–21.

[62] Jing, Y., Moore, L. R., Schneider, T., Williams, P. S., Chalmers, J. J., Farag, S. S., Bolwell, B. and Zborowski, M. (2007). Negative selection of hematopoietic progenitor cells by continuous magnetophoresis. Exp. Hematol. *35*, 662–72.

[63] Chalmers, J. J., Zborowski, M., Moore, L., Mandal, S., Fang, B. B. and Sun, L. (1998). Theoretical analysis of cell separation based on cell surface marker density. Biotechnol. Bioeng. *59*, 10–20.

[64] Chalmers, J. J., Zborowski, M., Sun, L. and Moore, L. (1998). Flow through, immunomagnetic cell separation. Biotechnol. Prog. *14*, 141–8.

[65] Moore, L. R., Rodriguez, A. R., Williams, P. S., McCloskey, K. E., Bolwell, B. J., Nakamura, M., Chalmers, J. J. and Zborowski, M. (2001). Progenitor cell isolation with a high-capacity quadrupole magnetic flow sorter. J. Magn. Magn. Mater. *225*, 277–84.

[66] Moore, L. R., Milliron, S., Williams, P. S., Chalmers, J. J., Margel, S. and Zborowski, M. (2004). Control of magnetophoretic mobility by susceptibility-modified solutions as evaluated by cell tracking velocimetry and continuous magnetic sorting. Anal. Chem. *76*, 3899–907.

[67] Giddings, J. C. (1985). A system based on split-flow lateral-transport thi (SPLITT) separation cells for rapid and continuous particle fractionatio Sep. Sci. Technol. *20*, 749–68.
[68] Williams, P. S., Zborowski, M. and Chalmers, J. J. (1999). Flow rat optimization for the quadrupole magnetic cell sorter. Anal. Chem. 7 3799–807.
[69] Sun, L., Zborowski, M., Moore, L. R. and Chalmers, J. J. (1998). Con tinuous, flow-through immunomagnetic cell sorting in a quadrupole fiel Cytometry *33*, 469–75.
[70] Williams, P. S., Zborowski, M. and Chalmers, J. J. (1999). Flow rat optimization for the quadrupole magnetic cell sorter. Anal. Chem. 7 3799–807.
[71] Gross, H. J., Verwer, B., Houck, D. and Recktenwald, D. (1993). Detec tion of rare cells at a frequency of one per million by flow cytometry Cytometry *14*, 519–26.
[72] Chalmers, J. J., Zhao, Y., Nakamura, M., Melnik, K., Lasky, L Moore, L. R. and Zborowski, M. (1999). An instrument to determin the magnetophoretic mobility of labeled biological cells and paramagneti particles. J. Magn. Magn. Mater. *194*, 231–41.
[73] Moore, L. R., Zborowski, M., Nakamura, M., McCloskey, K., Gura, S Zuberi, M., Margel, S. and Chalmers, J. J. (2000). The use of magnetite doped polymeric microspheres in calibrating cell tracking velocimetry J. Biochem. Biophys. Methods *44*, 115–30.
[74] Hoyos, M., McCloskey, K., Moore, L., Nakmura, M., Bolwell, B Chalmers, J. J. and Zborowski, M. (2002). Pulse-injection studies o blood progenitor cells in a quadrupole magnetic flow sorter. Sep. Sci Technol. *37*, 1–23.
[75] Williams, P. S., Moore, L. R., Chalmers, J. J. and Zborowski, M. (2003) Splitter imperfections in annular split-flow thin separation channels Effect on nonspecific crossover. Anal. Chem. *75*, 1365–73.
[76] Williams, P. S., Decker, K., Nakamura, M., Chalmers, J. J., Moore, L. R and Zborowski, M. (2003). Splitter imperfections in annular split-flov thin separation channels: Experimental study of nonspecific crossover Anal. Chem. *75*, 6687–95.
[77] McCloskey, K. E., Zborowski, M. and Chalmers, J. J. (2000). Magneto phoretic mobilities correlate to surface antigen binding capacities. Cyto metry *40*, 307–15.
[78] McCloskey, K. E., Comella, K., Chalmers, J. J., Margel, S. an Zborowski, M. (2001). Mobility measurements of immunomagneticall labeled cells allow quantitation of secondary antibody binding amplifica tion. Biotechnol. Bioeng. *75*, 642–55.

[79] Chalmers, J. J., Zhao, Y., Nakamura, M., Melnik, K., Lasky, L., Moore, L. and Zborowski, M. (1999). An instrument to determine the magnetophoretic mobility of labeled, biological cells and paramagnetic particles. J. Magn. Magn. Mater. *194*, 231–41.

[80] McCloskey, K. E., Moore, L. R., Hoyos, M., Rodriguez, A., Chalmers, J. J. and Zborowski, M. (2003). Magnetophoretic cell sorting is a function of antibody binding capacity. Biotechnol. Prog. *19*, 899–907.

[81] Giddings, J. C. (1966). A new separation concept based on a coupling of concentration and flow nonuniformities. Sep. Sci. *1*, 123–5.

[82] Thompson, G. H., Myers, M. N. and Giddings, J. C. (1967). An observation of a field-flow fractionation effect with polystyrene samples. Sep. Sci. *2*, 797–900.

[83] Giddings, J. C. (1993). Field-flow fractionation: Separation and characterization of macromolecular, colloidal, and particulate materials. Science *260*, 1456–65.

[84] Schimpf, M. E., Caldwell, K. and Giddings, J. C. (Eds). (2000). Field-Flow Fractionation Handbook. John Wiley & Sons, Inc., New York, NY.

[85] Giddings, J. C. (1985). A system based on split-flow lateral-transport thin (SPLITT) separation cells for rapid and continuous particle fractionation. Sep. Sci. Technol. *20*, 749–68.

[86] Springston, S. R., Myers, M. N. and Giddings, J. C. (1987). Continuous particle fractionation based on gravitational sedimentation in split-flow thin cells. Anal. Chem. *59*, 344–50.

[87] Gao, Y., Myers, M. N., Barman, B. N. and Giddings, J. C. (1991). Continuous fractionation of glass microspheres by gravitational sedimentation in split-flow thin (SPLITT) cells. Part Sci. Technol. *9*, 105–18.

[88] Giddings, J. C. (1992). Optimization of transport-driven continuous SPLITT fractionation. Sep. Sci. Technol. *27*, 1489–504.

[89] Thompson, G. H., Myers, M. N. and Giddings, J. C. (1969). Thermal field-flow fractionation of polystyrene samples. Anal. Chem. *41*, 1219–22.

[90] Giddings, J. C. and Myers, M. N. (1978). Steric field-flow fractionation: A new method for separating 1 to 100 μm particles. Sep. Sci. Technol. *13*, 637–45.

[91] Giddings, J. C., Myers, M. N., Caldwell, K. D. and Pav, J. W. (1979). Steric field-flow fractionation as a tool for the size characterization of chromatographic supports. J. Chromatogr. A. *185*, 261–71.

[92] Caldwell, K. D., Nguyen, T. T., Myers, M. N. and Giddings, J. C. (1979). Observations on anomalous retention in steric field-flow fractionation. Sep. Sci. Technol. *14*, 935–46.

[93] Giddings, J. C. (1968). Nonequilibrium theory of field-flow fractionation. J. Chem. Phys. *49*, 81–5.

[94] Giddings, J. C., Yoon, Y. H., Caldwell, K. D., Myers, M. N. and Hovingh, M. E. (1975). Nonequilibrium plate height for field-flow fractionation in ideal parallel plate columns. Sep. Sci. *10*, 447–60.

[95] Williams, P. S., Koch, T. and Giddings, J. C. (1992). Characterization of near-wall hydrodynamic lift forces using sedimentation field-flow fractionation. Chem. Eng. Commun. *111*, 121–47.

[96] Williams, P. S., Moon, M. H. and Giddings, J. C. (1992). Fast separation and characterization of micron size particles by sedimentation/steric field-flow fractionation: Role of lift forces. In: Particle Size Analysis (Stanley-Wood, N. G. and Lines, R. W., eds.). Royal Society of Chemistry, Cambridge, UK, pp. 280–9.

[97] Williams, P. S., Lee, S. and Giddings, J. C. (1994). Characterization of hydrodynamic lift forces by field-flow fractionation: Inertial and near-wall lift forces. Chem. Eng. Commun. *130*, 143–66.

[98] Williams, P. S., Moon, M. H., Xu, Y. and Giddings, J. C. (1996). Effect of viscosity on retention time and hydrodynamic lift forces in sedimentation/steric field-flow fractionation. Chem. Eng. Sci. *51*, 4477–88.

[99] Williams, P. S., Moon, M. H. and Giddings, J. C. (1996). Influence of accumulation wall and carrier solution composition on lift force in sedimentation/steric field-flow fractionation. Colloids and Surfaces A: Physicochemical and Engineering Aspects *113*, 215–28.

[100] Koch, T. and Giddings, J. C. (1986). High speed separation of large ($> 1\ \mu$m) particles by steric field-flow fractionation. Anal. Chem. *58*, 994–7.

[101] Giddings, J. C., Chen, X., Wahlund, K.-G. and Myers, M. N. (1987). Fast particle separation by flow/steric field-flow fractionation. Anal. Chem. *59*, 1957–62.

[102] Ratanathanawongs, S. K. and Giddings, J. C. (1989). High-speed size characterization of chromatographic silica by flow/hyperlayer field-flow fractionation. J. Chromatogr. A. *467*, 341–56.

[103] Jiang, Y., Miller, M. E., Hansen, M. E., Myers, M. N. and Williams, P. S. (1999). Fractionation and size analysis of magnetic particles using FFF and SPLITT technolodies. J. Magn. Magn. Mater. *194*, 53–61.

[104] Jiang, Y., Kummerow, A. and Hansen, M. (1997). Preparative particle separation by continuous SPLITT fractionation. J. Microcolumn. Sep. *9*, 261–73.

[105] Contado, C., Dondi, F., Beckett, R. and Giddings, J. C. (1997). Separation of particulate environmental samples by SPLITT fractionation using different operating modes. Anal. Chim. Acta *345*, 99–110.

[106] Dondi, F., Contado, C., Blo, G. and Garçia Martin, S. (1998). SPLITT cell separation of polydisperse suspended particles of environmental interest. Chromatographia. *48*, 643–54.

[107] Levin, S., Myers, M. N. and Giddings, J. C. (1989). Continuous separation of proteins in electrical split-flow thin (SPLITT) cell with equilibrium operation. Sep. Sci. Technol. *24*, 1245–59.

[108] Fuh, C. B. and Giddings, J. C. (1997). Isoelectric split-flow thin (SPLITT) fractionationation of proteins. Sep. Sci. Technol. *32*, 2945–67.

[109] Fuh, C. B. and Giddings, J. C. (1995). Isolation of human blood cells, platelets, and plasma proteins by centrifugal SPLITT fractionation. Biotechnol. Prog. *11*, 14–20.

[110] Fuh, C. B. and Giddings, J. C. (1997). Separation of submicron pharmaceutic emulsions with centrifugal split-flow thin (SPLITT) fractionation. J. Microcolumn. Sep. *9*, 205–11.

[111] Fuh, C. B., Myers, M. N. and Giddings, J. C. (1994). Centrifugal SPLITT fractionation: New technique for separation of colloidal particles. Ind. Eng. Chem. Res. *33*, 355–62.

[112] Fuh, C. B. and Chen, S. Y. (1998). Magnetic split-flow thin fractionation: new technique for separation of magnetically susceptible particles. J. Chromatogr. A *813*, 313–24.

[113] Fuh, C. B. and Chen, S. Y. (1999). Magnetic split-flow thin fractionation of magnetically susceptible particles. J. Chromatog. A *857*, 193–204.

[114] Fuh, C. B., Lai, J. Z. and Chang, C. M. (2001). Particle magnetic susceptibility determination using analytical split-flow thin fractionation. J. Chromatogr. A *923*, 263–70.

[115] Fuh, C. B., Tsai, H. Y. and Lai, J. Z. (2003). Development of magnetic split-flow thin fractionation for continuous particle separation. Anal. Chim. Acta *497*, 115–22.

[116] Wingo, R. M., Prenger, F. C., Johnson, M. D., Waynert, J. A., Worl, L. A. and Ying, T.-Y. (2004). High-gradient magnetic field split-flow thin channel (HGMF-SPLITT) fractionation of nanoscale paramagnetic particles. Sep. Sci. Technol. *39*, 2769–83.

[117] Moore, L. R., Rodriguez, A. R., Williams, P. S., McCloskey, K., Bolwell, B. J., Nakamura, M., Chalmers, J. J. and Zborowski, M. (2001). Progenitor cell isolation with a high-capacity quadrupole magnetic flow sorter. J. Magn. Magn. Mater. *225*, 277–84.

[118] Zborowski, M., Williams, P. S., Sun, L., Moore, L. R. and Chalmers, J. J. (1997). Cylindrical SPLITT and quadrupole magnetic

field in application to continuous-flow magnetic cell sorting. J. Liq Chromatogr. Related Technol. *20*, 2887–905.
[119] Chalmers, J. J., Zborowski, M., Sun, L. and Moore, L. (1998). Flow through, immunomagnetic cell separation. Biotechnol. Prog. *14*, 141–8
[120] Sun, L., Zborowski, M., Moore, L. R. and Chalmers, J. J. (1998) Continuous, flow-through immunomagnetic cell sorting in a quadrupole field. Cytometry *33*, 469–75.
[121] Zborowski, M., Sun, L., Moore, L. R. and Chalmers, J. J. (1999). Rapic cell isolation by magnetic flow sorting for applications in tissue engineering. ASAIO J. *45*, 127–30.
[122] Zborowski, M., Sun, L., Moore, L. R., Williams, P. S. and Chalmers, J. J. (1994, 1999). Continuous cell separation using nove magnetic quadrupole flow sorter. J. Magn. Magn. Mater. *194* 224–30.
[123] Hoyos, M., McCloskey, K. E., Moore, L. R., Nakamura, M. Bolwell, B. J., Chalmers, J. J. and Zborowski, M. (2002). Pulse-injection studies of blood progenitor cells in a quadrupole magnet flow sorter Sep. Sci. Technol. *37*, 745–67.
[124] Chiang, A. S., Kmiotek, E. H., Langan, S. M., Noble, P. T., Reis, J. F. G and Lightfoot, E. N. (1979). Preliminary experimental survey of hollow-fiber electro-polarization chromatography (electrical field-flow fractionation) for protein fractionation. Sep. Sci. Technol. *14*, 453–74.
[125] Carlshaf, A. and Jönsson, J. Å. (1989). Gradient elution in hollow-fibre flow field-flow fractionation. J. Chromatogr. A *461*, 89–93.
[126] Jonsson, J. A. and Carlshaf, A. (1989). Flow field flow fractionation in hollow cylindrical fibers. Anal. Chem. *61*, 11–8.
[127] Carlshaf, A. and Jönsson, J. Å. (1991). Effects of ionic strength of eluent on retention behavior and on the peak broadening process in hollow fiber flow field-flow fractionation. J. Microcolumn. Sep. *3*, 411–6.
[128] Carlshaf, A. and Jönsson, J. Å. (1993). Properties of hollow fibers used for flow field-flow fractionation. Sep. Sci. Technol. *28*, 1031–42.
[129] Carlshaf, A. and Jönsson, J. Å. (1993). Perturbations of the retention parameter due to sample overloading in hollow-fiber flow field-flow fractionation. Sep. Sci. Technol. *28*, 1191–201.
[130] Wijnhoven, J. E. G. J., Koorn, J. P., Poppe, H. and Kok, W. T. (1995) Hollow-fibre flow field-flow fractionation of polystyrene sulphonates J. Chromatogr. A *699*, 119–29.
[131] Wijnhoven, J. E. G. J., Koorn, J.-P., Poppe, H. and Kok, W. T. (1996) Influence of injected mass and ionic strength on retention of water-soluble polymers and proteins in hollow-fibre flow field-flow fractionation. J. Chromatogr. A *732*, 307–15.

[132] Lee, W. J., Min, B.-R. and Moon, M. H. (1999). Improvement in particle separation by hollow fiber flow field-flow fractionation and the potential use in obtaining particle size distribution. Anal. Chem. *71*, 3446–52.
[133] Moon, M. H., Lee, K. H. and Min, B. R. (1999). Effect of temperature on particle separation in hollow fiber flow field-flow fractionation. J. Microcolumn. Sep. *11*, 676–81.
[134] Giddings, J. C. and Brantley, S. L. (1984). Shear field-flow fractionation: Theoretical basis of a new highly selective technique. Sep. Sci. Technol. *19*, 631–51.
[135] Davis, J. M. and Giddings, J. C. (1986). Feasibility study of dielectrical field-flow fractionation. Sep. Sci. Technol. *21*, 969–89.
[136] Zolotarev, P. P., Ugrozov, V. V. and Skornyakov, E. P. (1988). Theory of the separation of mixtures in a coaxial column with a transverse temperature gradient. Zh. Fizi. Khimii. *62*, 1896–903.
[137] Ugrozov, V. V., Maksimycheva, M. A. and Zolotarev, P. P. (1989). Theory of flow fractionation in a coaxial channel. Russian J. Phys. Chem. *63*, 575–8.
[138] Davis, J. M., Fan, F.-R. F. and Bard, A. J. (1987). Retention by electrical field-flow fractionation of anions in a new apparatus with annular porous glass channels. Anal. Chem. *59*, 1339–48.
[139] Vickrey, T. M. and Garcia-Ramirez, J. A. (1980). Magnetic field-flow fractionation: Theoretical basis. Sep. Sci. Technol. *15*, 1297–304.
[140] Schunk, T. C., Gorse, J. and Burke, M. F. (1984). Parameters affecting magnetic field-flow fractionation of metal oxide particles. J. Sep. Sci. Technol. *19*, 653–66.
[141] Gorse, J., Schunk, T. C. and Burke, M. F. (1984–85). The study of liquid suspensions of iron oxide particles with a magnetic field-flow fractionation device. Sep. Sci. Technol. *19*, 1073–85.
[142] Semenov, S. N. and Kuznetsov, A. A. (1986). Flow fractionation in a transverse high-gradient magnetic field. Russian J. Phys. Chem. *60*, 247–50.
[143] Semenov, S. N. (1986). Flow fractionation in a strong transverse magnetic field. Russian J. Phys. Chem. *60*, 729–31.
[144] Mori, S. (1986). Magnetic field-flow fractionation using capillary tubing. Chromatographia. *21*, 642–4.
[145] Ohara, T., Mori, S., Oda, Y., Yamamoto, K., Wada, Y. and Tsukamoto, O. (1994). FFF using high gradient and high intensity magnetic field: Process analysis presented at Fourth International Symposium on Field-Flow Fractionation (FFF94), Lund, Sweden, June 13–5.
[146] Tsukamoto, O., Ohizumi, T., Ohara, T., Mori, S. and Wada, Y. (1995). Feasibility study on separation of several tens nanometer scale particles

by magnetic field-flow-fractionation technique using superconducting magnet. IEEE Trans. Appl. Supercond. *5*, 311–4.

[147] Ohara, T., Mori, S., Oda, Y., Wada, Y. and Tsukamoto, O. (1995) Feasibility of using magnetic chromatography for ultra-fine particle separation. Proceedings of the IEE Japan-Power & Energy *95*, 161–6.

[148] Ohara, T., Mori, S., Oda, Y., Wada, Y. and Tsukamoto, O. (1996) Feasibility of magnetic chromatography for ultra-fine particle separation. Trans. IEE Japan *116—B*, 979–86.

[149] Wang, X., Whitby, E. R., Karki, K. C. and Winstead, C. H. (1997) Computer simulation of magnetic chromatography system for ultra-fine particle separation. Trans. IEE Japan *117-B*, 1466–74.

[150] Ohara, T. (1997). Feasibility of using magnetic chromatography for ultra-fine particle separation. In: High Magnetic Fields: Applications, Generations, Materials (Schneider-Muntau, H. J., ed.). World Scientific, New Jersey, pp. 43–55.

[151] Ohara, T., Wang, X., Wada, H. and Whitby, E. R. (2000). Magnetic chromatography: Numerical analysis in the case of particle size distribution. Trans. IEE Japan *120-A*, 62–7.

[152] Karki, K. C., Whitby, E. R., Patankar, S. V., Winstead, C., Ohara, T and Wang, X. (2001). A numerical model for magnetic chromatography. Appl. Math. Model *25*, 355–73.

[153] Mitsuhashi, K., Yoshizaki, R., Ohara, T., Matsumoto, F., Nagai, H. and Wada, H. (2002). Retention of ions in a magnetic chromatograph using high-intensity and high-gradient magnetic fields. Sep. Sci. Technol. *37*, 3635–45.

[154] Berthier, J., Pham, P. and Massé, P. (2001). Numerical modeling of magnetic field flow fractionation in microchannels: A two-fold approach using particle trajectories and concentration, presented at 2001 International Conference on Modeling and Simulation of Microsystems, Hilton Head Island, South Carolina, USA, in Nanotech. (Vol. 1), pp. 202–5.

[155] Latham, A. H., Freitas, R. S., Schiffer, P. and Williams, M. E. (2005). Capillary magnetic field flow fractionation and analysis of magnetic nanoparticles. Anal. Chem. *77*, 5055–62.

[156] Giddings, J. C., Williams, P. S. and Beckett, R. (1987). Fractionating power in programmed field-flow fractionation: Exponential sedimentation field decay. Anal. Chem. *59*, 28–37.

[157] Davis, J. M. and Giddings, J. C. (1985). Retention theory for field-flow fractionation in annular channels. J. Phys. Chem. *89*, 3398–405.

[158] Bird, R. B., Stewart, W. E. and Lightfoot, E. N. (1960). Transport Phenomena. John Wiley & Sons, New York, NY.

[159] Martin, M. (1997). Time-based retention ratio for curved separation channels: Application to sedimentation field-flow fractionation. J. Microcolumn. Sep. *9*, 225–32.
[160] Davis, J. M. (1989). Nonequilibrium theory for field-flow fractionation in annular channels. Sep. Sci. Technol. *24*, 219–45.
[161] Williams, P. S., Giddings, M. C. and Giddings, J. C. (2001). A data analysis algorithm for programmed field-flow fractionation. Anal. Chem. *73*, 4202–11.
[162] Carpino, F., Moore, L. R., Zborowski, M., Chalmers, J. J. and Williams, P. S. (2005). Analysis of magnetic nanoparticles using quadrupole magnetic field-flow fractionation. J. Magn. Magn. Mater. *293*, 546–52.
[163] Carpino, F., Moore, L. R., Chalmers, J. J., Zborowski, M. and Williams, P. S. (2005). Quadrupole magnetic field-flow fractionation for the analysis of magnetic nanoparticles. J. Physics: Conference Series *17*, 174–80.
[164] Carpino F., Zborowski M. and Williams P. S. (2007). Quadrupole magnetic field-flow fractionation: A novel technique for the characterization of magnetic nanoparticles. J. Magn. Magn. Mater. *311*, 383–7.
[165] Williams, P. S. and Giddings, J. C. (1987). Power programmed field-flow fractionation: A new program form for improved uniformity of fractionating power. Anal. Chem. *59*, 2038–44.
[166] Williams, P. S. and Giddings, J. C. (1991). Comparison of power and exponential field programming in field-flow fractionation. J. Chromatogr. A *550*, 787–97.
[167] Williams, P. S. and Giddings, J. C. (1994). Theory of field programmed field-flow fractionation with corrections for steric effects. Anal. Chem. *66*, 4215–28.
[168] Williams, P. S., Giddings, J. C. and Beckett, R. (1987). Fractionating power in sedimentation field-flow fractionation with linear and parabolic field decay programming. J. Liq. Chromatogr. *10*, 1961–98.
[169] Williams, P. S., Kellner, L., Beckett, R. and Giddings, J. C. (1988). Comparison of experimental and theoretical fractionating power for exponential field decay sedimentation field-flow fractionation. Analyst *113*, 1253–9.
[170] Yamaura, M., Camilo, R. L., Sampaio, L. C., Macêdo, M. A., Nakamura, M. and Toma, H. E. (2004). Preparation and characterization of (3-aminopropyl) triethoxysilane-coated magnetite nanoparticles. J. Magn. Magn. Mater. *279*, 210–7.
[171] Hoyos, M. and Martin, M. (1994). Retention theory of sedimentation field-flow fractionation at finite concentrations. Anal. Chem. *66*, 1718–30.

[172] Martin, M. (1998). Theory of field-flow fractionation. In: Advances in Chromatography (Brown, P. R. and Grushka, E., eds.), Vol. 39. (Chapter 1). Marcel Dekker, Inc., New York, NY, pp. 1–138.
[173] Rosensweig, R. E. (1997). Ferrohydrodynamics. Dover Publications, Inc., Mineola, NY. (Chapter 2).
[174] Blums, E., Cebers, A. and Maiorov, M. M. (1996). Magnetic Fluids. Walter de Gruyter & Co., Berlin, Germany. (Chapter 5).

Laboratory Techniques in Biochemistry and Molecular Biology, Volume 32
Magnetic Cell Separation
M. Zborowski and J. J. Chalmers (Editors)

APPENDIX A

Nomenclature, abbreviations, units, and conversion factors

A.1. Nomenclature

Symbols are *italicized* in the text. The normal **bold** type indicates a vector quantity (a string of two or three components, in two or three dimensions, respectively)

Symbol	Description	Reference in text
$\langle x \rangle$	Root-mean-square displacement of Brownian motion	Equation (1.25)
$\langle \mu \rangle$	Ensemble average magnetic moment	Equation (2.2)
∇	Operator nabla, or gradient	Appendix B
α	Experimentally determined constant	Equation (8.23)
χ	Volume magnetic susceptibility	Equation (1.1)
ψ	Ratio of secondary antibodies bound to each primary antibody: $n_2\theta_2\lambda_2$	Section 8.5
δ	Transport lamina thickness	Equation (12.8)
ρ	Mass density	Equation (1.4)
μ	Magnetic dipole moment	Equation (8.2)
η	Viscosity	Equation (1.28)
μ_0	Magnetic permeability of free space	Equation (1.11)
θ_1	Fraction of antigens bound by primary antibody	Equation (8.4)
λ_1	Valence of primary antibody binding	Equation (8.4)
θ_2	Fraction of binding sites on primary antibody bound by secondary antibody	Equation (8.4)
λ_2	Valence of secondary antibody binding	Equation (8.4)
μ_A	Atomic magnetic moment	Equation (1.11)

(*continued*)

DOI: 10.1016/S0075-7535(06)32014-1

NOMENCLATURE (*continued*)

Symbol	Description	Reference in text
μ_B	Bohr magneton	Equation (2.1)
μ_{eff}	Effective magnetic moment	Equation (1.13)
μ_m	Absolute magnetic permeability	Equation (2.18)
μ_r	Relative magnetic permeability	Equation (2.18)
χ_{eff}	Effective, volume magnetic susceptibility	Equation (3.15)
χ_f	Volume susceptibility of a fluid	Equation (5.11)
χ_g	Mass (or specific) magnetic susceptibility	Equation (1.3)
$\chi_{g,globin}$	Mass (or specific) magnetic susceptibility of globin	Equation (1.4), and the following text
χ_{globin}	Volume magnetic susceptibility of globin	Equation (1.4), and the following text
χ_i	Volume magnetic susceptibility of component i	Equation (1.10)
χ_m	Molar magnetic susceptibility, or 1-g formula weight susceptibility	Equation (1.5)
$\chi_{m,ferro}$	Molar susceptibility of four ferro-heme groups of deoxyhemoglobin	Equation (1.18)
χ_p	Volume susceptibility of a particle	Equation (5.11)
χ_{RBC}	Volume susceptibility of the red blood cell	Equation (1.16)
ϕ_i	Volume fraction of ith component	Equation (1.10), and the following text
ρ_{ISS}	Inlet splitting surface radius (normalized)	Equation (12.5)
ρ_{OSS}	Outlet splitting surface radius (normalized)	Equation (12.6)
Ab	Antibody	Section 8.6
Ag	Antigen	Section 8.6
$Ag \cdot Ab$	Antigen-antibody complex	Equation (8.10)
B	Magnetic field intensity, or magnetic induction, or magnetic flux density	Equation (1.1)
CGS	Centimeter gram second system of units	Equation (1.2)
c_{Hb}	Intracellular hemoglobin concentration (molar)	Equation (1.19)
D	Diffusion coefficient	Equation (1.28)
D_c	Cell diameter	Equation (8.3)
e	Electron charge	Equation (2.1)
E	Electric field	Equation (5.8)
EMU	Electromagnetic units	Equation (1.2), and the following text
ESU	Electrostatic units	Equation (1.2), and the following text

NOMENCLATURE (*continued*)

Symbol	Description	Reference in text
F	Force	Equation (1.1)
F_d	Viscous drag force	Equation (1.37)
F_m	Magnetic force	Equation (1.37)
g	Gravitational acceleration	Equation (1.30)
H	Magnetic field strength	Equation (1.1)
h	Planck constant	Equation (2.1)
I	Electric current	Equation (2.12), and the following text
k	Boltzmann constant	Equation (1.11)
k_a, k_d	Association and dissociation reaction rate constants	Equation (8.8)
K_A, K_D	Association and dissociation chemical equilibrium constants	Equation (8.9)
L	Ligand	Equation (8.8)
L	Flow channel length adjacent to magnet	Equation (12.9)
l	Characteristic length, such as length of the magnetic pathline	Equation (5.12)
m_g	Mass	Equation (1.3)
M	Volume magnetization	Equation (3.25)
m	Magnetophoretic mobility	Equation (5.5)
m_{DEP}	Dielectrophoretic mobility	Equation (5.8)
m_e	Electron mass	Equation (2.1)
M_g	Mass magnetization	
M_m	Molar magnetization	Equation (2.3)
M_R	Remnant magnetization	Section 6.3
M_s	Saturation magnetization	Section 2.1
M_w	Molar mass, molecular weight	Equation (1.7), and the following text
N	Number of cells	Equation (9.2)
N	Number of moles	Equation (1.5)
$\mathbf{n}$	Surface unit normal vector	Appendix C
n_1	Number of antigen binding sites per cell	Equation (8.4)
n_2	Number of secondary antibody binding sites per primary antibody	Equation (8.4)
n_3	Number of magnetic particles conjugated to each secondary antibody	Equation (8.4)
N_A	Avogadro number	Equation (1.11)
n_B	Number of Bohr magnetons	Equation (1.17)
N_{mp}	Magnetic particle binding capacity	Section 8.6

(*continued*)

NOMENCLATURE (*continued*)

Symbol	Description	Reference in text
p	Pressure	Appendix C
P_t, $P(a)$, $P(b)$	Purity of target cells, or in outlets a or b	Equations (9.1), (12.12), (12.13)
Q	Heat of magnetization	Equation (3.17)
Q	Total flow rate	Equation (12.5)
$Q(a)$, $Q(a')$, $Q(b)$, $Q(b')$	Partial flow rates	Equations (12.5), (12.6)
R	Receptor	Equation (8.8)
R	Particle radius	Equation (1.28)
$R \cdot L$	Complexation of receptor and ligand	Equation (8.8)
r_0	Bohr radius	Equation (2.9)
r_i	Annular flow channel inner wall radius	Equation (12.1)
r_o	Annular flow channel outer wall radius	Equation (12.1)
S	Oxygen saturation of oxyhemoglobin	Equation (1.16)
s_g	Sedimentation coefficient	Equation (1.31)
SI	International System of Units	Equation (1.2)
S_m	Magnetophoretic driving force, or magnetic pressure gradient, or Maxwell pressure gradient	Equations (5.6), (12.3)
T	Absolute temperature (in kelvins, K)	Equation (1.11)
t	Time	Equation (1.26)
$\mathbf{T}$	Maxwell stress tensor	Appendix C
T_B	Blocking temperature	Equation (1.13)
T_C	Curie temperature	Equation (1.11)
U	Potential energy in the magnetostatic field	Equation (3.17)
u	Particle velocity	Equation (8.3)
$u_{settling}$	Terminal settling velocity	Equation (8.3)
V	Volume	Equation (1.1)
v	Particle velocity	Equation (1.38)
V_m	Molar volume	Equation (1.6), and the following text
x_g	Displacement due to gravitational sedimentation	Equation (1.29)
x_m	Displacement induced by the magnetic field	Equation (1.25)
Z	Fraction of hemoglobin converted to methemoglobin	Equation (1.23)

A.2. Abbreviations

Abbreviation	Description	Reference in text
AAD	Amino-actinomysin	Section 7.4.2.3
ABC	Antibody binding capacity	Section 8.6
ADME	Adsorption, distribution, metabolism, and excretion	Section 7.6.1
AM	Acetoxymethyl	Section 7.4.2.3
ASA	anti-sperm antibodies	Section 6.4.1
ATP	Adenosine triphosphate	Section 7.4.2.2
ATRP	Atomic transfer radical polymerization	Section 6.1
BM	Bone marrow	Table 9.1
BP	Benzyoyl peroxide	Section 6.5
CD	Cluster of differentiation (cell marker)	Section 7.5.1
CTV	Cell tracking velocimetry	Section 8.3
CL	Cell line	Table 9.1
DTPA	Diethylenetriamine pentaacetic acid	Section 7.3.2
EHS	Environmental Health and Safety (database at Rice University)	Section 7.7
ELISA	Enzyme-linked immunosorbent assay	Section 6.2.3.1
emu/g	Mass magnetization, in electromagnetic units of magnetic dipole moment per gram	Section 6.5
EPA	Environmental Protection Agency (U.S.)	Section 7.6.1
EPR	Electron paramagnetic resonance	Section 7.3.3
ESCA	Electron spectroscopy for chemical analysis	Section 7.3.2
FACS	Fluorescence-activated cell sorter	Section 7.4.2.3
Fc	Constant fragment of the IgG	Section 6.4.2.1
FCM	Flow cytometry	Table 9.1
FDA	Food and Drug Administration (in the U.S.)	Section 7.1
FITC	Fluorescein isothiocyanate	Section 7.4.2.4
FSE	Fast-spin echo	Section 6.2.4.2
GαH IgG	Goat anti-human IgG	Section 6.4.2.1
HSA	Human serum albumin	Section 6.2.3.1
HEA	Human epithelial antigen	Section 8.5
HGMS	High-gradient magnetic separator	Section 10.3
HPLC	High-performance liquid chromatography	Section 6.1
HRTEM	High-resolution TEM	Section 6.2.1
HSC	Hematopoietic stem cell	Table 9.1
IgG	Immunoglobulin G	Section 6.4.2.1
ICC	Immunocytochemistry	Table 9.1
IMS	Immunomagnetic separation	Table 9.1
INT	2-*p*-(iodophenyl)-3-(*p*-nitrophenyl)-5-phenyltetrazolium chloride	Section 7.4.2.3

(continued)

ABBREVIATIONS (*continued*)

Abbreviation	Description	Reference in text
IP	Intraperitoneal	Section 6.2.3.1
ISO	International Standardization Organization	Section 7.6
ISS	Inlet splitting surface	Equation (12.8)
OSS	Outlet splitting surface	Equation (12.8)
JNK	Kinase	Section 7.1
LD_{50}	Lethal dose, 50%	Section 7.2
LDA	Limited dilution assay	Table 9.3
LDH	Lactate dehydrogenase	Section 7.4.2.2
LHS	Left-hand side (of an equation)	
MAC	Methacryloyl chloride	Section 6.2.2.1
MACS	Magnetically activated cell sorter; trade name of magnetic separator and magnetic nanoparticles	Section 8.5
MAOETIB	Monomer, 2-methacryloyloxyethyl [2,3,5-triiodobenzoate]	Section 6.2.2.2
MCP-1	Monocyte chemoattractant protein-1	Section 7.5.2
METI	Ministry of Economics, International Trade and Industry (Japan)	Section 7.6.3
MHC	Major histocompatibility complex	Section 7.5.2
MMA	methyl methacrylate	Section 6.5
MNC	Mononuclear cells	Table 9.1
MPC	Magnetic Particle Concentrator	Section 10.1
MRI	(Nuclear) magnetic resonance imaging	Section 6.2.4.1
MTS	3-(4,5-dimethylthiazol-2-yl)-5-(3-carboxy-methoxyphenyl)-2-(4-sulfophenyl)-2H-tetrazolium	Section 7.4.2.2
MTT	3-(4,5 dimethyl thiazoloyl-2)-2,5-diphenyl tetrazolium bromide	Section 7.4.2.2
N&N	Nanosciences and nanotechnologies	Section 7.6.2
NAD	Nicotinamide adenine dinucleotide	Section 7.4.2.3
NADH	Nicotinamide adenine dinucleotide hydride	Section 7.4.2.3
NK	Natural killer (cell)	Section 7.5.1
NNI	National Nanotechnology Initiative (U.S.)	Section 7.6.1
NO	Nitric oxide	Section 7.3.2
NR	Neutral Red	Section 7.4.2.3
NSET	Nanoscale Science, Engineering and Technology Subcommittee (U.S.)	Section 7.6.1
P(PEGMA)	poly(poly(ethylene glycol) monomethacrylate)	Section 7.3.2
PB	Peripheral blood	Table 9.1
PBL	Peripheral blood lymphocytes	Table 9.1
PBPC	Peripheral blood progenitor cells	Table 9.1

ABBREVIATIONS (*continued*)

Abbreviation	Description	Reference in text
PBS	Phosphate-buffered saline	Section 6.2.3.1
PDVB	Poly(divinyl benzene)	Section 6.2.2.2
PE	Phycoerythrin	Section 8.4
PEG	Polyethylene glycol	Section 7.3.2
PLA	Polylactic alcohol	Section 7.3.2
PLGA	poly(lactide-co-glycolide)	Section 7.4.2.5
PMMA	Poly(methyl methacrylate)	Section 6.5
PMS	Phenazine methosulphate	Section 7.4.2.2
polyMAOETIB	Poly(2-methacryloyloxyethyl [2,3,5-triiodobenzoate])	Section 6.2.2.2
PS	Polystyrene	Section 6.3
PVA	Polyvinylalcohol	Section 7.3.2
RES	Reticuloendothelial system	Section 7.3.1
QMS	Quadrupole magnetic sorter	Section 9.4
RHS	Right-hand side (of an equation)	
ROS	Reactive oxygen species	Section 7.4.3.1
RT-PCR	Reverse transcription polymerase chain reaction	Section 9.4
SCC	Squamous cell carcinoma	Table 9.4
SDS	Sodium dodecyl sulfate	Section 6.2.2.2
SEM	Scanning electron microscopy	Section 6.4.1
SIMS	Secondary ion mass spectroscopy	Section 7.3.2
SPIO	Superparamagnetic iron oxides	Section 7.4.3
SPION	Superparamagnetic iron oxide nanoparticle	Section 7.3.2
SQUID	Superconducting quantum interference device	Section 8.2
SRBC	Sheep red blood cells	Section 6.4.2.2
STM	Scanning tunneling microscopy	Section 7.3.2
TBARS	Thiobarbituric acid-reactive substances	Section 7.3.3
TCC	Terminal complement complex	Section 7.5.2
TE	Time to echo	Section 6.2.4.2
TEM	Transmission electron microscopy	Section 6.2.1
TR	Repetition time	Section 6.2.4.2
TUNEL	Terminal transferase dUTP nick end labeling	Section 7.4.2.4
USPIO	Ultrasmall superparamagnetic iron oxide (particles)	Section 7.4.3.1
XPS	X-ray photoelectron spectroscopy	Section 6.3
XRD	X-ray diffraction	Section 6.3
XTT	Sodium(2,3-bis(2-methoxy-4-nitro-5-sulpho-phenyl)-2H-tetrazolium-5-carboxanilide	Section 7.4.2.2

A.3. Selected units and conversion factors. For symbol designation, see Table A.1. For susceptibility conversion, see Chapter 1 and Table 1.1

Symbol	Selected formulas (derived quantities)	EMU CGS units (abbreviation)	SI units (abbreviation)	To obtain value in SI units, multiply value in CGS units by
η	$\frac{1}{6\pi} F R^{-1} v^{-1}$	cm^{-1} g s^{-1}	m^{-1} kg s^{-1}	10^{-5}
μ	Il^2	$cm^{5/2}$ $g^{1/2}$ s^{-1}(emu)	A m^2	10^{-3}
μ_B		9.274×10^{-21}erg G^{-1}	9.274×10^{-24}J T^{-1}	10^{-3}
μ_0		1	$4\pi \times 10^{-7}$T m A^{-1}	$4\pi \times 10^{-7}$ T m A^{-1}
v	lt^{-1}	cm s^{-1}	m s^{-1}	10^{-2}
ρ		g cm^{-3}	kg m^{-3}	10^{-3}
χ		1	1	4π
B		gauss (G)	tesla (T)	10^{-4}
D	$\frac{1}{2} l^2 t^{-1}$	cm^2 s^{-1}	m^2 s^{-1}	10^{-4}
e			1.602×10^{-19} A s	
F	IlB_0	dyne	newton (N)	10^{-5}
g		981 cm s^{-2}	9.81 m s^{-2}	10^{-2}
H	Il^{-1}	oersted (Oe)	A m^{-1}	$1000 \times (4\pi)^{-1}$
I		$cm^{1/2}$ $g^{1/2}$ s^{-1}	ampere (A)	10^{-1}
k		1.381×10^{-16} erg K^{-1}	1.381×10^{-23}J K^{-1}	10^{-7}
l		centimeter (cm)	meter (m)	10^{-2}
m	$v{S_m}^{-1}$		m^3 T^{-1} A^{-1} s^{-1} = m^5 J^{-1} s^{-1} = m^3 s kg^{-1}	
m_e			9.11×10^{-31} kg	
M	μV^{-1}	$cm^{-1/2}$ $g^{1/2}$ s^{-1} (emu cm^{-3})	A m^{-1}	10^3
m_g		gram (g)	kilogram (kg)	10^{-3}
N_A		6.022×10^{23} mol^{-1}	6.022×10^{23} mol^{-1}	1
r_0			0.53×10^{-10} m	
s_g		s	s	1
S_m	$\frac{d}{dx}(\frac{1}{2} HB_0)$		T A m^{-2} = J m^{-4}	
[illegible]			[illegible]	10^{-7}

Laboratory Techniques in Biochemistry and Molecular Biology, Volume 32
Magnetic Cell Separation
M. Zborowski and J. J. Chalmers (Editors)

APPENDIX B

Vector notation

The notation follows the style in Becker [1]. The vector and tensor quantities are indicated by use of bold face (**A**), the scalar quantities and the vector magnitudes are indicated by italics (A). The unit normal vectors of the Cartesian coordinate system are indicated by $\mathbf{e}_x$, $\mathbf{e}_y$, $\mathbf{e}_z$. A good introduction to vector algebra and vector analysis can be found in any number of physics textbooks, such as Becker or Stratton [1, 2]. The formulas provided in this section are selected for their relevance to the material of the book and they are by no means exhaustive. In particular, the formulas are limited to Cartesian coordinates.

A vector is an ordered list of three numbers A_x, A_y, A_z (in three-dimensional space):

$$\mathbf{A} = [A_x, A_y, A_z] = A_x\mathbf{e}_x + A_y\mathbf{e}_y + A_z\mathbf{e}_z \tag{B.1}$$

where

$$\begin{aligned} \mathbf{e}_x &= [1, 0, 0] \\ \mathbf{e}_y &= [0, 1, 0] \\ \mathbf{e}_z &= [0, 0, 1] \end{aligned} \tag{B.2}$$

are unit normal vectors along (Cartesian) coordinate axes. Vector magnitude:

$$A = \sqrt{A_x^2 + A_y^2 + A_z^2} \tag{B.3}$$

DOI: 10.1016/S0075-7535(06)32015-3

An example of a vector is the particle position vector, **r**:

$$\mathbf{r} = [x, y, z] = x\mathbf{e}_x + y\mathbf{e}_y + z\mathbf{e}_z$$
$$r = \sqrt{x^2 + y^2 + z^2} \tag{B.4}$$

A vector that is a function of the position coordinates is also called a vector-valued function or a vector field.

Product of a scalar, ϕ, and a vector, **A**:

$$\phi\mathbf{A} = [\phi A_x, \phi A_y, \phi A_z] \tag{B.5}$$

Scalar product of two vectors, **A** and **B** (also known as the inner product or the dot product), is commutative:

$$\mathbf{A} \cdot \mathbf{B} = A_x B_x + A_y B_y + A_z B_z = B_x A_x + B_y A_y + B_z A_z = \mathbf{B} \cdot \mathbf{A} \tag{B.6}$$

In particular, a scalar product of a vector with itself

$$\mathbf{A} \cdot \mathbf{A} = A_x A_x + A_y A_y + A_z A_z = A_x^2 + A_y^2 + A_z^2 = AA = A^2 \tag{B.7}$$

Geometrical interpretation of the scalar product

$$\mathbf{A} \cdot \mathbf{B} = AB\cos(\measuredangle(\mathbf{A}, \mathbf{B})) \tag{B.8}$$

where $\measuredangle(\mathbf{A}, \mathbf{B})$ is the angle between vectors **A** and **B**. In particular, for orthogonal vectors

$$\mathbf{A} \cdot \mathbf{B} = 0 \text{ for } \mathbf{A} \perp \mathbf{B} \tag{B.9}$$

In particular:

$$\begin{aligned} \mathbf{e}_x \cdot \mathbf{e}_y = \mathbf{e}_y \cdot \mathbf{e}_z = \mathbf{e}_z \cdot \mathbf{e}_x = 0 \\ \mathbf{e}_x \cdot \mathbf{e}_x = \mathbf{e}_y \cdot \mathbf{e}_y = \mathbf{e}_z \cdot \mathbf{e}_z = 1 \end{aligned} \tag{B.10}$$

Definition of a gradient (grad) of the scalar quantity, ϕ:

$$\mathrm{grad}\phi = \nabla\phi = \lim_{V \to 0} \frac{\oiint_S \phi \mathrm{d}\mathbf{S}}{V} \tag{B.11}$$

where the right-hand side (RHS) denotes a ratio of closed surface integral of a function ϕ times directed surface element $\mathrm{d}\mathbf{S} = \mathbf{n}\mathrm{d}S$, to the volume, V, enclosed by the surface, S, in the limit for the vanishingly small volume, V. The vector $\mathbf{n}$ is the unit vector normal to the surface element dS, directed to the outside of the surface S. The illustration of such surfaces and directed surface elements are provided in Appendix C (Figs. C.1 and C.6). The symbol for the gradient, ∇, is called a nabla operator. In Cartesian coordinates, for a function ϕ that is differentiable in V, Eq. (B.11) reduces to

$$\nabla\phi = \left[\frac{\partial\phi}{\partial x}, \frac{\partial\phi}{\partial y}, \frac{\partial\phi}{\partial z}\right] \equiv \frac{\partial\phi}{\partial x}\mathbf{e}_x + \frac{\partial\phi}{\partial y}\mathbf{e}_y + \frac{\partial\phi}{\partial z}\mathbf{e}_z \tag{B.12}$$

where symbols $\frac{\partial}{\partial x}$ and the like denote partial differentiation with respect to variable x and the like. Equation (B.12) can be treated formally as a product of a scalar, ϕ, and a vector, ∇, in a manner analogous to Eq. (B.5), so that

$$\nabla \equiv \left[\frac{\partial}{\partial x}, \frac{\partial}{\partial y}, \frac{\partial}{\partial z}\right] \tag{B.13}$$

The nabla symbol should not be treated as an ordinary vector, however (it is a vector operator, also called a "del" operator). In particular, it does not follow the commutative rule under the scalar product operation, as the ordinary vectors do, Eq. (B.6). For instance

$$\mathbf{A}\cdot\nabla = A_x\frac{\partial}{\partial x} + A_y\frac{\partial}{\partial y} + A_z\frac{\partial}{\partial z} \tag{B.14}$$

and the result is a partial differentiation operator, which, when applied to a function, ϕ, returns a scalar value that is a gradient ϕ magnitude along the vector $\mathbf{A}$:

$$(\mathbf{A}\cdot\nabla)\phi = A_x\frac{\partial\phi}{\partial x} + A_y\frac{\partial\phi}{\partial y} + A_z\frac{\partial\phi}{\partial z} \tag{B.15}$$

On the other hand

$$\nabla\cdot\mathbf{A} = \frac{\partial A_x}{\partial x} + \frac{\partial A_y}{\partial y} + \frac{\partial A_z}{\partial z} \equiv \mathrm{div}\mathbf{A} \tag{B.16}$$

where the RHS is the divergence operator acting on a vector, $\mathbf{A}$.

The result of the multiplication of a vector operator, $\mathbf{A} \cdot \nabla$, and a vector, $\mathbf{B}$:

$$(\mathbf{A} \cdot \nabla)\mathbf{B} = \left[A_x \frac{\partial B_x}{\partial x} + A_y \frac{\partial B_x}{\partial y} + A_z \frac{\partial B_x}{\partial z}, A_x \frac{\partial B_y}{\partial x} + A_y \frac{\partial B_y}{\partial y} + A_z \frac{\partial B_y}{\partial z}, A_x \frac{\partial B_z}{\partial x} + A_y \frac{\partial B_z}{\partial y} + A_z \frac{\partial B_z}{\partial z} \right] \neq A(\nabla B) \tag{B.17}$$

which is a consequence of another property of the vector operator ∇ multiplication, that is not associative under the scalar operation. The inequality on the RHS changes to equality, however, for conservative vector fields $\mathbf{A}$ and $\mathbf{B}$, that is, the fields that can be expressed as a gradient of a certain scalar function, ϕ:

$$\mathbf{A} = \nabla\phi \quad \text{for a conservative field } \mathbf{A} \tag{B.18}$$

In those instances, the function ϕ is referred to as the field potential. Therefore

$$(\mathbf{A} \cdot \nabla)\mathbf{B} = A\nabla B \quad \text{conservative fields } \mathbf{A} \text{ and } \mathbf{B} \tag{B.19}$$

The exercise of proving that this is indeed so, is left to the reader. Definition of a divergence (div) of the vector quantity, $\mathbf{A}$:

$$\text{div}\mathbf{A} \equiv \nabla \cdot \mathbf{A} = \lim_{V \to 0} \frac{\oiint_S \mathbf{A} \cdot \text{d}\mathbf{S}}{V} \tag{B.20}$$

where the RHS denotes a ratio of a closed surface integral of a field $\mathbf{A}$ in a scalar product with the directed surface element $\text{d}\mathbf{S} = \mathbf{n}\text{d}S$, over volume, V, enclosed by the surface, S, in the limit for the vanishingly small volume, V. Here, as in Eq. (B.11), the vector $\mathbf{n}$ is the unit vector normal to the surface element $\text{d}S$, directed to the outside of the surface S (Figs. C.1 and C.6). In Cartesian coordinates

$$\text{div}\mathbf{A} \equiv \nabla \cdot \mathbf{A} = \frac{\partial A_x}{\partial x} + \frac{\partial A_y}{\partial y} + \frac{\partial A_z}{\partial z} \tag{B.21}$$

as already shown in Eq. (B.16).

A tensor quantity, **T**, is an array of components that follow the same transformation rules as the position vectors under the operation of coordinate transformation (such as translation or rotation). In particular, a tensor may have a form of a matrix

$$\mathbf{T} = \begin{pmatrix} T_{xx} & T_{xy} & T_{xz} \\ T_{yx} & T_{yy} & T_{yz} \\ T_{zx} & T_{zy} & T_{zz} \end{pmatrix} \tag{B.22}$$

Here the Cartesian coordinates are used as indices of the tensor components (as in the Maxwell stress tensor). An inner product of a tensor, **T** and a vector, **n**, is defined as follows:

$$\mathbf{T}\cdot\mathbf{n} = \begin{pmatrix} T_{xx} & T_{xy} & T_{xz} \\ T_{yx} & T_{yy} & T_{yz} \\ T_{zx} & T_{zy} & T_{zz} \end{pmatrix} \cdot \begin{pmatrix} n_x \\ n_y \\ n_z \end{pmatrix} = \begin{pmatrix} T_{xx}n_x + T_{xy}n_y + T_{xz}n_z \\ T_{yx}n_x + T_{yy}n_y + T_{yz}n_z \\ T_{zx}n_x + T_{zy}n_y + T_{zz}n_z \end{pmatrix} \tag{B.23}$$

The result is a vector, or a vector field, which obeys the same rules as described above for vectors. In general, the inner product of a tensor and a vector (or a tensor and a tensor, not shown here) is not commutative.

The divergence operation on a tensor is defined as follows:

$$\begin{aligned} \mathrm{div}\mathbf{T} \equiv \nabla\cdot\mathbf{T} &= \left[\frac{\partial}{\partial x}, \frac{\partial}{\partial y}, \frac{\partial}{\partial z}\right] \cdot \begin{pmatrix} T_{xx} & T_{xy} & T_{xz} \\ T_{yx} & T_{yy} & T_{yz} \\ T_{zx} & T_{zy} & T_{zz} \end{pmatrix} \\ &= \begin{pmatrix} \dfrac{\partial T_{xx}}{\partial x} + \dfrac{\partial T_{yx}}{\partial y} + \dfrac{\partial T_{zx}}{\partial z} \\ \dfrac{\partial T_{xy}}{\partial x} + \dfrac{\partial T_{yy}}{\partial y} + \dfrac{\partial T_{zy}}{\partial z} \\ \dfrac{\partial T_{xz}}{\partial x} + \dfrac{\partial T_{yz}}{\partial y} + \dfrac{\partial T_{zz}}{\partial z} \end{pmatrix} \end{aligned} \tag{B.24}$$

Again, the result is a vector.

References

[1] Becker, R. (1982). Electromagnetic fields and interactions. Dover Publications, Inc., New York.
[2] Stratton, J. A. (1941). Electromagnetic theory. McGraw-Hill Book Company, Inc., New York.

Laboratory Techniques in Biochemistry and Molecular Biology, Volume 32
Magnetic Cell Separation
M. Zborowski and J. J. Chalmers (Editors)

APPENDIX C

Magnetic body force and the Maxwell stresses

C.1. Magnetic body force and the Maxwell stresses

The total magnetic force acting on a magnetically susceptible body of volume V is equal to the integral of the force density over volume:

$$\mathbf{F} = \int_V \mathbf{f} \mathrm{d}V \tag{C.1}$$

It has been shown by Maxwell [1, 2] that the magnetic force density is equal to a divergence of a certain tensor, $\mathbf{T}$:

$$\mathbf{f} = \mathrm{div}\mathbf{T} \equiv \lim_{V \to 0} \frac{\oint_A \mathbf{T} \cdot \mathbf{n} \mathrm{d}S}{V} \tag{C.2}$$

(see Appendix B for additional information about vector analysis). In free space, the Maxwell stress tensor $\mathbf{T}$ has the form:

$$\mathbf{T} = \begin{pmatrix} H_x B_x - \frac{1}{2}\mathbf{H} \cdot \mathbf{B} & H_x B_y & H_x B_z \\ H_y B_x & H_y B_y - \frac{1}{2}\mathbf{H} \cdot \mathbf{B} & H_y B_z \\ H_z B_x & H_z B_y & H_z B_z - \frac{1}{2}\mathbf{H} \cdot \mathbf{B} \end{pmatrix} \tag{C.3}$$

In matter, $\mathbf{B} = \mu_0(\mathbf{M} + \mathbf{H})$.

DOI: 10.1016/S0075-7535(06)32016-5

In the mechanics of continuous media and in hydrostatics, the stress tensor plays an important role because it determines local pressure and stress components, such as tension [1–4]. Similarly, the magnetostatic stress exerted on a surface bounding the magnetic body is equal to a scalar product of the Maxwell stress tensor, $\mathbf{T}$ and a unit vector normal to the surface, $\mathbf{n}$, directed away from the body [1]

$$\mathbf{T} \cdot \mathbf{n} \tag{C.4}$$

Substituting for $\mathbf{T}$ the expression for the Maxwell stress tensor, Eq. (C.3), one obtains the following expression for the Maxwell stress, or the magnetic stress:

$$\begin{aligned}\mathbf{T} \cdot \mathbf{n} &= \begin{pmatrix} (H_x B_x - \frac{1}{2}\mathbf{H} \cdot \mathbf{B}) n_x + H_x B_y n_y + H_x B_z n_z \\ H_y B_x n_x + (H_y B_y - \frac{1}{2}\mathbf{H} \cdot \mathbf{B}) n_y + H_y B_z n_z \\ H_z B_x n_x + H_z B_y n_y + (H_z B_z - \frac{1}{2}\mathbf{H} \cdot \mathbf{B}) n_z \end{pmatrix} \\ &= \mathbf{H}(\mathbf{B} \cdot \mathbf{n}) - \tfrac{1}{2}(\mathbf{H} \cdot \mathbf{B})\mathbf{n} \end{aligned} \tag{C.5}$$

Elements of vector and tensor calculus are provided in Appendix B.

C.2. Maxwell stress on a sphere

The Maxwell stress on a sphere is calculated using Eq. (C.5) and the coordinate system as in Fig. C.1 The simplest case is a sphere made of a homogeneous, isotropic, and linearly polarizable (regular) material of magnetization M_2, suspended in a regular medium of magnetization M_1 when exposed to an applied magnetic field H. It follows that $\mathbf{B} \parallel \mathbf{H}$ and therefore $\mathbf{H} \cdot \mathbf{B} = HB$ both inside the sphere and in the suspending medium. For the inside leaf of the sphere surface (Fig. C.1) one obtains from Eq. (C.5):

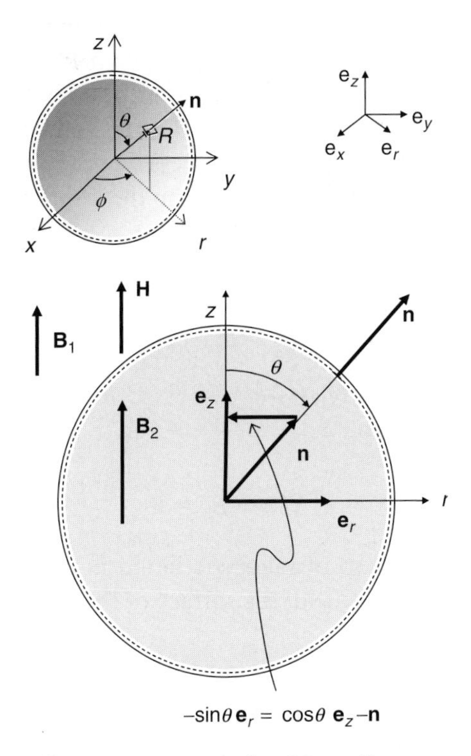

Fig. C.1. The coordinate system to calculate Maxwell stress on the surface of a sphere of radius R. The unit normal vectors directed along Cartesian coordinate axes are designated $\mathbf{e}_x$, $\mathbf{e}_y$, $\mathbf{e}_z$. The unit vector along polar coordinate r is $\mathbf{e}_r$, and the unit normal vector to the sphere surface is $\mathbf{n}$. The inside of the sphere is indicated by gray shading, the boundary of the outside medium by a solid line circle, the interface by a broken line circle. The diagram in the lower panel serves the purpose of demonstrating the equality: $-\sin\theta\mathbf{e}_r = \cos\theta\mathbf{e}_z - \mathbf{n}$.

$$
\begin{aligned}
(\mathbf{T}\cdot\mathbf{n})_2 &= \mathbf{H}(\mathbf{B}_2\cdot\mathbf{n}) - \tfrac{1}{2}(\mathbf{H}\cdot\mathbf{B}_2)\mathbf{n} \\
&= \mathbf{H}B_2\cos\theta - \tfrac{1}{2}HB_2\mathbf{n} \\
&= \tfrac{1}{2}\mathbf{H}B_2\cos\theta + \tfrac{1}{2}\mathbf{H}B_2\cos\theta - \tfrac{1}{2}HB_2\mathbf{n} \\
&= \tfrac{1}{2}HB_2\cos\theta\mathbf{e}_z + \tfrac{1}{2}HB_2(\cos\theta\mathbf{e}_z - \mathbf{n}) \\
&= \tfrac{1}{2}HB_2(\cos\theta\mathbf{e}_z - \sin\theta\mathbf{e}_r) \qquad \text{(C.6)}
\end{aligned}
$$

The last equation results from the geometrical analysis illustrated in Fig. C.1. A similar expression obtains for the outside leaf of the sphere surface, except that the magnetic induction in the surrounding medium is B_1. The net stress on the sphere surface is a difference between the magnetic stress on the inside and the outside leaves of the sphere surface, and therefore

$$\begin{aligned}\Delta(\mathbf{T}\cdot\mathbf{n}) &\equiv (\mathbf{T}\cdot\mathbf{n})_2 - (\mathbf{T}\cdot\mathbf{n})_1 \\ &= \tfrac{1}{2}H(B_2 - B_1)(\cos\theta\mathbf{e}_z - \sin\theta\mathbf{e}_r) \\ &= \tfrac{1}{2}\mu_0 H\Delta M(\cos\theta\mathbf{e}_z - \sin\theta\mathbf{e}_r) \qquad (C.7)\end{aligned}$$

where

$$\mu_0\Delta M = B_2 - B_1 \qquad (C.8)$$

For a spherical drop of magnetic liquid in equilibrium, in the absence of the magnetic field, the surface of the sphere corresponds to the stress of the form:

$$\mathbf{t}_0 = t_0\cos\theta\mathbf{e}_z + t_0\sin\theta\mathbf{e}_r \qquad (C.9)$$

where t_0 is the surface stress magnitude at equilibrium. The application of the magnetic field introduces an excess stress that causes deformation of the liquid drop, so that at the surface the following equation holds:

$$\mathbf{t}_0 + \Delta(\mathbf{T}\cdot\mathbf{n}) = (t_0 + \tfrac{1}{2}\mu_0 H\Delta M)\cos\theta\mathbf{e}_z + (t_0 - \tfrac{1}{2}\mu_0 H\Delta M)\sin\theta\mathbf{e}_r \qquad (C.10)$$

Equation (C.10) is illustrated for a number of cases that may help visualize the Maxwell stress effects in heterogeneous and magnetically susceptible media. In the case of biological materials, such effects become noticeable only for high fields (and gradients) as the magnetic susceptibility of most such materials is very small and are less regular than illustrated here because of the magnetic anisotropy of cellular structures.

C.2.1. Rigid sphere, uniform field, $\Delta M > 0$

The Maxwell stress distribution on a surface of a sphere of a constant magnetization, **M**, in a uniform magnetic field, **H** = **const**., as calculated from Eq. (C.10) is illustrated in Fig. C.2A. Note that the magnitude of the Maxwell stress is constant over the surface of the sphere and equal to $\frac{1}{2}\mu_0 \Delta MH$, but that the direction changes with the polar angle, θ, so that it acts as a pure tension at the poles ($\theta = \pi/2$ and $-\pi/2$) and a pure compression at the equator ($\theta = 0$). In between, it has a nonvanishing component tangent to the surface, and therefore acts as a shear. For a rigid sphere in a uniform magnetic field, the net Maxwell stress over the sphere surface is zero and therefore the field has no effect on the sphere motion.

C.2.2. Rigid sphere, constant field gradient, $\Delta M > 0$

The effect of the nonuniform magnetic field is illustrated here for the simplest case of a constant field gradient along z direction, so that

$$\begin{aligned} H_1 &\equiv \frac{\mathrm{d}H}{\mathrm{d}z} = \text{const} \\ H &= H_0 + H_1 z = H_0 + H_1 R \sin\theta \end{aligned} \tag{C.11}$$

where R is the radius of the sphere (Fig. C.1). The resulting Maxwell stress distribution on the sphere surface, calculated by substituting Eq. (C.11) into Eq. (C.10), is illustrated in Fig. C.2B One notes a shift in the tension at the poles in the direction of the gradient when compared with the case of the uniform field, resulting in the net stress acting in the direction of the field gradient, and the motion of the sphere following the direction of the gradient. The field gradient indicates the direction toward the field source, and thus the net effect is the sphere moving toward the magnetic field source (magnetic attraction).

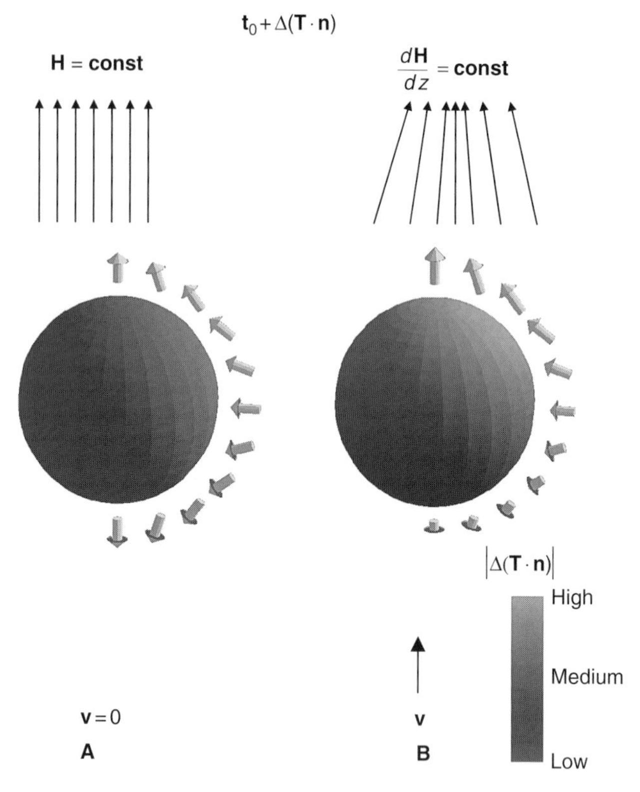

Fig. C.2. Magnetic stress distribution on a surface of rigid sphere that is more magnetic than the medium ($\Delta M > 0$, Eq. (C.10)). (A) Uniform applied field. Note purely tensile contribution at the poles and purely compressive stress contribution at the equator, as indicated by the arrows, that cancel out when integrated over the sphere surface (the sphere velocity, $\mathbf{v} = 0$). (B) In the field gradient, the tension at the pole exposed to a stronger field overcomes tension of the pole exposed to a weaker field, with the net effect of the sphere being drawn toward the field source ($\mathbf{v} \neq 0$). (See Color Insert.)

C.2.3. Rigid sphere, uniform field, $\Delta M < 0$

The direction of the Maxwell stress reverses, Fig. C.3A, as compared to the case of the uniform field and $\Delta M > 0$, discussed above, however, without affecting the motion of the sphere as a whole.

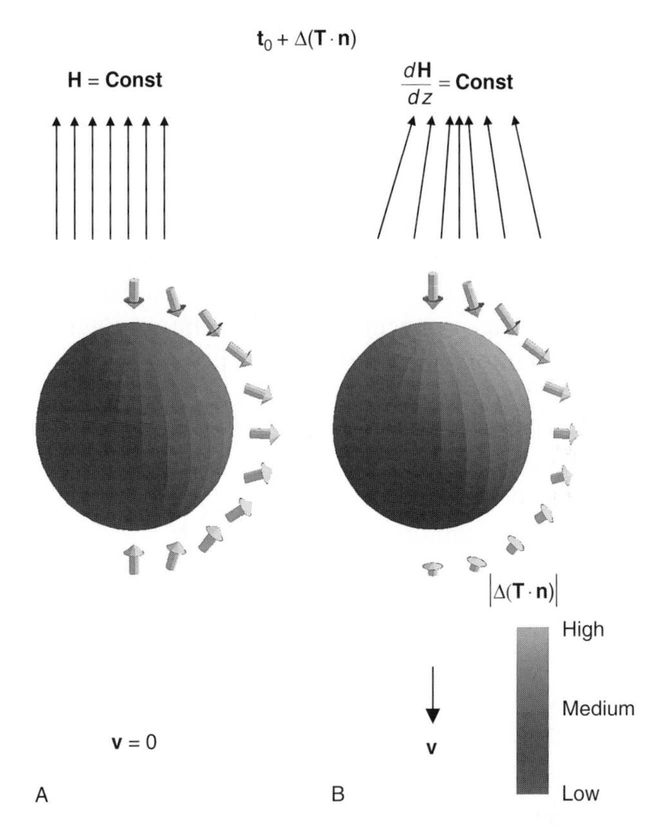

Fig. C.3. Magnetic stress distribution on a surface of rigid sphere that is less magnetic than the medium (ΔM <0, Eq. (C.10)). (A) Uniform applied field. Note change in the direction of stress arrows as compared to Fig. C.2, with no change in the net effect on the sphere ($\mathbf{v} = 0$). (B) In the field gradient, the compressive stress at the pole exposed to a stronger field overcomes stress at the pole exposed to a weaker field, with the net effect of the sphere being pushed away from the field source ($\mathbf{v} \neq 0$). (See Color Insert.)

C.2.4. Rigid sphere, constant field gradient, $\Delta M < 0$

Here also the direction of the Maxwell stress reverses, Fig. C.3B, as compared to the case of constant field gradient and $\Delta M > 0$, with the net effect of the sphere being pushed away from the field source (magnetic repulsion).

C.2.5. Elastic sphere, uniform field, $\Delta M > 0$

The Maxwell stress formulation is particularly instructive when applied to visco-elastic magnetic media, such as ferrofluids, extensively studied by Rosensweig and others [5]. It's relevance to the magnetic cell separation becomes apparent only in the case of very high fields and gradients. With the rapid development of new, stronger permanent magnets and permanent magnet configurations (such as the Halbach configurations), and the continuing search for low temperature, superconducting materials, such high field applications to the magnetic cell separation become increasingly economically viable and likely in the future. The expression for the Maxwell stress, Eq. (C.10), provides a convenient means to illustrate the field effects on a simple model of a homogeneous, viscoelastic media of uniform magnetization, M.

Here the surface normal determines a paraboloid of revolution with the major half-axis equal to $a \equiv p_0 + \frac{1}{2}\mu_0 H \Delta M$ (directed along the field vector H if $\Delta M > 0$) and the minor half-axis equal to $b \equiv p_0 - \frac{1}{2}\mu_0 H \Delta M$ (lying in the plane, perpendicular to the field vector **H** if $\Delta M > 0$), as illustrated in Fig. C.4A The half-distance between the foci is equal to $c = \sqrt{a^2 - b^2} = \sqrt{2 p_0 \mu_0 H \Delta M}$. The effect of the field is the elongation of the elastic body along the field lines leading to a prolate spheroid, however, with no net stress to effect the translational motion.

C.2.6. Elastic sphere, constant field gradient, $\Delta M > 0$

As in the case of the rigid sphere, the simplest case of the field gradient directed along the z axis is considered, and the total stress is calculated by inserting the expression for H as a function of z (or θ), Eq. (C.11), into Eq. (C.10). The resulting Maxwell stress distribution and the shape of the elastic body are illustrated in Fig. C.4B. Note a pear-like shape with the pointed end directed toward the magnetic field source. Also, there is now a net stress pushing the body in the direction of the field gradient, with the pointed end

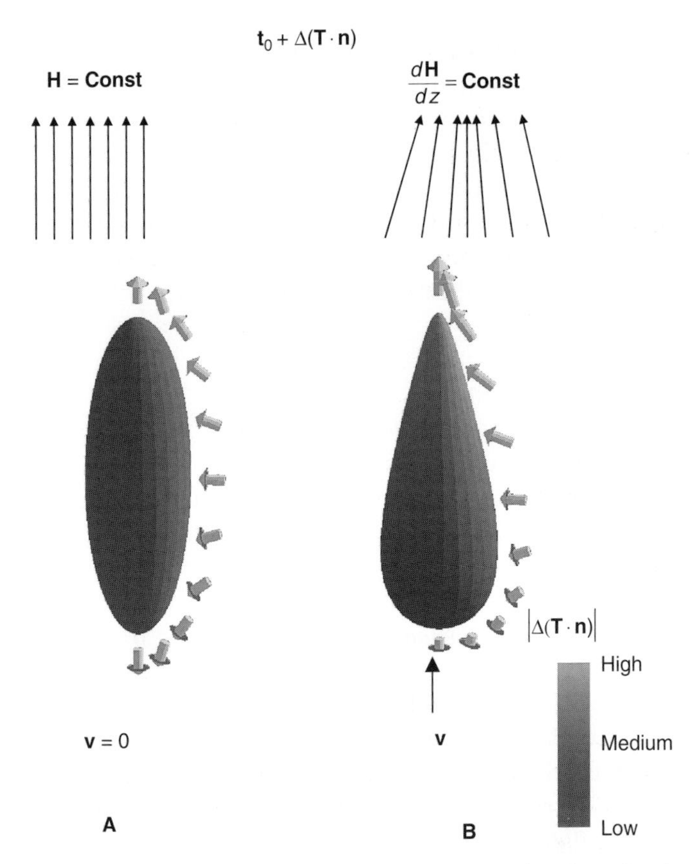

Fig. C.4. Magnetic stress distribution applied to an elastic sphere that is more magnetic than the medium ($\Delta M > 0$, Eq. (C.10)). (A) Uniform applied field. Note deformation as a result of purely tensile contribution at the poles and purely compressive stress contribution at the equator, as indicated by the arrows, that cancel out over the sphere surface ($\mathbf{v} = 0$). (B) In the field gradient, the tension at the pole exposed to a stronger field overcomes tension of the pole exposed to a weaker field, resulting in a characteristic, pear-like deformation with the net effect of the body being drawn toward the field source ($\mathbf{v} \neq 0$). (See Color Insert.)

being pushed the strongest. This is reminiscent of the shape of an inverted hot air balloon and suggests analogies with the buoyancy of heterogeneous media in the gravitational field, except that the buoyancy due to gravity is related to the field intensity alone, unlike the case of the magnetic buoyancy, which requires the field gradient. In the limit of high fields and gradients and large magnitude of the term ΔM, one observes separation of the pointed end of the visco-elastic body and the subsequent beading [5].

C.2.7. Elastic sphere, uniform field, $\Delta M < 0$

Here the Maxwell stress causes the deformation to an oblate spheroid, Fig. C.5A, with no net stress and no motion of the body as a whole.

C.2.8. Elastic sphere, constant field gradient, $\Delta M < 0$

One observes flattening of the side of the oblate spheroid exposed to the source of the magnetic field and the net stress pushing the body away from the field source, Fig. C.5B. The shape is reminiscent of that of an air bubble rising in water [6], except that it is inverted in relation to the direction of the net stress, and that the magnetic field gradient is required to effect the motion. The repulsion of the magnetic body whose "magnetic density" (magnetization M) is less than that of the surrounding medium (M) gave rise to the term of the "magnetic Archimedes effect."

C.3. Magnetic force on a particle in the limit of small particle size

The total magnetic force exerted on a body is equal to a closed surface integral of the Maxwell stress

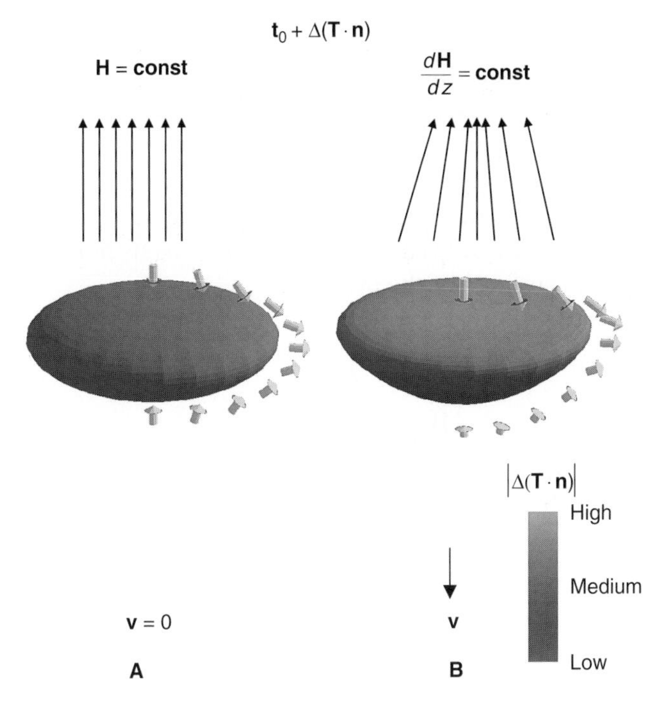

Fig. C.5. Magnetic stress distribution applied to an elastic sphere that is less magnetic than the medium ($\Delta M < 0$, Eq. (C.10)). (A) Uniform applied field. Note change in the direction of stress arrows as compared to Fig. C.4 resulting in the change in deformation from prolate to oblate ellipsoid of revolution ($\mathbf{v} = 0$). (B) In the field gradient, the stress at the pole exposed to a stronger field overcomes stress of the weaker field, resulting in an asymmetrical flattening of the oblate ellipsoid and the motion of the spheroid away from the field source ($\mathbf{v} \neq 0$). (See Color Insert.)

$$\mathbf{F} = \int_V \mathbf{f} \mathrm{d}V = \oint_A \mathbf{T} \cdot \mathbf{n} \mathrm{d}S \tag{C.12}$$

Interestingly, the surface may not correspond to the surface of the magnetic body, as long as it completely encloses the body [1]. An example of the Maxwell surface is that of the cylinder with its axis directed along the major component of the magnetic field, Fig. C.6. The difference in the magnetic stress between the inner leaf, $(\mathbf{T}\cdot\mathbf{n})'$, and outer leaf, $(\mathbf{T}\cdot\mathbf{n})$, of the Maxwell surface is calculated separately for the top, bottom, and the side walls of the cylinder, taking advantage of the nearly cylindrical symmetry of the field in the limit of small particle size. The Maxwell surface representation convention in Fig. C.6 is the same as that in Fig. C.1 (the gray shade indicates inside of the magnetic body).

The surface integral is a sum of contributions from the bottom, side, and top surfaces of the cylinder:

$$\oint_A \mathbf{T}\cdot\mathbf{n}\mathrm{d}S = \left(\iint \mathbf{T}\cdot\mathbf{n}\mathrm{d}S\right)_{\mathrm{bot}} + \left(\iint \mathbf{T}\cdot\mathbf{n}\mathrm{d}S\right)_{\mathrm{side}} + \left(\iint \mathbf{T}\cdot\mathbf{n}\mathrm{d}S\right)_{\mathrm{top}} \tag{C.13}$$

One notes that because of the axial symmetry, the stress contribution from the side of the cylinder vanishes, that is $\left(\iint \mathbf{T}\cdot\mathbf{n}\mathrm{d}S\right)_{\mathrm{side}} = 0$. Therefore, the total magnetic force acting on the body is the result of difference of net stresses on top and bottom sides of the cylinder.

In the limit of a vanishingly small volume of the magnetic body, the stress difference between the top and bottom sides of the cylinder, divided by its volume equals to

$$\lim_{V\to 0} \frac{\oint_A \mathbf{T}\cdot\mathbf{n}\mathrm{d}S}{V} = \nabla\left(\tfrac{1}{2}\mu_0 HM\right) \tag{C.14}$$

where ∇ is the gradient operator (see Appendix B for vector analysis notation). The derivation of Eq. (C.14) is illustrated in Fig. C.6 As it happens, the left-hand side of Eq. (C.14) is the definition of the divergence of a tensor, div**T**, Eq. (C.2), and therefore in the limit of small particle size Eqs. (C.2) and (C.14) lead to

$$\mathbf{f} = \nabla\left(\tfrac{1}{2}\mu_0 HM\right) \tag{C.15}$$

For a small particle suspended in a medium of magnetization, M_1, the net body force acting on the particle equals difference of body forces acting on the particle and on the volume of medium

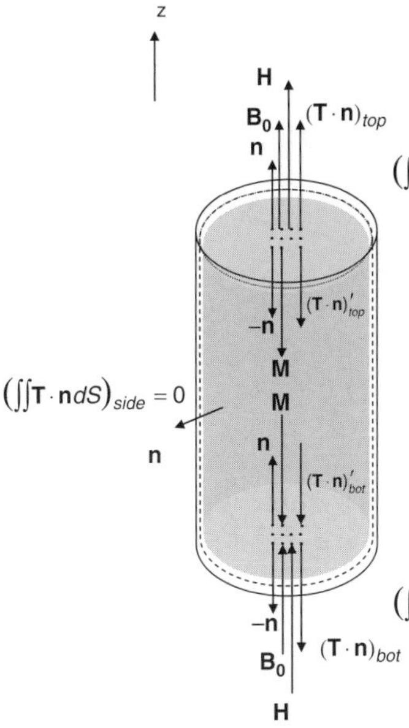

$$(\mathbf{T}\cdot\mathbf{n})_{top} = \mathbf{H}(B_0\cdot\mathbf{n}) - \tfrac{1}{2}(\mathbf{H}\cdot B_0)\mathbf{n} = \mathbf{H}B_0 - \tfrac{1}{2}HB_0\mathbf{n} = \tfrac{1}{2}HB_0\mathbf{n}$$

$$(\mathbf{T}\cdot\mathbf{n})'_{top} = \tfrac{1}{2}HB(-\mathbf{n}) = \tfrac{1}{2}H\mu_0(-M+H)(-\mathbf{n}) = -\tfrac{1}{2}H\mu_0M(-\mathbf{n}) + \tfrac{1}{2}HB_0(-\mathbf{n})$$

$$\left(\iint \mathbf{T}\cdot\mathbf{n}dS\right)_{top} = (\mathbf{T}\cdot\mathbf{n})_{top} + (\mathbf{T}\cdot\mathbf{n})'_{top} = \left(\tfrac{1}{2}HB_0\mathbf{n} - \tfrac{1}{2}H\mu_0M(-\mathbf{n}) + \tfrac{1}{2}HB_0(-\mathbf{n})\right)_{top} = \left(\tfrac{1}{2}H\mu_0M\mathbf{n}\right)_{top}$$

$$(\mathbf{T}\cdot\mathbf{n})_{bot} = \mathbf{H}(B_0\cdot(-\mathbf{n})) - \tfrac{1}{2}(\mathbf{H}\cdot B_0)(-\mathbf{n}) = -\mathbf{H}B_0 + \tfrac{1}{2}HB_0\mathbf{n} = -\tfrac{1}{2}HB_0\mathbf{n}$$

$$\left(\iint \mathbf{T}\cdot\mathbf{n}dS\right)_{side} = 0$$

$$(\mathbf{T}\cdot\mathbf{n})'_{bot} = \tfrac{1}{2}HB'\mathbf{n} = \tfrac{1}{2}H\mu_0(-M+H)\mathbf{n} = -\tfrac{1}{2}H\mu_0M\mathbf{n} + \tfrac{1}{2}HB_0\mathbf{n}$$

$$\left(\iint \mathbf{T}\cdot\mathbf{n}dS\right)_{bot} = (\mathbf{T}\cdot\mathbf{n})_{bot} + (\mathbf{T}\cdot\mathbf{n})'_{bot} = \left(-\tfrac{1}{2}HB_0\mathbf{n} - \tfrac{1}{2}H\mu_0M\mathbf{n} + \tfrac{1}{2}HB_0\mathbf{n}\right)_{bot} = \left(-\tfrac{1}{2}H\mu_0M\mathbf{n}\right)_{bot}$$

$$\mathbf{f} = div\mathbf{T} = \lim_{V\to 0}\frac{\int_A \mathbf{T}\cdot\mathbf{n}dS}{V} = \lim_{V\to 0}\frac{\left(\iint \mathbf{T}\cdot\mathbf{n}ds\right)_{top} + \left(\iint \mathbf{T}\cdot\mathbf{n}ds\right)_{bot}}{V} = \lim_{V\to 0}\frac{(\mathbf{T}\cdot\mathbf{n})_{top}S + (\mathbf{T}\cdot\mathbf{n})_{bot}S}{V}$$

$$= \lim_{\Delta z\to 0}\frac{\left((\tfrac{1}{2}H\mu_0M\mathbf{n})_{top} + (-\tfrac{1}{2}H\mu_0M\mathbf{n})_{bot}\right)S}{S\,\Delta z} = \lim_{\Delta z\to 0}\frac{\Delta\left(\tfrac{1}{2}H\mu_0M\right)}{\Delta z}\mathbf{n} = \nabla\left(\tfrac{1}{2}\mu_0HM\right)$$

Fig. C.6. In the limit of a small volume, divergence of the Maxwell stress tensor, div**T**, equals gradient of the dipole energy density in the magnetic field, grad $(\frac{1}{2}MB)$ (see text). Here the magnetization of the volume $V = S\Delta z$, indicated in grey, is $\mathbf{M}_2 = \mathbf{M}$, and the magnetization of the outside medium is $\mathbf{M}_1 = 0$.

displaced by the particle (this is sometimes referred to as a "magnetic Archimedes law"):

$$\mathbf{f} = \nabla(\tfrac{1}{2}\mu_0HM_2) - \nabla(\tfrac{1}{2}\mu_0HM_1) = \nabla(\tfrac{1}{2}\mu_0H\Delta M) \tag{C.16}$$

where

$$\Delta M = M_2 - M_1 \tag{C.17}$$

For linearly polarizable media with the magnetic susceptibilities, χ_2 (particle) and χ_1 (fluid):

$$\Delta M = \chi_2 H - \chi_1 H = \Delta\chi H \tag{C.18}$$

where

$$\Delta\chi = \chi_2 - \chi_1 \tag{C.19}$$

and therefore

$$\mathbf{f} = \nabla(\tfrac{1}{2}\Delta\chi H B_0) \tag{C.20}$$

where $B_0 = \mu_0 H$ is the applied magnetic field induction. For homogeneous, incompressible, and linearly polarizable media (i.e., paramagnetic and diamagnetic substances), $\Delta\chi = \text{const}$ (independent of field and spatial coordinates), and the expression for the body force takes the form:

$$\mathbf{f} = \Delta\chi\nabla(\tfrac{1}{2}HB_0) \tag{C.21}$$

where H and B_0 are the field strength and field induction (or field intensity) in free space, respectively.

C.4. Elastic body, discrete surface magnetization

Perhaps of the most immediate interest to magnetic separation is the behavior of an elastic body whose magnetization is essentially the same as that of the surrounding medium (the cell), and whose surface is dotted with small (discrete) particles that are highly magnetizable (superparamagnetic nanobeads). The motion of such a body can be illustrated by using the Maxwell stress and the magnetic pathlines, Fig. C.7. The magnetic particles bound to the surface of the cell are reduced to point-like masses that are randomly distributed over the cell surface. Here, it is assumed that their trajectories in the magnetic field are unaffected by the presence of the cell to which they are bound. Such a simplified picture applies only to a quasi-static motion of a perfectly elastic body. The real dynamics of the cell motion in the magnetic is much more complex, in particular, the friction of the viscous fluid medium, the elasticity of the cell membrane, and the highly localized force exerted by the magnetic particles at the points of contact with the cell surface may cause stresses and strains that locally may be even higher than suggested in Fig. C.7. In the extreme cases of the large magnetic particles, the high local forces exerted by the bound bead cause tearing and loss of integrity of the cell membrane. For sufficiently high gradients and magnetization of the magnetic nano-particles, the particles distal to the magnetic field source may puncture the cell membrane and enter the cytosol (magnetoporation).

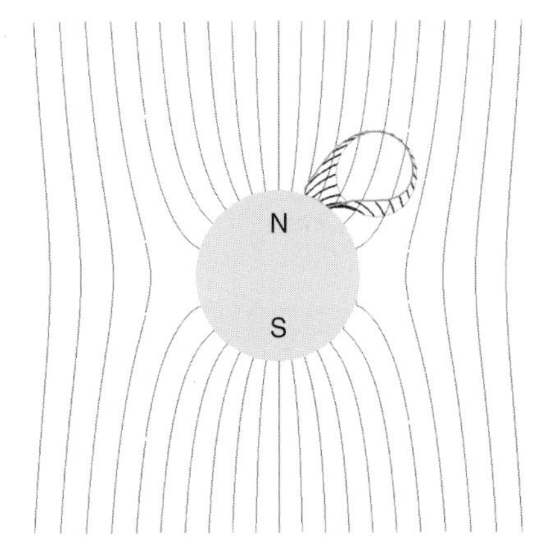

Fig. C.7. Deformation of an elastic body in the high gradient magnetic field of a high permeability wire of circular cross-section (marked in gray). The initial position of the elastic body is indicated by an open circle. The final position is indicated by a pear-shaped oval. The thick black lines indicate pathlines of magnetic particles attached to the surface of the elastic body as it is dragged towards the wire. The figure is an idealized representation of type of strains that a magnetically labeled cell undergoes in the high gradient magnetic field.

References

[1] Becker, R. (1982). Electromagnetic fields and interactions. Dover Publications, Inc., New York.
[2] Melcher, J. R. (1981). Continuum electromechanics. The MIT Press, Cambridge, MA.
[3] Longair, M. S. (1994). Theoretical concepts in physics. Cambridge University Press, Cambridge.
[4] Stratton, J. A. (1941). Electromagnetic theory. McGraw-Hill Book Company, Inc., New York.
[5] Rosensweig, R. E. (1985). Ferrohydrodynamics. MA: Cambridge University Press, Cambridge.
[6] Brodkey, R. S. (1995). The phenomena of fluid motions. Dover Publications, Inc., Mineola NY.

Laboratory Techniques in Biochemistry and Molecular Biology, Volume 32
Magnetic Cell Separation
M. Zborowski and J. J. Chalmers (Editors)

APPENDIX D

Volume magnetic susceptibilities of selected substances

Table of volume magnetic susceptibilities
(dimensionless, SI system of units)

Substance	Formula	B_0, T	Temperature (K)	$\chi \times 10^6$	References
Aluminum	Al		293	20.8	[1]
Benzene	C_6H_6		305	−7.73	[1]
Bismuth	Bi		293	−165	[1]
Carbohydrates	—			−10.4	[2]
Deltamax	—	1×10^{-5}		1.4×10^{11}	[3]
deoxyHB RBC	—			−8.77	[4]
Dysprosium oxide	Dy_2O_3		287.2	23,600	[1]
Erbium oxide	Er_2O_3		286	21,000	[1]
Ferritin, calculated	—			239	Ch. 1
Ferritin, native horse spleen	—			87.9	Ch. 1
Gadolinium chloride	$GdCl_3$		293	6,020	[1]
Graphite	C		293	−14.1	[1]
Hematite	Fe_2O_3		1033	1,480	[1]
Iron (II) chloride	$FeCl_2$		293	4,620	[1]
Magnetite	Fe_3O_4	0.2	293	1,790,000	[1]
Magnetite	Fe_3O_4	0.002	293	10,910,000	[5]
Iron (II) oxide (wuestite)	FeO		293	7,180	[1]
Lipids	—			−8.71	[2]
Manganese chloride	$MnCl_2$		293	4,390	[1]

(*continued*)

DOI: 10.1016/S0075-7535(06)32017-7

(*continued*)

Substance	Formula	B_0, T	Temperature (K)	$\chi \times 10^6$	References
metHb RBC	—			−8.73	[4]
Mumetal	—			2.0×10^{10}	[3]
Nucleic acids	—			−10.5	[2]
oxyHb RBC	—			−9.05	[4], Ch. 1
Platinum	Pt		290.3	279	[1]
Supermalloy	—	2×10^{-7}		9.0×10^{10}	[3]
Water	H_2O		293	−9.03	[1]

References

[1] Weast, R. C. (1981). CRC handbook of chemistry and physics, 62nd ed. CRC Press, Inc., Boca Raton, Fl.

[2] Chikov, V., Kuznetsov, A. and Schutt, W. (1991). Analytical cell magnetophoresis. In: Physical Characteristics of Biological Cells (Schütt, W., Klinkmann, H., Lamprecht, I. and Wilson, T., eds.). Verlag Gesundheit GmbH, Berlin, pp. 381–389.

[3] Cullity, B. D. (1972). Introduction to magnetic materials. Addison-Wesley, Reading, Mass.

[4] Zborowski, M., Ostera, G. R., Moore, L. R., Milliron, S., Chalmers, J. J. and Schechter, A. N. (2003). Red blood cell magnetophoresis. Biophys. J. *84*, 2638–45.

[5] Yamaura, M., Camilo, R. L., Sampaio, L. C., Macedo, M. A., Nakamura, M. and Toma, H. E. (2004). Preparation and characterization of (3-aminopropyl)triethoxysilane-coated magnetite nanoparticles. J. Magn. Magn. Mater. *279*, 210–7.

Index

D

E

F

M

N

O

P

Q

R

S

T

U

V

X

Z

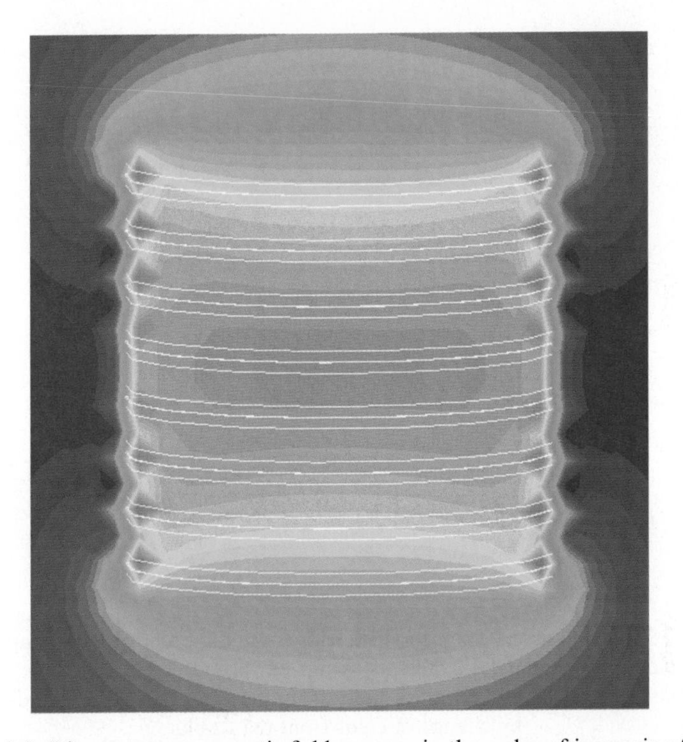

Fig. 2.8. Direct current magnetic field sources, in the order of increasing field inhomogeneity and field gradients that they produce. Solenoid: Note uniformly high field inside the solenoid, $B_0 \approx$ const., as indicated by red color. Note also field gradient at the rim of the solenoid, as indicated by color change with distance.

Fig. 2.9. Direct current magnetic field sources, in the order of increasing field inhomogeneity and field gradients that they produce. Current line: Note decrease of field intensity with distance from the wire, $B_0 \propto 1/r$, as indicated by change of color from red to blue.

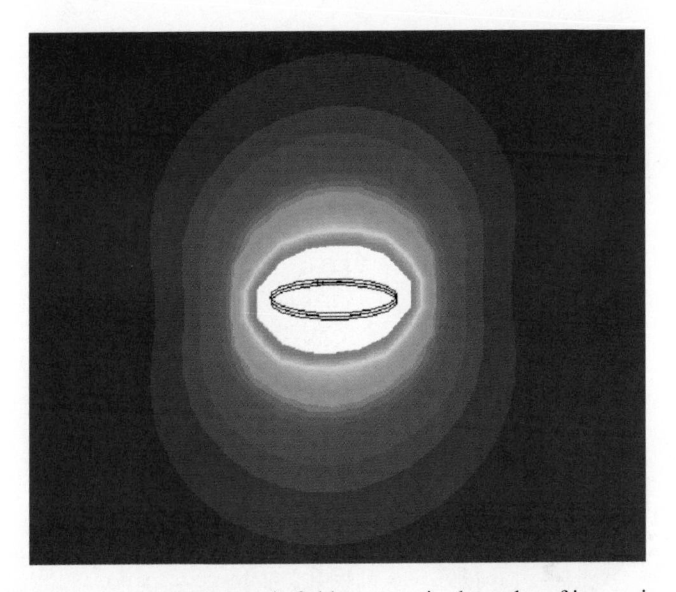

Fig. 2.10. Direct current magnetic field sources, in the order of increasing field inhomogeneity and field gradients that they produce. Current ring: Note decrease of field intensity with distance from the ring, that is much faster than that seen for the current line. The far field of the ring current is equivalent to the field of a magnetic dipole $B_0 \propto 1/r^3$.

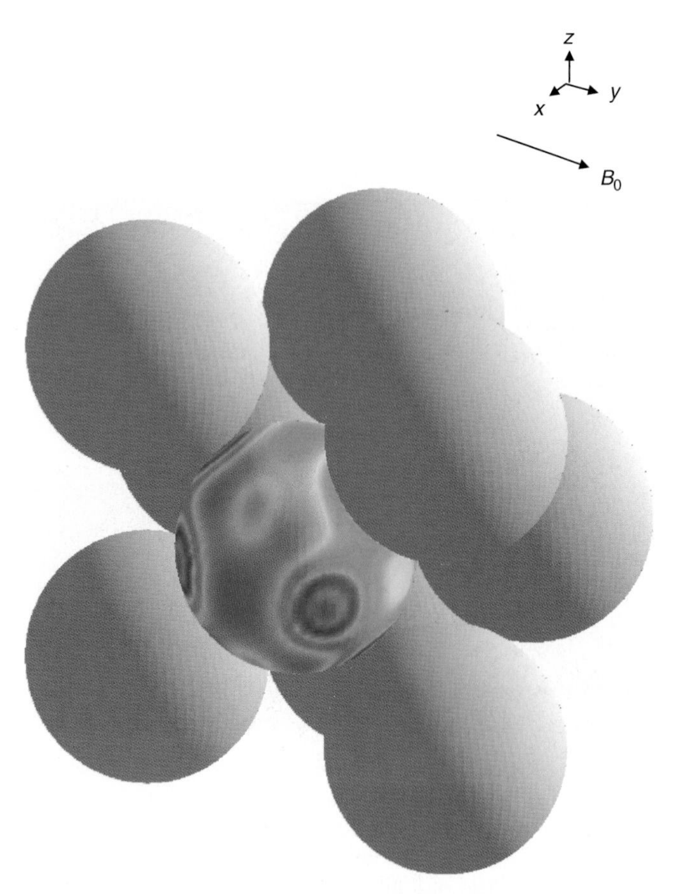

Fig. 4.6. Field of a stack of ferromagnetic spheres. False color representation of the magnetic pressure (Maxwell stress) distribution at the surface of a sphere (violet—high, red—low). The nearest neighbors are rendered in gray. Only 8 of 12 nearest neighbors are shown for clarity. Note magnetic stress concentration at points of contact between spheres.

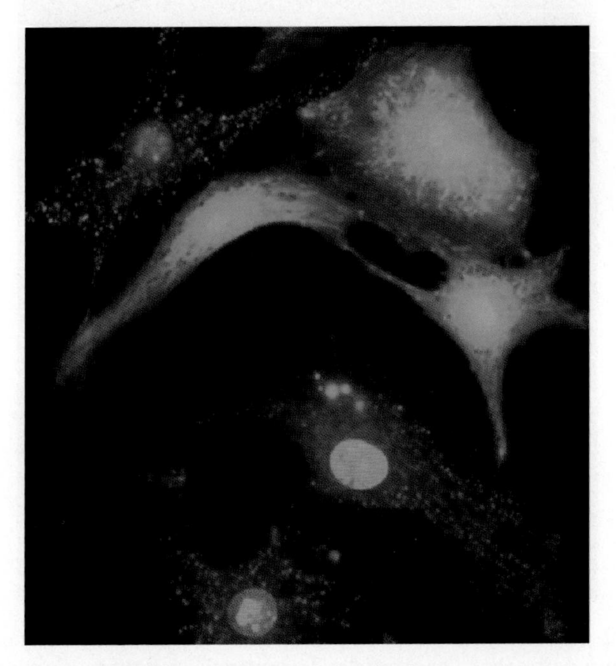

Fig. 7.8. Epithelial cells stained with the Live/Dead® assay. Live cells fluoresce bright green, whereas dead cells with compromised membranes show red-fluorescing nuclei.

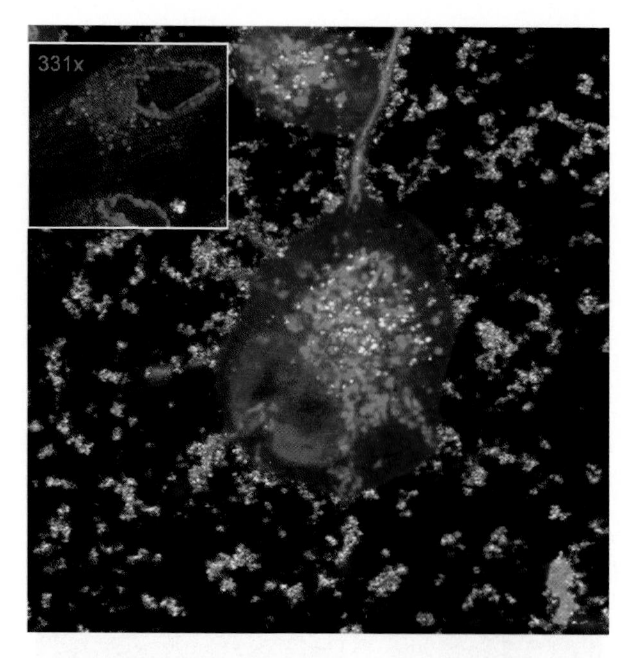

Fig. 7.12. Confocal microscopy of prostate cancer C4–2 cells after incubation for 8 hours with cobalt nanoparticles. Large amounts of the particles are taken up into the cell plasma, some are also seen adhering to the outside. The insert shows the same cells after incubation with magnetic nanoparticles coated with a cytotoxic copolymer. The cells are dying, as seen by the nuclei showing the typical picture of condensing chromatin while undergoing apoptosis.

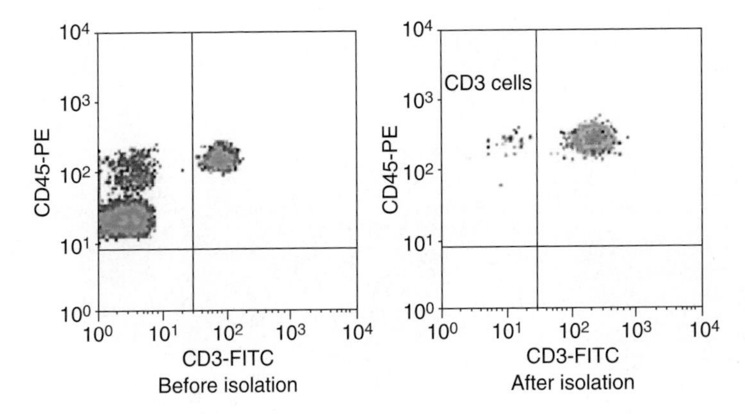

Fig. 11.1. Untouched CD3+ T cells negatively isolated from MNC with the Dynal T Cell Negative Isolation Kit. The figure shows the CD3+ population before and after negative isolation (from the Company website [1], used with permission).

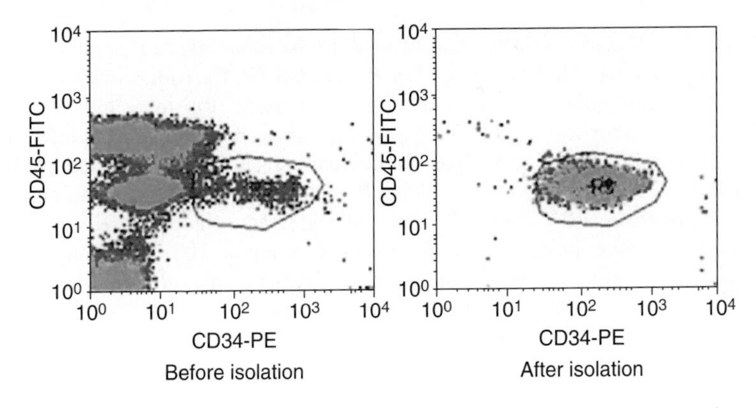

Fig. 11.2. Isolation results: CD34+ cells positively isolated from bone marrow-derived MNC with the Dynal® CD34 Progenitor Cell Selection System (from the Company website [2], used with permission).

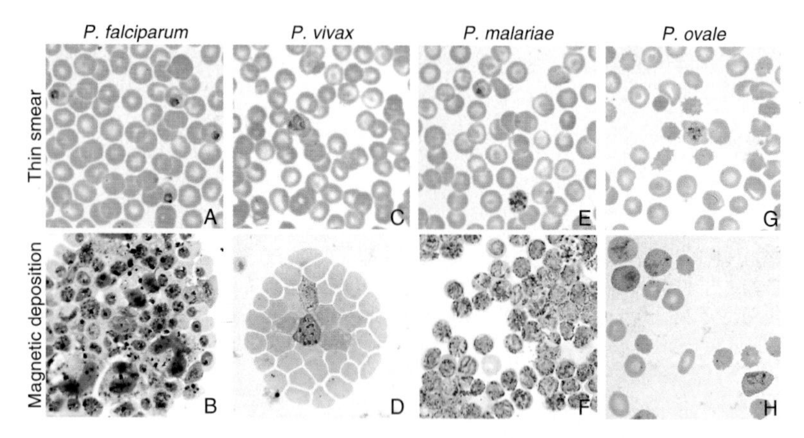

Fig. 12.5. Malaria MDM concentrates *Plasmodium*-infected erythrocytes. *P. falciparum* (A, B), *P. vivax* (C, D), *P. malariae* (E, F), and *P. ovale* (G, H) infections comparing conventional thin blood smear (top row) and malaria MDM (bottom row) for each *Plasmodium* species were prepared from infected nonhuman primate blood samples. Individual parasitemias determined by the Earle and Perez method were 2.7% for *P. falciparum*, 0.1% for *P. vivax*, 0.4% for *P. malariae*, and 0.2% for *P. ovale*. All slides were stained using standard Giemsa staining procedures and examined using a 100× oil immersion objective. In part B, "M" = macrophage (with permission from *American Journal of Tropical Medicine and Hygiene*).

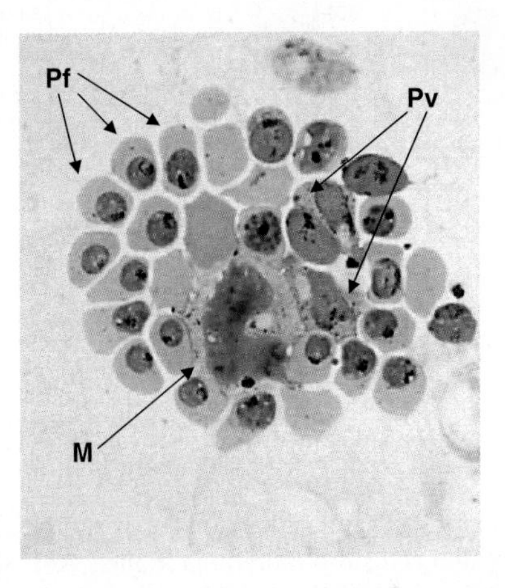

Fig. 12.6. MDM detection of *P. falciparum* and *P. vivax* from a mixed blood sample. Equal volumes of blood from *P. falciparum* (initial parasitemia of 2.7%) and *P. vivax* (initial parasitemia of 0.1%) infected monkeys were mixed and then subjected to MDM analysis. Giemsa stained slides show MDM concentration of *P. falciparum* (Pf), *P. vivax* (Pv), and macrophages (M) containing hemozoin.

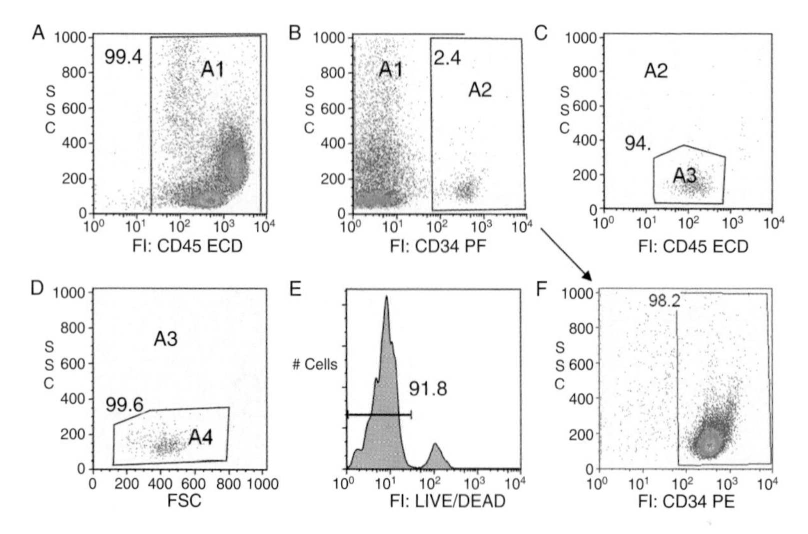

Fig. 12.10. FCM analysis of clinical leukapheresis sample. Numbers in panels are percentages of gated cell subpopulations. Panels A–E show sample before QMS sorting. Panel F shows QMS sorted fraction. (A) Dot plot of side scatter versus CD45-ECD FI. A1: CD45+ leukocytes in the total population. (B) Dot plot of side scatter versus CD34-PE FI. A2: the CD34+ cells in the subpopulation A1. (C) Dot plot of side scatter versus CD45-ECD FI; A3: CD45 low to medium and SSC low in the subpopulation A2. (D) Dot plot of side scatter versus forward scatter to gate A4 in the subpopulation A3. (E) Dot plot of side scatter versus FI of LIVE/DEAD Reduced Biohazard Cell Viability Kit No. 4 to check the viability of the cells in A4. Here the purity of CD34+ is defined as: (number of events gated in A4)/(number of events gated in A1) × 100%. (F) Dot plot of side scatter versus CD34-PE FI. A2: the CD34+ cells in the subpopulation A1 (after QMS separation).

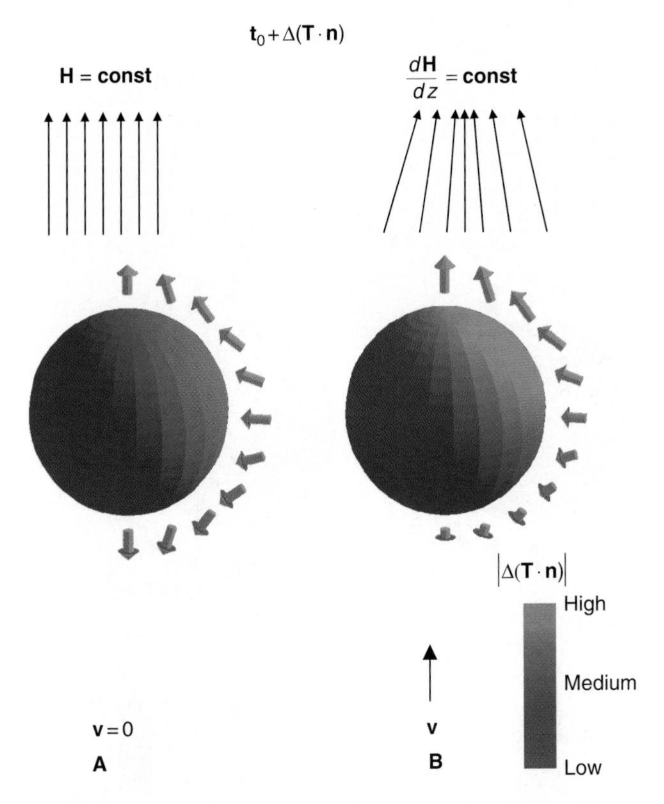

Fig. C.2. Magnetic stress distribution on a surface of rigid sphere that is more magnetic than the medium ($\Delta M > 0$, Eq. (C.10)). (A) Uniform applied field. Note purely tensile contribution at the poles and purely compressive stress contribution at the equator, as indicated by the arrows, that cancel out when integrated over the sphere surface (the sphere velocity, $\mathbf{v} = 0$). (B) In the field gradient, the tension at the pole exposed to a stronger field overcomes tension of the pole exposed to a weaker field, with the net effect of the sphere being drawn toward the field source ($\mathbf{v} \neq 0$).

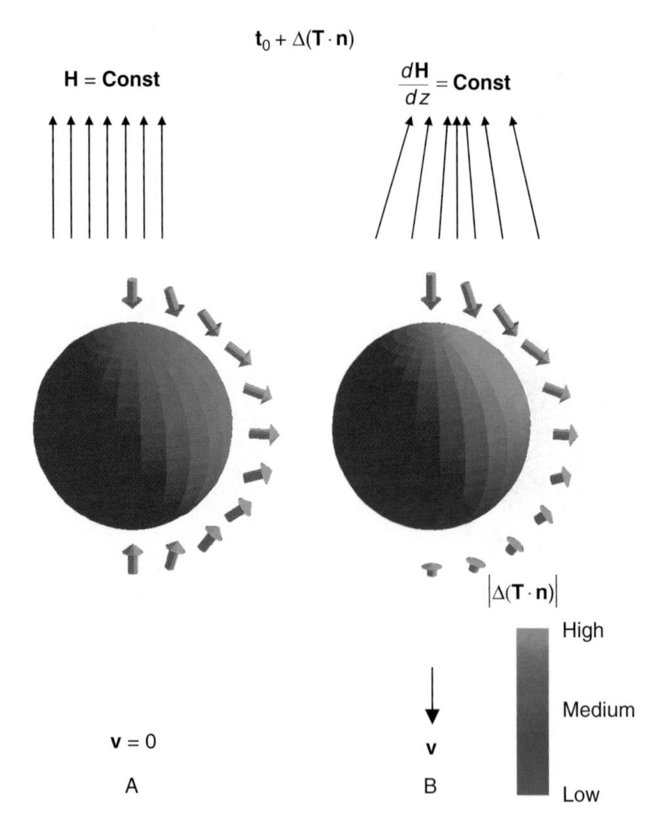

Fig. C.3. Magnetic stress distribution on a surface of rigid sphere that is less magnetic than the medium ($\Delta M < 0$, Eq. (C.10)). (A) Uniform applied field. Note change in the direction of stress arrows as compared to Fig. C.2, with no change in the net effect on the sphere ($\mathbf{v} = 0$). (B) In the field gradient, the compressive stress at the pole exposed to a stronger field overcomes stress at the pole exposed to a weaker field, with the net effect of the sphere being pushed away from the field source ($\mathbf{v} \neq 0$).

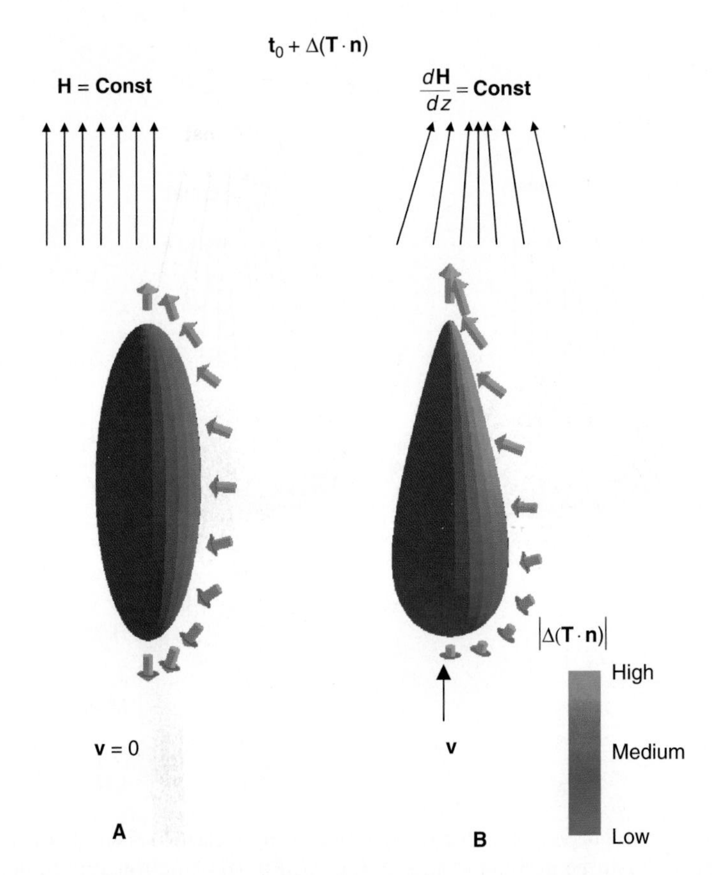

Fig. C.4. Magnetic stress distribution applied to an elastic sphere that is more magnetic than the medium ($\Delta M > 0$, Eq. (C.10)). (A) Uniform applied field. Note deformation as a result of purely tensile contribution at the poles and purely compressive stress contribution at the equator, as indicated by the arrows, that cancel out over the sphere surface ($\mathbf{v} = 0$). (B) In the field gradient, the tension at the pole exposed to a stronger field overcomes tension of the pole exposed to a weaker field, resulting in a characteristic, pear-like deformation with the net effect of the body being drawn toward the field source ($\mathbf{v} \neq 0$).

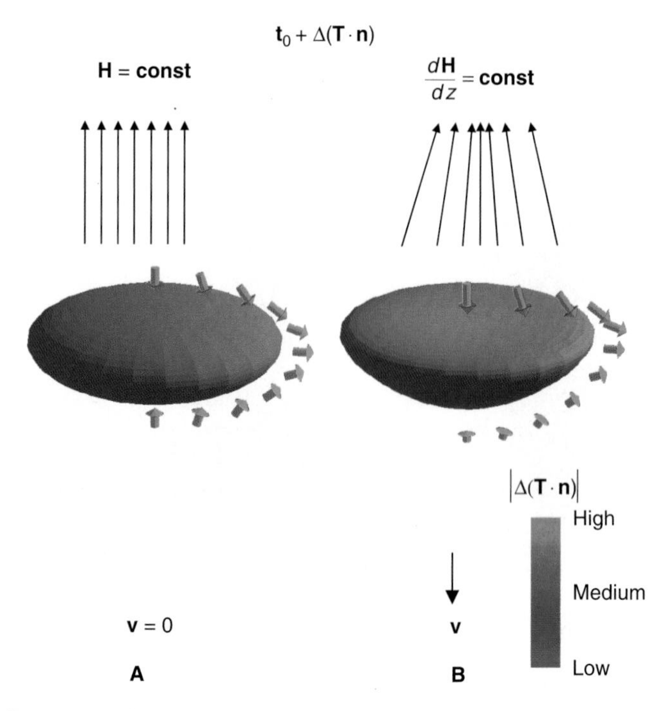

Fig. C.5. Magnetic stress distribution applied to an elastic sphere that is less magnetic than the medium ($\Delta M < 0$, Eq. (C.10)). (A) Uniform applied field. Note change in the direction of stress arrows as compared to Fig. C.4 resulting in the change in deformation from prolate to oblate elispoid of revolution ($\mathbf{v} = 0$). (B) In the field gradient, the stress at the pole exposed to a stronger field overcomes stress of the weaker field, resulting in an asymmetrical flattening of the oblate ellipsoid and the motion of the spheroid away from the field source ($\mathbf{v} \neq 0$).